Applications of Nanomaterials in Biomedical Imaging and Cancer Therapy: 3rd Edition

Applications of Nanomaterials in Biomedical Imaging and Cancer Therapy: 3rd Edition

Guest Editor

James C. L. Chow

MDPI

Basel • Beijing • Wuhan • Barcelona • Belgrade • Novi Sad • Cluj • Manchester

Guest Editor
James C. L. Chow
Department of Radiation Oncology
University of Toronto
Toronto, ON
Canada

Editorial Office
MDPI AG
Grosspeteranlage 5
4052 Basel, Switzerland

This is a reprint of the Special Issue, published open access by the journal *Nanomaterials* (ISSN 2079-4991), freely accessible at: https://www.mdpi.com/journal/nanomaterials/special_issues/Z0U20L95FF.

For citation purposes, cite each article independently as indicated on the article page online and as indicated below:

Lastname, A.A.; Lastname, B.B. Article Title. *Journal Name* **Year**, *Volume Number*, Page Range.

ISBN 978-3-7258-6300-6 (Hbk)
ISBN 978-3-7258-6301-3 (PDF)
https://doi.org/10.3390/books978-3-7258-6301-3

Contents

About the Editor

James C. L. Chow

James C. L. Chow is a Professor in the Department of Radiation Oncology and is cross-appointed to the Department of Materials Science and Engineering at the University of Toronto. He is also a Medical Physicist at Princess Margaret Cancer Center, University Health Network, Toronto. Dr. Chow is internationally recognized for his contributions to computer simulation, nanomedicine, radiation dosimetry, and AI applications in oncology.

He has published over 220 peer-reviewed articles and 20 book chapters and has edited multiple books and Special Issues including the acclaimed "Applications of Nanomaterials in Biomedical Imaging and Cancer Therapy" series. His research focuses on Monte Carlo simulation, nanoparticle-enhanced radiotherapy, biomedical imaging, and the integration of artificial intelligence in cancer care.

He serves on editorial boards of leading journals and has been honored with numerous distinctions, including Outstanding Reviewer Awards and recognition among MDPI's Top 1000 Reviewers. His work has advanced precision oncology through innovations in imaging contrast agents, theranostic nanomaterials, and AI-driven clinical tools, influencing both research and clinical practice worldwide.

nanomaterials

MDPI

Editorial

Applications of Nanomaterials in Biomedical Imaging and Cancer Therapy: 3rd Edition

James C. L. Chow [1,2,3]

[1] Radiation Medicine Program, Princess Margaret Cancer Centre, University Health Network, Toronto, ON M5G 1X6, Canada; james.chow@uhn.ca
[2] Department of Radiation Oncology, University of Toronto, Toronto, ON M5T 1P5, Canada
[3] Department of Materials Science and Engineering, University of Toronto, Toronto, ON M5S 3E4, Canada

Following the success of our first two editions of "Applications of Nanomaterials in Biomedical Imaging and Cancer Therapy", this third edition continues to highlight the rapid evolution of nanotechnology and its transformative impact on biomedical imaging, targeted therapy, and cancer theranostics [1,2]. The series has consistently provided a forum for researchers worldwide to showcase innovative nanomaterial-based approaches addressing some of oncology's greatest challenges—treatment selectivity, biocompatibility, and clinical translation.

Building upon the foundations laid in the first and second edition [3,4], which emphasized nanoparticle-based imaging modalities, multifunctional composites, and radiation dose enhancement, this new collection extends the discussion to emerging trends in tumor microenvironment modulation, responsive nanocarriers, and computational modeling in nanomedicine. Collectively, the ten contributions in this issue present significant advancements that continue to drive the field toward precision, safety, and clinical readiness.

For innovative nanomaterials in cancer therapy and imaging, Chkair et al. [5] report the development of hydrophilic, fluorescent organic nanoparticles (FONPs) functionalized with purpurin-18 for photodynamic therapy (PDT) in colorectal cancer. Their study demonstrates efficient induction of apoptosis via reactive oxygen species (ROS) generation, highlighting a promising approach for minimally invasive treatment. In a complementary approach, Snyder et al. [6] explore the synergistic use of silver nanoparticles (AgNPs) with proteotoxic stress inhibitors to selectively induce death in triple-negative breast cancer (TNBC) cells. This work highlights how stress pathway modulation can refine nanotoxicity into a therapeutic advantage. The paper by Kitamura et al. [7] introduces a tumor microenvironment-responsive liposomal system incorporating iRGD-conjugated peptides to enhance tissue penetration. These nanoparticles exhibited Neuropilin-1-mediated tumor targeting, representing a promising advancement for siRNA and drug delivery deep within tumors. Cakir et al. [8] design chrysin-loaded micelles synthesized via RAFT polymerization, enabling dual apoptotic pathway activation in ovarian cancer cells. This multi-functional system combines hydrophobic drug encapsulation with precise endosomal escape and improved biocompatibility.

Focusing on nanoparticle engineering and mechanistic insights, Quispe Cohaila et al. [9] describe a green synthesis of biogenic zinc oxide (ZnO) nanoparticles using Bacillus licheniformis. These nanoparticles exhibit selective cytotoxicity against glioblastoma cells while maintaining normal retinal cell viability. Their findings highlight the promise of microbial nanomanufacturing as a safe and effective strategy for anticancer agent development. Zimina et al. [10] investigate the magnetically controlled transport of nanoparticles using a tumor-on-a-chip model. Their findings reveal how particle size and coating influence migration dynamics, offering insights for controlled nanoparticle delivery in dense tumor tissues. Moreover, the use of carbon-encapsulated iron nanoparticles for molecular MRI imaging of glioma, developed by Stawarska et al. [11], demonstrates targeted imaging through integrin $\alpha v\beta 3$ receptors. This study contributes to non-invasive diagnosis and image-guided therapy development.

In the realm of modeling, simulation, and quantum-based nanomedicine, Monte Carlo modeling remains a powerful computational tool. FLASH radiotherapy (FLASH-RT) is an emerging modality that delivers radiation at ultrahigh dose rates (UHDRs), typically greater than 40 Gy per second. This technique has attracted significant attention for its ability to spare normal tissues while maintaining or even enhancing tumor control [12]. Chow [13] offers a comprehensive review of its applications in nanoparticle-enhanced radiotherapy, nanodosimetry, and the optimization of FLASH-RT. This review bridges computational and clinical nanoscience, emphasizing patient-specific and quantum modeling strategies for next-generation treatment planning. Kim and Chow [14] further explore this frontier by quantifying ROS yields around gold nanoparticles under UHDR electron beams. Their study elucidates nanoparticle–radiation interactions at the nanoscale, providing a computational foundation for combining gold nanoparticles with FLASH-RT. Moreover, Sarwat et al. [15] extend nanotechnology applications to ophthalmology through the development of hydrophobic silicon quantum dots (Si-QDs) for imaging the tear film lipid layer. This innovation exemplifies the versatility of nanomaterials beyond oncology, addressing the imaging challenges in dry eye disease.

The papers featured in this third edition collectively advance our understanding of how nanomaterials can revolutionize both diagnostic and therapeutic modalities. From photodynamic and radiotherapy enhancement to computational simulation and bio-inspired synthesis, these studies showcase the interdisciplinary depth and translational potential of nanotechnology in medicine.

Together with the first and second editions, this ongoing series emphasizes a central theme: the convergence of nanotechnology, biology, and computation is pivotal to advancing precision oncology. As emerging platforms such as tumor microenvironment-responsive nanoparticles, FLASH-RT optimization, and machine learning-driven nanomaterial design continue to mature, the boundary between simulation and clinical application grows increasingly narrow.

We thank all authors, reviewers, and collaborators for their exceptional contributions that make this Special Issue a vibrant reflection of nanomedicine's promise.

Funding: This research received no external funding.

Data Availability Statement: Not applicable.

Acknowledgments: The author expresses heartfelt appreciation to all contributors and editors of this Special Issue for their exceptional commitment and invaluable contributions.

Conflicts of Interest: The author declares no conflicts of interest.

Abbreviations

The following abbreviations are used in this manuscript:

FONPs	Fluorescent Organic Nanoparticles
PDT	Photodynamic Therapy
ROS	Reactive Oxygen Species
AgNPs	Silver Nanoparticles

TNBC	Triple-Negative Breast Cancer
iRGD	Internalizing Arginine–Glycine–Aspartic Peptide
siRNA	Small Interfering RNA
RAFT	Reversible Addition–Fragmentation Chain Transfer
ZnO	Zinc Oxide
MRI	Magnetic Resonance Imaging
FLASH-RT	FLASH Radiotherapy
UHDR	Ultrahigh Dose Rate
Si-QDs	Silicon Quantum Dots

References

1. Siddique, S.; Chow, J.C.L. Application of Nanomaterials in Biomedical Imaging and Cancer Therapy. *Nanomaterials* **2020**, *10*, 1700. [CrossRef] [PubMed]
2. Siddique, S.; Chow, J.C.L. Recent Advances in Functionalized Nanoparticles in Cancer Theranostics. *Nanomaterials* **2022**, *12*, 2826. [CrossRef] [PubMed]
3. Chow, J.C.L. Special Issue: Application of Nanomaterials in Biomedical Imaging and Cancer Therapy. *Nanomaterials* **2022**, *12*, 726. [CrossRef] [PubMed]
4. Chow, J.C.L. Application of Nanomaterials in Biomedical Imaging and Cancer Therapy II. *Nanomaterials* **2024**, *14*, 1627. [CrossRef] [PubMed]
5. Chkair, R.; Couvez, J.; Brégier, F.; Diab-Assaf, M.; Sol, V.; Blanchard-Desce, M.; Liagre, B.; Chemin, G. Activity of Hydrophilic, Biocompatible, Fluorescent, Organic Nanoparticles Functionalized with Purpurin-18 in Photodynamic Therapy for Colorectal Cancer. *Nanomaterials* **2024**, *14*, 1557. [CrossRef] [PubMed]
6. Snyder, C.M.; Mateo, B.; Patel, K.; Fahrenholtz, C.D.; Rohde, M.M.; Carpenter, R.; Singh, R.N. Enhancement of Triple-Negative Breast Cancer-Specific Induction of Cell Death by Silver Nanoparticles by Combined Treatment with Proteotoxic Stress Response Inhibitors. *Nanomaterials* **2024**, *14*, 1564. [CrossRef] [PubMed]
7. Kitamura, K.; Matsui, R.; Itagaki, N.; Takeuchi, Y.; Fukuda, H.; Tanaka, K.-I.; Hama, S. Development of Tumor Microenvironment-Responsive Nanoparticles with Enhanced Tissue Penetration. *Nanomaterials* **2025**, *15*, 1695. [CrossRef]
8. Cakir, S.; Yildiz, U.; Yildirim, T.; Aydin, O. Chrysin-Loaded Micelles Regulate Cell Cycle and Induce Intrinsic and Extrinsic Apoptosis in Ovarian Cancer Cells. *Nanomaterials* **2025**, *15*, 1362. [CrossRef] [PubMed]
9. Quispe Cohaila, A.B.; Fora Quispe, G.d.L.; Cáceda Quiroz, C.J.; Mamani Anccasi, R.; Mejía García, T.A.; Tamayo Calderón, R.M.; Gamarra Gómez, F.; Sacari Sacari, E.J. Biogenic ZnO Nanoparticles Synthesized by *B. licheniformis*: A Selective Cytotoxicity Against NG-108 Glioblastoma Cells. *Nanomaterials* **2025**, *15*, 1338. [CrossRef] [PubMed]
10. Zimina, T.; Sitkov, N.; Brusina, K.; Fedorov, V.; Mikhailova, N.; Testov, D.; Gareev, K.; Samochernykh, K.; Combs, S.; Shevtsov, M. Magnetically Controlled Transport of Nanoparticles in Solid Tumor Tissues and Porous Media Using a Tumor-on-a-Chip Format. *Nanomaterials* **2024**, *14*, 2030. [CrossRef] [PubMed]
11. Stawarska, A.; Bamburowicz-Klimkowska, M.; Szeszkowski, W.; Grudzinski, I.P. Dynamic Susceptibility Contrast Magnetic Resonance Imaging with Carbon-Encapsulated Iron Nanoparticles Navigated to Integrin Alfa V Beta 3 Receptors in Rat Glioma. *Nanomaterials* **2025**, *15*, 1277. [CrossRef] [PubMed]
12. Chow, J.C.L.; Ruda, H.E. Mechanisms of Action in FLASH Radiotherapy: A Comprehensive Review of Physicochemical and Biological Processes on Cancerous and Normal Cells. *Cells* **2024**, *13*, 835. [CrossRef] [PubMed]
13. Chow, J.C.L. Monte Carlo Simulations in Nanomedicine: Advancing Cancer Imaging and Therapy. *Nanomaterials* **2025**, *15*, 117. [CrossRef] [PubMed]
14. Kim, C.D.; Chow, J.C.L. Reactive Oxygen Species Yield near Gold Nanoparticles Under Ultrahigh-Dose-Rate Electron Beams: A Monte Carlo Study. *Nanomaterials* **2025**, *15*, 1303. [CrossRef] [PubMed]
15. Sarwat, S.; Stapleton, F.; Willcox, M.D.P.; O'Mara, P.B.; Roy, M. Hydrophobic Silicon Quantum Dots for Potential Imaging of Tear Film Lipid Layer. *Nanomaterials* **2025**, *15*, 552. [CrossRef] [PubMed]

nanomaterials

MDPI

Article

Development of Tumor Microenvironment-Responsive Nanoparticles with Enhanced Tissue Penetration

Karin Kitamura [1], Ryo Matsui [2], Nagisa Itagaki [2], Yuka Takeuchi [1], Hana Fukuda [1], Ken-Ichiro Tanaka [3,4] and Susumu Hama [1,4,*]

1 Laboratory of Pharmaceutical Technology, Faculty of Pharmacy, Musashino University, Tokyo 202-8585, Japan
2 Department of Biophysical Chemistry, Kyoto Pharmaceutical University, Kyoto 607-8414, Japan
3 Laboratory of Bio-Analytical Chemistry, Faculty of Pharmacy, Musashino University, Tokyo 202-8585, Japan
4 Research Institute of Pharmaceutical Sciences, Musashino University, Tokyo 202-8585, Japan
* Correspondence: s-hama@musashino-u.ac.jp; Tel.: +81-42-468-8679

Abstract

Liposomes modified with slightly acidic pH-sensitive peptides (SAPSp-lipo) are effectively delivered to tumor tissues, followed by cellular uptake in the tumor microenvironment. Although SAPSp-lipo can penetrate tumor tissues via the interspace route between cancer cells and the extracellular matrix (ECM), penetration needs to be enhanced to deliver liposomes into tumor cores comprising malignant cancer cells. To enhance the intratumoral penetration of SAPSp-lipo, we focused on the internalizing RGD peptide (iRGD), which can penetrate tumor tissue, differing from the penetration mechanism of SAPSp. In this study, we developed liposomes modified with iRGD-conjugated SAPSp (SAPSp-iRGD-lipo). Compared with SAPSp-lipo, SAPSp-iRGD-lipo was delivered to deeper regions within both spheroids and tumor tissues. The enhanced penetration was suppressed by a co-treatment with a Neuropilin-1 inhibitor, and the fluorescence signals from intratumorally injected SAPSp-iRGD-lipo were localized in Neuropilin-1-expressing regions, indicating a Neuropilin-1-mediated tumor penetration. Moreover, SAPSp-iRGD-lipo reduced F-actin formation in monolayered cells and was not localized in F-actin-rich regions in tumors, suggesting that SAPSp-iRGD-lipo facilitates tumor penetration through actin depolymerization. In addition, anticancer siRNA delivered by SAPSp-iRGD-lipid nanoparticles effectively induced apoptosis in cells under slightly acidic conditions. Taken together, SAPSp-iRGD-modified nanoparticles represent a novel class of tumor-penetrable and microenvironment-responsive drug carriers capable of efficient intratumoral delivery and therapeutic activity.

Keywords: tumor microenvironment; tissue penetration; liposomes

1. Introduction

To achieve effective chemotherapy for solid tumors, it is imperative to develop drug carriers that can modulate systemic pharmacokinetics and intratumoral and intracellular drug dynamics [1]. Currently, polyethylene glycol (PEG)-modified carriers, which exhibit extended circulation times, are extensively employed to passively deliver drugs to tumors through the enhanced permeability and retention (EPR) effect [2]. However, PEGylated carriers have limitations, including a low cellular uptake efficiency [3] and the Accelerated Blood Clearance (ABC) phenomenon upon repeated administration [4]. This necessitates the development of alternative PEG-free delivery systems. In contrast, cationic carriers, which demonstrate a high cellular uptake efficiency, tend to interact with biological components in the bloodstream, leading to reduced circulation times and suboptimal tumor accumulation [5].

Liposomal formulations have been extensively investigated as versatile nanocarriers for improving drug delivery efficiency. Previous comprehensive reviews have summarized their composition, preparation methods, and physicochemical properties that influence therapeutic performance [6]. More recently, advances in liposome-based nanocarriers for targeted and controlled delivery were highlighted by Alavi et al. [7], underscoring the continuing evolution of liposome technologies. These studies provide the foundation for the development of pH- and peptide-responsive liposomal systems explored in the present work.

In our previous study, we developed an innovative liposomal carrier modified with a slightly acidic pH-sensitive peptide (SAPSp) [8–10]. This carrier is engineered to respond to a slightly acidic tumor microenvironment by altering its surface charge from negative to positive. At physiological pH, SAPSp-modified liposomes (SAPSp-lipo), which possess a negative surface charge, accumulate in tumors, similar to PEGylated liposomes (PEG-lipo) [8]. When exposed to the slightly acidic conditions present within tumors, the surface charge becomes positive, thereby facilitating an enhanced cellular uptake and the efficient cytoplasmic delivery of encapsulated drugs [8].

Furthermore, SAPSp-lipo infiltrates tumors through the cell–extracellular matrix (ECM) pathway by inducing actin depolymerization in cancer cells [10]. Nonetheless, its intratumoral penetration remains inadequate for the effective delivery to deeply situated cancer cells, suggesting the need for further enhancements. Recent studies have introduced various strategies to enhance tumor penetration through the physicochemical and biological modulation of nanocarriers, including charge conversion and the use of tissue-penetrating peptides. In particular, the conjugation of iRGD peptides or related CendR motifs can significantly enhance tumor penetration and drug delivery efficiency [11–13]. Incorporating these concepts, the present work aims to design SAPSp-iRGD-lipo as a dual-function carrier that integrates pH responsiveness and iRGD-mediated tissue penetration. Upon proteolytic cleavage within the tumor microenvironment, iRGD reveals its CendR motif, which subsequently binds to and activates neuropilin-1 (NRP-1), initiating trans-tissue transport [14–16]. The present study introduces the novel concept of a dual-functional peptide-modified liposome (SAPSp-iRGD-lipo) that integrates the pH-responsive behavior of SAPSp with the NRP-1-mediated tumor-penetrating ability of iRGD. This design enables a dynamic adaptation to the tumor microenvironment, allowing for both efficient blood circulation and deep tumor penetration—overcoming the inherent trade-off between stability and permeability observed in conventional liposomal systems.

In this study, we developed SAPSp-iRGD-modified liposomes (SAPSp-iRGD-lipo) and investigated their tissue-penetrating capabilities and pH-responsive characteristics. Furthermore, we encapsulated antitumor siRNAs within SAPSp-iRGD-modified lipid nanoparticles (SAPSp-iRGD-LNPs) and evaluated their cytotoxic efficacy against cancer cells.

2. Materials and Methods

2.1. Materials and Animals

The mouse melanoma cell line B16−F1 and human melanoma cell line A375 were obtained from DS Pharma Biomedical Co., Ltd. (Osaka, Japan). Human colorectal adenocarcinoma Caco-2 cells were obtained from the RIKEN BioResource Research Center (Ibaraki, Japan). Egg phosphatidylcholine (EPC; Product No. available upon request NOF Corporation) was purchased from the NOF Corporation (Tokyo, Japan).

1,2-Dioleoyl-3-trimethylammonium propane (DOTAP; Product No. 890890P), 1,2-dioleoyl-sn-glycero-3-phosphoethanolamine (DOPE; Product No. 850725P), and 1,2-dioleoyl-sn-glycero-3-phosphoethanolamine-N-(lissamine rhodamine B sulfonyl) (Rh-PE; Product No. 810150P) were obtained from Avanti Polar Lipids (Alabaster, AL, USA). Matrigel® Matrix Basement Membrane Phenol-Red Free was purchased from Corning Life Sciences (Tewksbury, MA, USA; Product No. 35623). Fluorescent dyes, including 3,3′-Dioctadecyl-5,5′-Di (4-Sulfophenyl) Oxacarbocyanine, Sodium Salt (DiO; product No. D7778), LysoTracker Green DND-26 (product No. L7526), Alexa Fluor® 594 F(ab′)$_2$ fragment of rabbit anti-goat IgG (H+L) (product No. A11080), and Alexa Fluor® 488 goat anti-rabbit IgG, were obtained from Invitrogen (product No.A11008) (Carlsbad, CA, USA). Goat anti-Neuropilin-1 antibody was purchased from R&D Systems (Minneapolis, MN, USA; product No. AF556), and rabbit anti-ZO-1 IgG was obtained from Santa Cruz Biotechnology (Dallas, TX, USA; product No. sc-10804). Rhodamine phalloidin was purchased from Cytoskeleton, Inc. (Denver, CO, USA; product No. PHDR1). DiR (1,1′-dioctadecyl-3,3,3′,3′-tetramethylindotricarbocyanine iodide) was purchased from PerkinElmer (XenoLight DiR, product No. 125964; Revvity Inc., Boston, MA, USA). Stearylated peptides, including SAPSp (stearyl-GGGGHGAH-EHAGHEHAAGEHHAHE-NH$_2$), iRGD (stearyl-GGGG(C)RGDKGD(C), circularized at cysteine), and SAPSp-iRGD (stearyl-GGGGHGAH-EHAGHEHAAGEHHAHEGGGG(C)RGDKGD(C), circularized at cysteine), were synthesized by Scrum, Inc. (Tokyo, Japan). Anti-luciferase siRNA (5′-GCGCUGCUGGUGCCAACCCTT-3′, 5′-GGGUUGGCACCAGCAGCGCTT-3′) and anti-human KIF11 siRNA (5′-CUGAAGACCUGAAGACAAUTT-3′, 5′-AUUGUCUUCAGGU CUUCAGTT-3′) were synthesized by Invitrogen. Five-week-old male Hos:HR-1 hairless mice and BALB/cSlc-nu/nu mice were purchased from Sankyo Labo Service Corporation, Inc. (Tokyo, Japan). All animal procedures were conducted in accordance with the institutional guidelines for the care and use of laboratory animals and were approved by the Institutional Animal Care and Use Committee (IACUC) of Musashino university (Approval No. [08-A-2024 and 08-A-2025]). Anesthesia was performed using sodium pentobarbital, and all efforts were made to minimize the suffering of the animals.

2.2. Preparation of Liposomes and Lipid Nanoparticles

Liposomes were prepared using a simple hydration technique, as described previously [10,17]. EPC and DOTAP, dissolved in ethanol, were combined at a molar ratio of 7.6:1 and subsequently dried under nitrogen gas to form lipid films. The films were hydrated with PBS(−) (final lipid concentration: 10 mM) and subjected to sonication using a bath-type sonicator. Surface modification was achieved by incubating the liposomes with 5 mol% stearylated SAPSp, iRGD, or SAPSp-iRGD for 30 min, resulting in SAPSp-lipo, iRGD-lipo, and SAPSp-iRGD-lipo, respectively. Lipid nanoparticles (LNPs) were prepared using the butanol dilution method, as described previously [18]. EPC, DOTAP, and DOPE were mixed at a molar ratio of 5.6:1:2. siRNA was diluted in RNase-free water, added under vortexing, and rapidly diluted in 2 mL citrate buffer. The resulting suspension was filtered using Amicon® Ultra-15 Centrifugal Filter Units (Ultracel®-100K, Merck Millipore Ltd., Burlington, MA, USA) with 7 mL PBS(−) and centrifuged at 1000× g for 15 min. The flow-through was discarded, and the retentate was washed with 10 mL PBS(−) through a second centrifugation (25 min, 1000× g). SAPSp-iRGD modification was accomplished by a 30 min incubation with 5 mol% stearylated SAPSp-iRGD. The encapsulation efficiency of siRNA was quantified using the RiboGreen® Assay (Thermo Fisher Scientific Inc., Waltham, MA, USA). The particle size and zeta potential were measured using a Zetasizer Nano (Malvern Instruments Ltd., Malvern, Worcestershire, UK).

2.3. Spheroid Penetration of Liposomes

A375 cells were cultured in DMEM supplemented with 10% FBS and 1% Matrigel® in 96-well NanoCulture Plates (SCIVAX Life Sciences, Kanagawa, Japan) for 96 h under humidified 5% CO_2 at 37 °C to facilitate spheroid formation [10]. Subsequently, the spheroids were transferred to PLL-coated glass-bottom dishes and incubated in DMEM with 10% FBS for an additional 24 h. Following washing with PBS(–), the spheroids were treated with DiD-labeled SAPSp-lipo, iRGD-lipo, or SAPSp-iRGD-lipo in a medium at pH 7.4 for 24 h and then fixed with 4% paraformaldehyde. Samples were mounted using VECTASHIELD® with DAPI (Vector Laboratories, Inc., Newark, CA, USA) and examined using a confocal laser scanning microscope (CLSM; Nikon Corporation, Tokyo, Japan). Fluorescence intensity profiles were generated across the x–y planes at a depth of 4 µm using the NIS-Elements software (version 5.42.01; Nikon Corporation, Tokyo, Japan) [10]. To evaluate Neuropilin-1-mediated penetration, spheroids were pretreated with 30 µM EG3287 prior to treatment with DiO-labeled liposomes.

2.4. Immunofluorescent Staining of Neuropilin-1 in Tumors

B16-F1 cells (3.5×10^6 cells) were combined with Matrigel at a 5:1 ratio and subcutaneously injected into the dorsal skin of 5-week-old hairless mice, as previously described [19,20]. Eight days after inoculation, when the mean tumor volume reached 280 mm^3, DiO-labeled liposomes (50 µL) were administered into the tumor. After a 5 h interval, the tumors were excised, embedded in an OCT compound, and cryosectioned at a thickness of 16 µm using a LEICA CM 1100 (Leica Biosystems, Nussloch, Germany). The sections were fixed with 4% paraformaldehyde, permeabilized with PBS containing 1% Triton X-100, and blocked with Protein Block ((Dako, Agilent Technologies, Santa Clara, CA, USA). Subsequently, the sections were incubated overnight at 4 °C with an anti-Neuropilin-1 antibody, followed by staining with an Alexa Fluor® 594 secondary antibody (Invitrogen™, Thermo Fisher Scientific Inc., Waltham, MA, USA) at 4 °C for 1 h. The sections were mounted using DAPI-containing VECTASHIELD® and visualized using CLSM.

2.5. Actin Staining in Cells and Tumor Tissue

A375 cells were seeded onto poly-L-lysine (PLL)-coated chamber slides at a density of 3×10^4 cells per well. Following a wash with PBS(–), the cells were exposed to SAPSp-lipo, iRGD-lipo, or SAPSp-iRGD-lipo in a pH 7.4 medium for one hour. For in vivo analysis, DiO-labeled liposomes were administered into the tumor tissue (50 µL), and the tumors were excised five hours post-injection, embedded in OCT, and cryosectioned to a thickness of 16 µm. Both cellular and tissue samples were fixed with 4% paraformaldehyde, permeabilized with 1% Triton X-100, and stained with rhodamine phalloidin according to the manufacturer's instructions. Samples were mounted with DAPI-containing VECTASHIELD® and analyzed using CLSM to observe actin polymerization.

2.6. ZO-1 Immunostaining

Caco-2 cells were seeded at a density of 2×10^5 cells/well onto 0.002% poly-L-lysine (PLL)-coated 8-well chamber slides and incubated at 37 °C for 5 days, with medium changes every other day using Caco-2 Monolayer Formation Medium (Oriental Yeast Co., Ltd., Tokyo, Japan). After washing with PBS(−), the cells were treated with SAPSp-iRGD-lipo and incubated at 37 °C for 3 h. Following treatment, the cells were fixed with cold methanol, permeabilized with 1% Triton X-100, and blocked with 5% fetal bovine serum (FBS) in 0.1% Triton X-100/PBS(−) for 1 h at room temperature. After washing, the cells were incubated overnight at room temperature with a primary antibody solution containing rabbit polyclonal anti-ZO-1 IgG (4 µg/mL). Cells were then washed three times with 0.1% Triton X-100/PBS(−) and incubated with the secondary antibody, Alexa Fluor® 488 goat anti-rabbit IgG (2 µg/mL), for 1 h at 37 °C in the dark. Finally, the samples were washed three times with 0.1% Triton X-100/PBS(−) and three times with ultrapure water, mounted with Vectashield® Mounting Medium with DAPI (Vector Laboratories, Burlingame, CA, USA), and observed using CLSM.

2.7. Measurement of Transepithelial Electrical Resistance (TEER) in Caco-2 Cell Monolayers

Caco-2 cells were seeded onto 24-well cell culture inserts at a density of 6×10^4 cells/insert and cultured with Caco-2 Monolayer Formation Medium (Oriental Yeast Co., Ltd., Tokyo, Japan). The culture medium was replaced every other day until the transepithelial electrical resistance (TEER) exceeded 500 Ω/cm^2, indicating the formation of a tight monolayer. Subsequently, SAPSp-iRGD-lipo was added to the apical side of the monolayers, and TEER values were measured at 0, 10, 30, 60, and 180 min using a Millicell® ERS-2 Volt-Ohm Meter (Merck Millipore, Darmstadt, Germany). As a positive control, EDTA treatment was used to induce tight junction disruption.

2.8. Biodistribution of Intravenously Administered Liposomes in Tumor-Bearing Mice

A suspension containing A375 cells (2×10^6 cells/mouse) mixed 1:1 with Matrigel® Basement Membrane Matrix (Corning Incorporated, Corning, NY, USA) was subcutaneously injected into the flank of BALB/cSlc-nu/nu mice. Eleven days after implantation, when the tumor volume had reached approximately 100 mm^3, DiR-labeled liposomes (1 mol% DiR; total lipid dose, 1 µmol/mouse) were administered via the tail vein. In vivo biodistribution at 1 h and 24 h post-injection was assessed using the IVIS Lumina LT In Vivo Imaging System (PerkinElmer, Inc./Revvity Inc., Boston, MA, USA). Fluorescence intensities of the whole body and tumor regions were quantified using the system software. In addition, plasma samples collected at 1 h and 24 h after injection were diluted at least 50-fold with ethanol, and the fluorescence intensity of DiR was quantified using a microplate reader (Infinite M Plex, Tecan Group Ltd., Männedorf, Switzerland).

2.9. Cell Association and Intracellular Trafficking of Liposomes

B16-F1 cells (5×10^4 cells/well) were incubated with rhodamine-labeled liposomes in serum-free medium at varying pH levels. Following a 1 h incubation at 4 °C, surface-bound liposomes were evaluated using flow cytometry. For trafficking analysis, cells were cultured in poly-L-lysine (PLL)-coated glass-bottom dishes and incubated with rhodamine-labeled liposomes at 37 °C. After 1 h, the endosomes/lysosomes and nuclei were stained with LysoTracker Green DND-26 and Hoechst 33342 (Invitrogen™, Thermo Fisher Scientific Inc., Waltham, MA, USA), respectively, and intracellular localization was analyzed using CLSM [8].

2.10. Evaluation of Cell Death

A375 cells (2.5×10^3 cells/well) were seeded into poly-L-lysine (PLL)-coated 8-well chamber slides and incubated for 24 h. Following a wash with PBS(–), DMEM(–) was adjusted to pH 7.4, 6.5, or 6.0. SAPSp-iRGD-LNPs encapsulating KIF11 siRNA (final concentration 100 nM; 30 pmol/well) were used, with Lipofectamine 2000 serving as a positive control. After a 4 h incubation period, the medium was replaced with complete DMEM(+), and the cells were incubated for an additional 92 h. The cells were then fixed with 4% paraformaldehyde, washed, and mounted using VECTASHIELD® with DAPI. Cell death was evaluated using CLSM.

2.11. Statistical Analysis

Statistical analyses were conducted using one-way ANOVA, followed by Tukey–Kramer HSD post hoc tests. Statistical significance was set at $p < 0.05$.

3. Results and Discussion

3.1. Comparison of Spheroid Penetration Between SAPSp- and SAPSp-iRGD-lipo

As shown in Table 1, the particle size of SAPSp-iRGD-lipo was approximately 100 nm. The surface charge of SAPSp-iRGD-lipo was measured as -15 mV at pH 7.4, which changed to −5.5 mV at pH 6.5. Similarly to SAPSp-lipo, the surface charge of SAPSp-iRGD-lipo transitioned from negative to near neutral under mildly acidic conditions. These findings indicate that the conjugation of the iRGD peptide sequence to SAPSp via a glycine linker does not compromise the pH-responsive charge-reversal capability of SAPSp [8].

Table 1. Physicochemical properties of SAPSp-iRGD liposomes.

pH	Particle Size (nm)	Polydispersity Index	ζ-potential (mV)
7.4	103 ± 12	0.262 ± 0.001	−15 ± 1.8
6.5	103 ± 11	0.341 ± 0.016	−5.5 ± 1.3

Given that the internal microenvironment of spheroids can vary depending on the cell line employed, we assessed the internal environment of spheroids composed of either A375 or B16-F1 cells using a Hypoxia Probe. This probe demonstrated oxygen-dependent fluorescence quenching, with the fluorescence intensity being indicative of the oxygen concentration [21]. As depicted in Supplementary Figure S1, spheroids composed of A375 cells were more compact than those composed of B16-F1 cells and exhibited a stronger fluorescence of the Hypoxia Probe in their cores than in their peripheries. In addition, although detailed quantitative data on the ECM density of A375 tumors have not been reported, our spheroid model revealed that A375 spheroids exhibited a larger hypoxic core compared with B16-F1 spheroids (Figure S1). Since hypoxia is known to promote ECM remodeling and stiffening through the upregulation of collagen crosslinking enzymes such as lysyl oxidase and prolyl hydroxylase [22,23], this finding suggests that A375 tumors may possess a denser ECM microenvironment than B16-F1 tumors. Consistently, previous histological studies have reported a prominent collagen I and fibronectin deposition in A375-derived melanoma tissues [24,25], supporting the notion that A375 tumors exhibit a highly fibrotic and mechanically stiff ECM.

Based on these observations, we used A375 cell-derived spheroids, which formed more compact structures, to evaluate the spheroid penetration of iRGD-lipo and SAPSp-lipo. In spheroids treated with iRGD-lipo, red fluorescence signals from the liposomes were observed even in the central region, suggesting that iRGD modification enhanced spheroid penetration (Figure 1A). Conversely, in spheroids treated with SAPSp-lipo, the liposomes remained localized near the peripheral region (Figure 1A). However, the spheroids treated with SAPSp-iRGD-lipo also exhibited red fluorescence signals in the central regions, similar to those observed following iRGD-lipo treatment (Figure 1A).

The semi-quantitative analysis based on CLSM images revealed that SAPSp-iRGD-lipo was more widely distributed in the core of the spheroid than SAPSp-lipo (Figure 1B). Even when evaluated using B16-F1 cell-derived spheroids, which form looser structures, SAPSp-iRGD-lipo penetrated deeper than SAPSp-lipo (Supplementary Figure S2). Collectively, these findings suggest that SAPSp-iRGD-lipo demonstrates superior spheroid penetration compared to SAPSp-lipo.

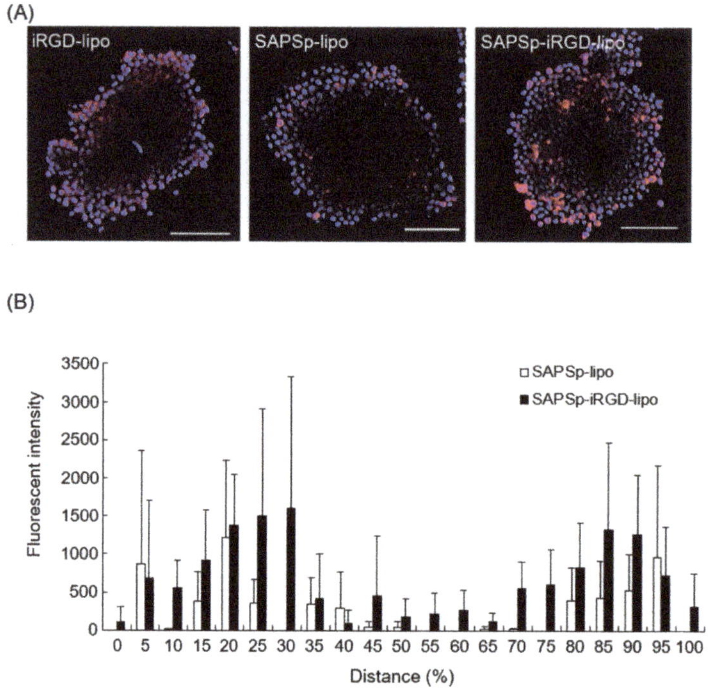

Figure 1. Spheroid penetration of iRGD-lipo and SAPSp-lipo. (**A**) Spheroids composed of A375 cells were treated with DiD-labeled iRGD-lipo, SAPSp-lipo, or SAPSp-iRGD-lipo and subsequently observed after 24 h using CLSM. Blue: nuclei (DAPI); red: liposomes (DiD). Scale bar: 100 μm. (**B**) A quantitative comparison of the spheroid penetration capability at pH 7.4 between SAPSp-lipo and SAPSp-iRGD-lipo was conducted using CLSM images of each sample in the x-y plane of three spheroids. The fluorescence distribution (%) of SAPSp-lipo and the total fluorescence intensity of SAPS-iRGD-lipo (white column) and SAPSp-iRGD-lipo (black column) in the x-y plane were determined at specified constant intervals (%) against the total fluorescence intensity on a line from a certain marginal region to the contralateral marginal region using a plot profile analysis. Data are presented as the mean ± SD from three spheroids. Distance (%) represents the relative position within the spheroid, where 50% corresponds to the spheroid core, and values smaller or larger than 50% indicate regions closer to the periphery.

3.2. *Involvement of Neuropilin-1 in the Penetration of SAPSp-iRGD-lipo*

The iRGD peptide facilitates tissue penetration by activating Neuropilin-1 [26]. To determine whether the enhanced penetration of SAPSp-iRGD-lipo was due to iRGD conjugation, we assessed spheroid penetration in the presence of a Neuropilin-1 inhibitor. A co-treatment with the NRP-1 inhibitor EG3287 markedly suppressed the penetration of both iRGD-lipo and SAPSp-iRGD-lipo into the spheroids (Figure 2A and Supplementary Figure S3). In contrast, the penetration of SAPSp-lipo was unaffected by the inhibitor. These findings indicate that the enhanced spheroid penetration of SAPSp-iRGD-lipo is mediated, at least in part, by an NRP-1-dependent mechanism. To further validate the involvement of NRP-1 in tissue penetration in vivo, we examined the correlation between the liposome distribution and NRP-1 expression following intratumoral injection in tumor-bearing mice. As shown in Figure 2B, both iRGD-lipo and SAPSp-iRGD-lipo exhibited green fluorescence signals that were predominantly localized in regions with a high NRP-1 expression. A quantitative co-localization analysis revealed Pearson's correlation coefficients of 0.464 and 0.442, respectively, indicating a moderate yet specific molecular interaction between the iRGD motif and NRP-1-expressing tumor regions. Collectively, these results demonstrate that the enhanced penetration of SAPSp-iRGD-lipo arises from its NRP-1-mediated, pH-responsive dual functionality, integrating active tumor targeting via iRGD with the pH-dependent charge conversion properties conferred by the SAPSp segment. This dual mechanism likely contributes to the improved intratumoral diffusion observed in both spheroid and tumor models.

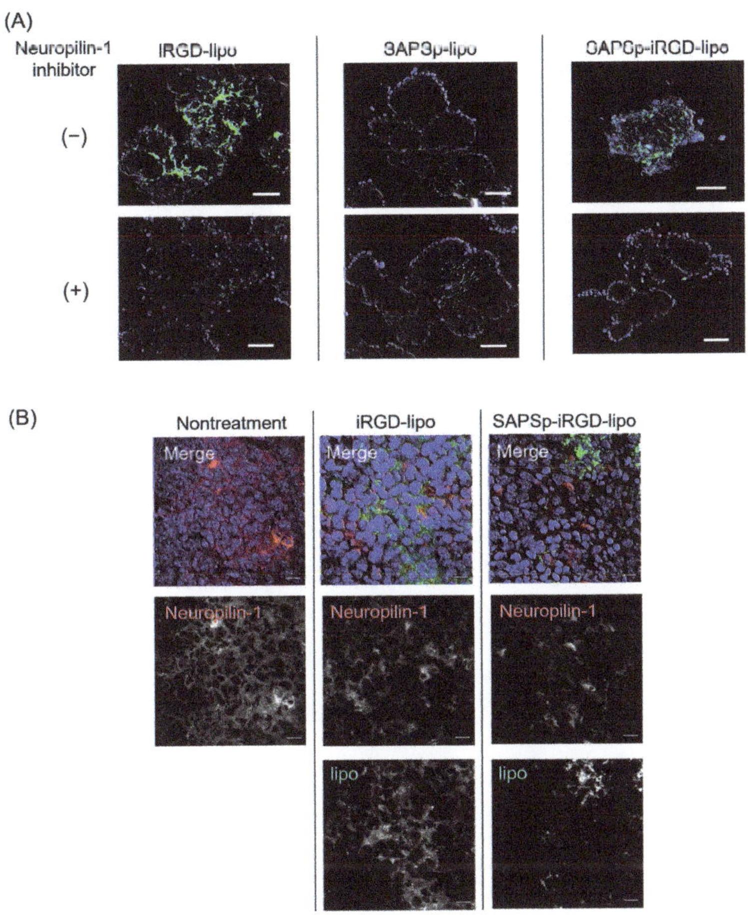

Figure 2. Relationship between the tissue penetration of SAPSp-iRGD-lipo and Neuropilin-1. (**A**) Effect of Neuropilin-1 inhibitor on the spheroid penetration of SAPSp-iRGD-lipo. Spheroids composed of A375 cells were pre-incubated with the Neuropilin-1 inhibitor EG3287 (final concentration: 30 μM) for 30 min. Subsequently, DiO-labeled iRGD-lipo, SAPSp-lipo, or SAPSp-iRGD-lipo was added, and spheroids were observed after 24 h using CLSM. Blue: nuclei (DAPI); green: liposomes (DiO). Scale bar: 100 μm. (**B**) Intratumoral distribution of Neuropilin-1 expression and liposomes. DiO-labeled iRGD-lipo or SAPSp-iRGD-lipo was locally administered into tumors. After 5 h, the tumors were excised, fixed with 4% formaldehyde, and subjected to immunostaining, followed by CLSM. Blue: nuclei (DAPI); green: liposomes (DiO); red: Neuropilin-1 (Alexa Fluor 594). Scale bar: 20 μm.

3.3. *Involvement of Actin Depolymerization in the Tissue Penetration of SAPSp-iRGD-lipo*

In our previous study, we demonstrated that SAPSp-lipo facilitates tissue penetration by inducing actin depolymerization [10]. To ascertain whether actin depolymerization contributed to the enhanced tissue penetration of SAPSp-iRGD-lipo, we investigated actin dynamics in a monolayer culture system.

Actin depolymerization was indicated by a reduction in the red fluorescence of F-actin staining. Consistent with prior findings [10], treatment with SAPSp-lipo resulted in a significant reduction in F-actin fluorescence compared to that in untreated cells, confirming its actin depolymerizing activity. Similarly, SAPSp-iRGD-lipo treatment led to a notable decrease in F-actin staining, suggesting that SAPSp-iRGD-lipo retained the actin depolymerizing activity of SAPSp-lipo. Furthermore, we quantified the fluorescence intensity of F-actin and found that SAPSp-iRGD-lipo treatment induced a stronger reduction in F-actin compared to SAPSp-lipo or iRGD-lipo alone (Figure 3B). The calculated Bliss Excess was 5.6%, indicating a mild synergistic effect between the SAPSp and iRGD motifs (Table 2). To distinguish whether this synergistic effect originated from the covalent (chimeric) linkage of SAPSp and iRGD or from a simple physical combination, we prepared additional liposomes co-modified with SAPSp (5 mol%) and iRGD (5 mol%) without covalent linkage. Their physicochemical properties are summarized in Supplementary Table S1. The ζ-potentials of these co-modified liposomes remained nearly neutral at both pH 7.4 and pH 6.5, contrasting with the pH-dependent charge reversal observed in SAPSp-iRGD-lipo, which shifts from negative (pH 7.4) to positive (pH 6.5). These results indicate that the chimeric linkage between SAPSp and iRGD is essential for integrating pH responsiveness and tumor-penetrating ability within a single molecular design.

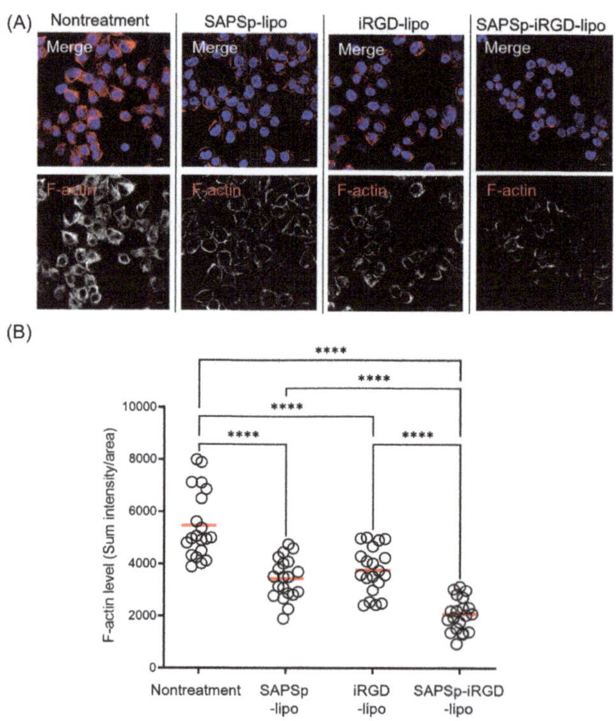

Figure 3. Actin depolymerization induced by SAPSp-iRGD in monolayered cells. A375 cells were exposed to liposomes for 1 h, followed by F-actin staining and observation using CLSM. (**A**) Representative fluorescence images. Blue: nuclei (DAPI); red: F-actin (rhodamine-phalloidin). Scale bar = 10 μm. (**B**) Semi-quantitative analysis of F-actin fluorescence intensity per cell. Each circle represents the fluorescence intensity of an individual cell, and the red bar indicates the mean value. $n = 20$, **** $p < 0.0001$.

Table 2. Comparison between observed and Bliss-predicted actin depolymerization efficiencies following liposome treatments.

Treatment	Mean %	Bliss-Predicted	Bliss Excess	Interaction
SAPSp-lipo	37.2	-	-	-
iRGD-lipo	31.1	-	-	-
SAPSp-iRD-lipo	62.3	56.7	+5.6	Mild synergy

To evaluate the role of actin depolymerization in the in vivo tissue penetration of SAPSp-iRGD-lipo, we analyzed its distribution in tumor-bearing mice following intratumoral injection. Similarly to SAPSp-lipo, SAPSp-iRGD-lipo exhibited a preferential diffusion into tumor regions characterized by a low F-actin expression (Figure 4). To further investigate the spatial relationship between liposomal localization and the actin cytoskeleton, we analyzed the co-localization between F-actin and each liposome formulation. The Pearson's correlation coefficients were 0.218 ± 0.021 for SAPSp-lipo, 0.247 ± 0.028 for iRGD-lipo, and 0.257 ± 0.029 for SAPSp-iRGD-lipo, confirming the reduced spatial overlap between F-actin and SAPSp-iRGD-lipo (Table 3). These results support the notion that actin depolymerization contributes to the enhanced tissue penetration of SAPSp-iRGD-lipo. In conjunction with the results presented in Figures 2–4, these findings indicate that SAPSp-iRGD-lipo achieves substantial tissue penetration through the synergistic mechanisms of SAPSp-mediated actin depolymerization and iRGD-mediated, Neuropilin-1-dependent transcytosis [15,16].

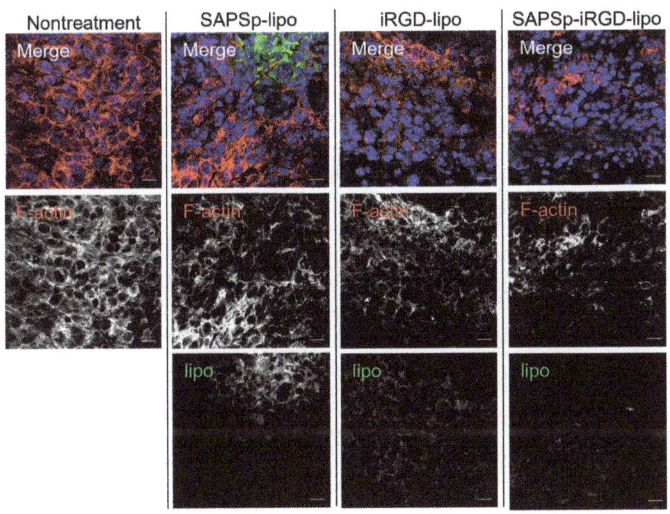

Figure 4. Relationship between intratumoral localization of liposomes and F-actin expression. SAPSp-lipo, iRGD-lipo, and SAPSp-iRGD-lipo were labeled with DiO and locally administered to tumor tissues. After 5 h, tumors were excised, fixed with 4% formaldehyde, and stained with rhodamine-labeled phalloidin for observation using CLSM. Blue: nuclei (DAPI); green: liposomes (DiO); red: F-actin (rhodamine-phalloidin). Scale bar = 20 μm.

Table 3. Evaluation of the co-localization between intratumoral distribution of various liposomes and F-actin-stained regions.

	SAPSp-lipo	iRGD-lipo	SAPSp-iRGD-lipo
Pearson's correlation coefficient	0.218 ± 0.021	0.247 ± 0.028	0.257 ± 0.029

3.4. *Effect of SAPSp-iRGD-lipo on the Intercellular Barrier*

The penetration of nanoparticles into tissues typically occurs via two principal routes: the paracellular pathway, which involves a movement through intercellular gaps, and transcytosis, which entails a passage through the cells [27,28]. Owing to the technical challenges associated with the direct and quantitative assessment of the contribution of transcytosis, our investigation focused on evaluating the role of the paracellular pathway in the penetration of SAPSp-iRGD-lipo. For nanoparticles to navigate tissues via the paracellular route, intercellular junctions must open. To determine whether SAPSp-iRGD-lipo induced the opening of intercellular junctions, we measured the transepithelial electrical resistance (TEER) in Caco-2 monolayers [29]. The results indicated no significant changes in TEER following treatment with either SAPSp-lipo or SAPSp-iRGD-lipo (Figure 5A), suggesting that neither formulation compromised the epithelial barrier integrity.

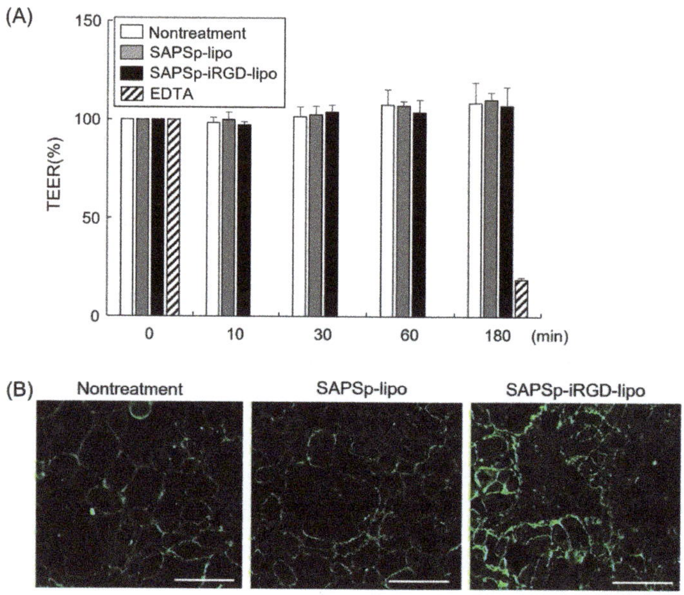

Figure 5. Effect of SAPSp-iRGD-lipo on intercellular barrier. (**A**) Caco-2 cell monolayers were treated with either SAPSp-lipo or SAPSp-iRGD-lipo, and alterations in transepithelial electrical resistance (TEER) were assessed. Data are expressed as percentages relative to the TEER value at 0 min. Data are presented as mean ± SD (n = 3). Treatment with 50 mM EDTA was used as a positive control. (**B**) Caco-2 cell monolayers were treated with either SAPSp-lipo or SAPSp-iRGD-lipo for 3 h. Following immunostaining with an anti-ZO-1 antibody, samples were examined using CLSM. Green: ZO-1, Scale bar: 20 μm.

Furthermore, the disruption of tight junctions correlates with an altered localization of the tight junction protein Zo-1 [30,31]. Consequently, we performed immunostaining using an anti-Zo-1 antibody to assess the tight junction morphology. The findings revealed that Zo-1 localization remained consistent after treatment with either SAPSp-lipo or SAPSp-iRGD-lipo compared to the untreated controls. These results imply that the paracellular pathway does not constitute a major route for the tissue penetration of SAPSp-iRGD-lipo nanoparticles.

3.5. Comparison of Biodistribution of Intravenously Administered Liposomes in Tumor-Bearing Mice

To examine whether SAPSp-iRGD-lipo retains its functional properties and systemic stability in vivo, DiR-labeled liposomes were intravenously administered to tumor-bearing mice, and their biodistribution was analyzed using an IVIS imaging system. As shown in Figure 6A–C, SAPSp-iRGD-lipo exhibited a biodistribution pattern similar to that of SAPSp-lipo. The total fluorescence intensity in mice treated with SAPSp-iRGD-lipo or SAPSp-lipo was lower than that of PEG-lipo, which served as a long-circulating control, and both formulations accumulated predominantly in the liver. A quantitative analysis of plasma samples (Figure 6D) revealed that the blood concentrations of SAPSp-iRGD-lipo and SAPSp-lipo were approximately 20% of the injected dose at 1 h post-injection. Previous pharmacokinetic studies have shown that anionic liposomes typically retain approximately 10–30% of the injected dose in the bloodstream at 1 h post-injection, whereas cationic liposomes exhibit much lower blood levels, often below 10% of the injected dose [32–34]. Consistent with these reports, the blood retention of SAPSp-iRGD-lipo and SAPSp-lipo falls within the range expected for anionic liposomes, suggesting that the surface-modified peptides remain stably associated with the liposomal surface under in vivo conditions. Importantly, both SAPSp-iRGD-lipo and SAPSp-lipo accumulated in tumor tissues as early as 1 h post-injection, whereas PEG-lipo showed a gradual tumor accumulation at 24 h. Taken together, these results demonstrate that SAPSp-iRGD-lipo maintains its in vivo stability and achieves a rapid tumor accumulation through active targeting driven by pH-responsive surface charge conversion, rather than passive accumulation via the enhanced permeability and retention (EPR) effect.

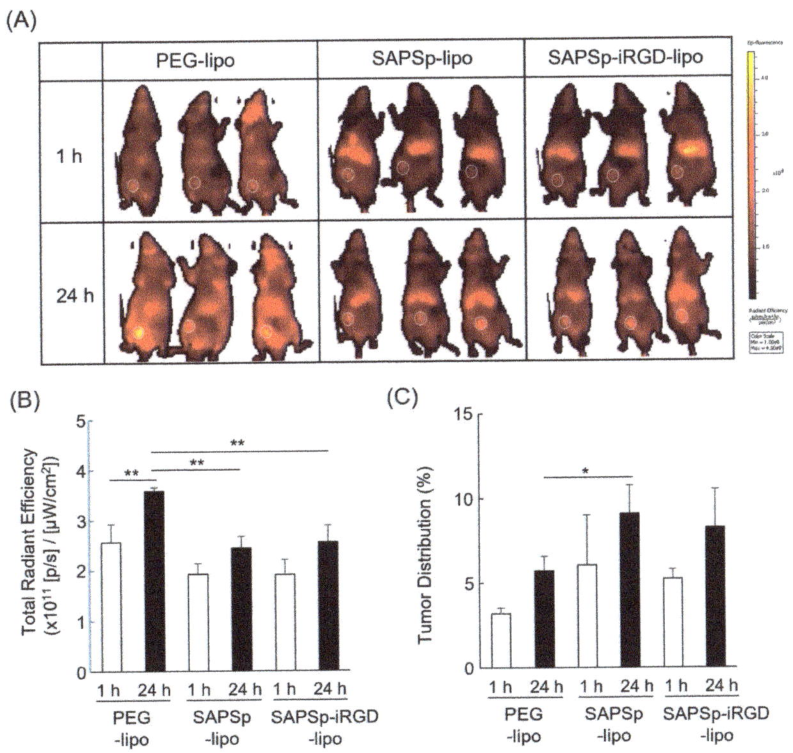

Figure 6. *Cont.*

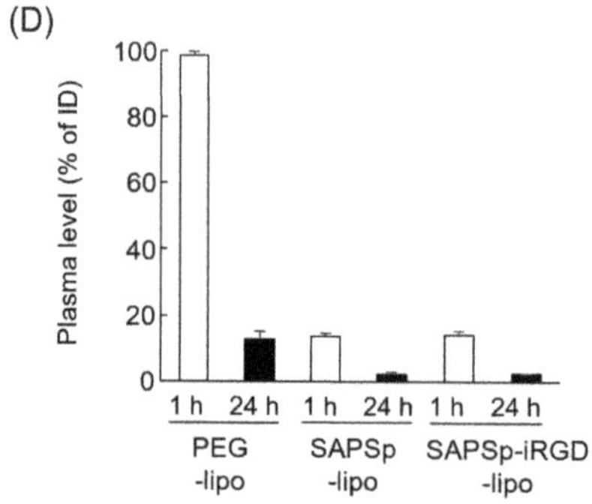

Figure 6. Biodistribution of intravenously administered liposomes in tumor-bearing mice. A375 tumor-bearing mice were intravenously injected with DiR-labeled liposomes. In vivo fluorescence images were acquired at 1 h and 24 h post-injection using an IVIS Lumina LT imaging system. (**A**) Representative fluorescence images. White circles indicate tumor regions. (**B**,**C**) Quantification of fluorescence intensities in whole body and tumor regions, respectively, using the system software. (**D**) Plasma fluorescence at 1 h and 24 h was measured using a microplate reader. n = 3, * $p < 0.05$, ** $p < 0.01$.

3.6. Cytosolic Delivery of Cargo by SAPSp-iRGD-Modified Nanoparticles

Our previous research demonstrated that SAPSp-modified liposomes respond to a mildly acidic tumor microenvironment, facilitating cargo delivery into the cytosol [8]. To ascertain whether SAPSp-iRGD-lipo exhibited a pH-responsive cytosolic delivery capability akin to that of SAPSp-lipo, we examined its cellular uptake and intracellular localization. As illustrated in Figure 7A, SAPSp-iRGD-lipo displayed a significantly enhanced cellular uptake at pH 6.5 compared to that at pH 7.4, with the magnitude of this increase comparable to that observed with SAPSp-lipo [8]. Furthermore, confocal laser scanning microscopy (CLSM) revealed a minimal red fluorescence (indicative of liposome presence) within cells at pH 7.4, whereas a pronounced intracellular red fluorescence was observed at pH 6.5, indicating an enhanced uptake under mildly acidic conditions. Importantly, the red fluorescent signal from SAPSp-iRGD-lipo did not colocalize with endosomal or lysosomal markers at pH 6.5, suggesting a successful delivery to the cytosol rather than a retention in endocytic compartments [8]. These findings suggest that SAPSp-iRGD-lipo, similar to SAPSp-lipo, possesses a pH-responsive cellular uptake and cytosolic delivery capabilities in mildly acidic conditions.

To further verify whether SAPSp-iRGD modification confers a cytosolic delivery ability, we prepared lipid nanoparticles (LNPs) encapsulating apoptosis-inducing siRNA and modified their surfaces with SAPSp-iRGD (SAPSp-iRGD-LNPs). Kinesin family member 11 (KIF11), a motor protein involved in mitosis, plays a significant role in cancer cells by influencing tumor progression, prognosis, and therapeutic responses [35]. The inhibition of KIF-11 activates the intrinsic apoptotic pathway [36]. In this study, anti-KIF11 siRNA was encapsulated in LNPs. As presented in Table 4, SAPSp-iRGD-LNPs encapsulating either anti-KIF11 or control siRNA exhibited a surface charge shift from negative to positive as the pH decreased from 7.4 to 6.5, similar to that of SAPSp-iRGD-lipo. Although the encapsulation efficiencies of anti-KIF11 siRNA (85.0%) and control siRNA (72.7%) differed slightly, both formulations showed comparable particle sizes and ζ-potentials, indicating that the physicochemical properties and pH-responsive charge-reversal behavior of SAPSp-iRGD-LNPs were unaffected by the type of encapsulated siRNA. To eliminate any influence of encapsulation efficiency on the cytotoxicity results, the siRNA dose was standardized to

30 pmol per well for both formulations. Therefore, the difference in cytotoxicity observed in Figure 7C can be attributed to the gene-specific knockdown of KIF11 rather than to variations in the encapsulation efficiency. Moreover, as depicted in Supplementary Figure S4, fluorescently labeled siRNA-loaded SAPSp-iRGD-LNPs demonstrated a greater spheroid penetration than SAPSp-LNPs, indicating that SAPSp-iRGD enhanced tissue penetration even when conjugated to structurally more rigid LNPs than liposomes. As shown in Figure 7C, under mildly acidic conditions (pH 6.5 and 6.0), the SAPSp-iRGD-LNP-mediated delivery of anti-KIF11 siRNA resulted in a significant induction of cell death, whereas the control siRNA delivery did not induce cell death under these conditions. To further compare the siRNA delivery performance between SAPSp-LNP and SAPSp-iRGD-LNP, fluorescently labeled siRNA-loaded nanoparticles were analyzed using CLSM. Both formulations exhibited a markedly enhanced cellular uptake under acidic conditions (pH $\leq$ 6.5; Supplementary Figure S5). Notably, the intracellular fluorescence of siRNA delivered by SAPSp-iRGD-LNP was slightly lower than that of SAPSp-LNP, suggesting a marginally reduced in vitro delivery efficiency and consequently a weaker gene-knockdown effect. However, considering that SAPSp-iRGD-LNP demonstrated a superior tissue penetration in spheroids (Supplementary Figure S4) and tumors, its enhanced intratumoral distribution is expected to compensate for this minor difference in cellular uptake, potentially yielding a higher overall antitumor gene-silencing effect in vivo. To further evaluate the therapeutic potential of SAPSp-iRGD-LNPs, we examined their biodistribution and tumor-delivery behavior following intravenous administration. As shown in Figure 6, both SAPSp-lipo and SAPSp-iRGD-lipo achieved an efficient tumor delivery, although their overall accumulation was lower than that of PEG-lipo. These results indicate that SAPSp-based systems exhibit active targeting properties mediated by their pH-responsive surface charge conversion, yet further optimization of their pharmacokinetic and tumor-delivery profiles will be necessary to translate their enhanced penetration into a measurable in vivo therapeutic efficacy. Given these findings, we consider the present results as a foundational step toward developing an optimized SAPSp-iRGD-LNP formulation capable of demonstrating improved antitumor outcomes in future studies. In comparison with previous reports, various nanoparticle systems have been developed to enhance intratumoral delivery, including pH-responsive charge-conversion systems [37,38], tumor-penetrating peptide-modified carriers [39,40], and integrin-targeting or ECM-degrading approaches [41,42]. These formulations typically rely on either (i) pH-triggered activation or (ii) receptor-mediated internalization to promote tumor penetration. In contrast, the present SAPSp-iRGD system integrates both mechanisms within a single chimeric peptide, combining the slightly acidic pH-responsive charge reversal of SAPSp with the Neuropilin-1-dependent tissue penetration of iRGD. This design differs from previously reported systems using mixed or co-administered peptides [43,44], where cooperative functionality can be limited by a heterogeneous surface presentation. In our study, SAPSp-iRGD-LNPs exhibited a coordinated pH-responsiveness and active tumor penetration without altering the particle size or peptide density, demonstrating that the covalent linkage of SAPSp and iRGD yields a synergistic functionality. Thus, SAPSp-iRGD-LNPs represent a distinct class of dual-functional, microenvironment-responsive nanocarriers that unify pH-triggered activation and receptor-mediated penetration, providing a mechanistic bridge to extend existing tumor penetration strategies.

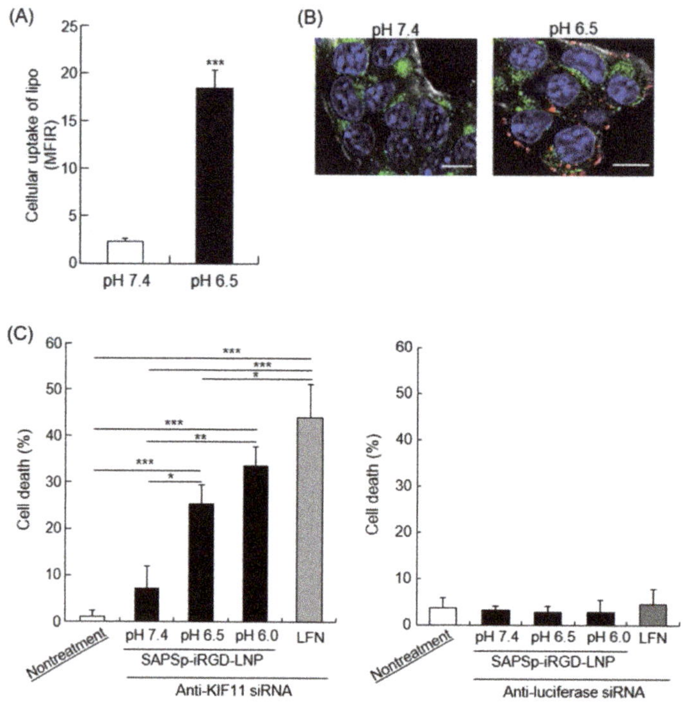

Figure 7. Intracellular trafficking and cargo delivery of SAPSp-iRGD-lipo. (**A**) Cells were exposed to rhodamine-labeled SAPSp-iRGD-lipo at pH 7.4 or 6.5 for 1 h at 4 °C, and the binding of liposomes to the cells was assessed via flow cytometry. (**B**) Cells were treated with rhodamine-labeled SAPSp-iRGD-lipo at pH 7.4 or 6.5 for 1 h at 37 °C, and the intracellular distribution of the liposomes was examined using CLSM. Blue: nuclei (Hoechst 33342); green: endosomes/lysosomes (LysoTracker Green DND-26); red: SAPSp-iRGD. Scale bar: 10 µm. (**C**) Induction of cell death through anti-KIF11 siRNA delivery by SAPSp-iRGD-LNPs. Cells were treated with SAPSp-iRGD-LNPs encapsulating anti-KIF11 siRNA at pH 7.4, 6.5, and 6.0 for 96 h. After fixation with 4% paraformaldehyde, the nuclei were stained with DAPI and observed using CLSM. Apoptotic cells in the CLSM images were quantified, and their percentages relative to the total number of cells were calculated. Anti-luciferase siRNA was used as a control siRNA. Data are presented as mean ± SD. (* $p < 0.05$; ** $p < 0.01$; *** $p < 0.001$; n = 3).

Table 4. Physicochemical properties of SAPSp-iRGD-LNP encapsulating siRNAs.

siRNA Encapsulated in LNPs	Encapsulation (%)	pH	Particle Size (nm)	Polydispersity Index	ζ-Potential (mV)
Anti-KIF11 siRNA	85.0	7.4	73.4 ± 9.2	0.353 ± 0.080	−16 ± 2.5
		6.5	74.5 ± 11	0.353 ± 0.053	−6.7 ± 1.9
Control siRNA	72.7	7.4	70.0 ± 5.1	0.297 ± 0.061	−13 ± 3.1
		6.5	72.9 ± 21	0.312 ± 0.119	−6.4 ± 3.2

4. Conclusions

To enhance the tissue penetration capability of the tumor pH-responsive peptide SAPSp, we developed an innovative peptide, SAPSp-iRGD, by conjugating the tissue-penetrating peptide sequence iRGD to the C-terminus of SAPSp. Nanoparticles modified with SAPSp-iRGD retained the pH-responsive properties of SAPSp and demonstrated a significantly improved tissue penetration. Moreover, our findings indicate that this enhanced penetration involves both neuropilin-1-mediated pathways and actin depolymerization. Owing to their combined tumor-specific tissue penetration and mildly acidic pH responsiveness, SAPSp-iRGD-modified nanoparticles represent a promising novel delivery system for targeted nucleic acid delivery into deep tumor regions. Although SAPSp-iRGD-modified nanoparticles demonstrated an enhanced tumor penetration and pH-responsive delivery, further studies are required to evaluate their pharmacokinetic profile and therapeutic efficacy in multiple tumor models. In addition, the optimization of peptide density and assessment of safety and large-scale production will be important for future clinical translation.

Supplementary Materials: The following supporting information can be downloaded at https://www.mdpi.com/article/10.3390/nano15221695/s1. Table S1: Physicochemical properties of liposomes co-modified with stearyl-SAPSp and stearyl-iRGD; Figure S1: Comparison of the hypoxic microenvironment within spheroids composed of A375 cells and those composed of B16-F1 cells; Figure S2: Quantitative comparison of the permeability of SAPSp-lipo and SAPSp-iRGD-lipo in B16-F1 cell-derived spheroids; Figure S3: Surface intensity plot showing the effect of Neuropilin-1 inhibition on the spheroid penetration of SAPSp-iRGD-lipo; Figure S4: Spheroid penetration of SAPSp-LNP and SAPSp-iRGD-LNP. Figure S5: Cellular uptake of SAPSp-LNP and SAPSp-iRGD-LNP.

Author Contributions: Conceptualization, S.H.; methodology, S.H.; validation, S.H.; formal analysis, S.H.; investigation, K.K., R.M., N.I., Y.T., H.F., K.-I.T., and S.H.; data curation, S.H.; writing—original draft preparation, S.H.; writing—review and editing, S.H.; visualization, S.H.; supervision, S.H.; funding acquisition, S.H. All authors have read and agreed to the published version of the manuscript.

Funding: This study was supported by a Grant-in-Aid for Scientific Research from the Japan Society for the Promotion of Science (Kakenhi Kiban (C) 23K11858) and a DAIGAKUTOKUBETSU KENKYUHI Grant (Musashino University).

Institutional Review Board Statement: The animal study protocol was approved by the Institutional Animal Care and Use Committee of Musashino University (protocol code 08-A-2024 and 08-A-2025, approval date not available).

Data Availability Statement: The original contributions presented in this study are included in the article/Supplementary Materials. Further inquiries can be directed to the corresponding author.

Conflicts of Interest: The authors declare no competing financial conflicts of interest.

Abbreviations

The following abbreviations are used in this manuscript:

lipo	Liposomes
SAPSp	Slightly acidic pH-sensitive peptides
ECM	Extracellular matrix
iRGD	Internalizing RGD
LNPs	Lipid nanoparticles
PEG	Polyethylene glycol
EPR	Enhanced permeability and retention
KIF11	Kinesin family member 11
CLSM	Confocal laser scanning microscopy
TEER	Transepithelial electrical resistance

References

1. Macewan, S.R.; Chilkoti, A. From Composition to Cure: A Systems Engineering Approach to Anticancer Drug Carriers. *Angew. Chem.* **2017**, *56*, 6712–6733. [CrossRef]

2. Wu, J. Enhanced Permeability and Retention (EPR) Effect: Significance of the Concept and Methods to Enhance Its Application. *J. Pers. Med.* **2021**, *11*, 771. [CrossRef]
3. Sebak, A.A. Limitations of Pegylated Nanocarriers: Unfavourable Physicochemical Properties, Biodistribution Patterns and Cellular and Subcellular Fates. *Int. J. Appl. Pharm.* **2018**, *10*, 6. [CrossRef]
4. Ishida, T.; Kiwada, H. Accelerated Blood Clearance (ABC) Phenomenon upon Repeated Injection of PEGylated Liposomes. *Int. J. Pharm.* **2007**, *354*, 56–62. [CrossRef]
5. Gautam, B.; Hsiao, J.-C.; Yu, H.-H.; Luo, C.-H.; Lee, H.-M.; Tseng, H.-R.; Gao, H.-D. Nano-On-Nano: Responsive Nanosubstrate-Mediated Liposome Delivery with High Cellular Uptake Efficiency. *ACS Appl. Bio Mater.* **2023**, *6*, 1611–1620. [CrossRef]
6. Madni, A.; Sarfraz, M.; Rehman, M.; Ahmad, M.; Akhtar, N.; Ahmad, S.; Tahir, N.; Ijaz, S.; Al-Kassas, R.; Löbenberg, R. Liposomal drug delivery: A versatile platform for challenging clinical applications. *J. Pharm. Pharm. Sci.* **2014**, *17*, 401–426. [CrossRef]
7. Basak, S.; Das, T.K. Liposome-Based Drug Delivery Systems: From Laboratory Research to Industrial Production—Instruments and Challenges. *ChemEngineering* **2025**, *9*, 56. [CrossRef]
8. Hama, S.; Itakura, S.; Nakai, M.; Nakayama, K.; Morimoto, S.; Suzuki, S.; Kogure, K. Overcoming the Polyethylene Glycol Dilemma via Pathological Environment-Sensitive Change of the Surface Property of Nanoparticles for Cellular Entry. *J. Control. Release* **2015**, *206*, 67–74. [CrossRef]
9. Itakura, S.; Hama, S.; Matsui, R.; Kogure, K. Effective Cytoplasmic Release of SiRNA from Liposomal Carriers by Controlling the Electrostatic Interaction of SiRNA with a Charge-Invertible Peptide in Response to Cytoplasmic PH. *Nanoscale* **2016**, *8*, 10649–10658. [CrossRef]
10. Suzuki, S.; Nakayama, K.; Kogure, K.; Hama, S.; Matsui, R.; Nishi, T.; Itakura, S.; Nishimoto, A. Tumor Microenvironment-Sensitive Liposomes Penetrate Tumor Tissue via Attenuated Interaction of the Extracellular Matrix and Tumor Cells and Accompanying Actin Depolymerization. *Biomacromolecules* **2017**, *18*, 535–543. [CrossRef]
11. Sugahara, K.N.; Teesalu, T.; Karmali, P.P.; Kotamraju, V.R.; Agemy, L.; Greenwald, D.R.; Ruoslahti, E. Tissue-Penetrating Delivery of Compounds and Nanoparticles into Tumors. *Biomaterials* **2014**, *35*, 6829–6838. [CrossRef]
12. Teesalu, T.; Sugahara, K.N.; Ruoslahti, E. Tumor-Penetrating Peptides. *Adv. Healthc. Mater.* **2015**, *4*, 2699–2707. [CrossRef]
13. Cun, X.; Chen, J.; He, Q.; Gao, H.; Zhang, L.; Ruan, S.; Wan, J. A Novel Strategy Combining IRGD Peptide with Tumor-Microenvironment-Responsive and Multistage Nanoparticles for Deep Tumor Penetration. *ACS Appl. Mater. Interfaces* **2015**, *7*, 27458–27466. [CrossRef]
14. Kang, S.; Lee, S.; Park, S. IRGD Peptide as a Tumor-Penetrating Enhancer for Tumor-Targeted Drug Delivery. *Polymers* **2020**, *12*, 1906. [CrossRef]
15. Liu, X.; Jiang, J.; Nel, A.E.; Meng, H. Major Effect of Transcytosis on Nanodrug Delivery to Pancreatic Cancer. *Mol. Cell. Oncol.* **2017**, *4*, e1335273. [CrossRef]
16. Mamnoon, B.; Loganathan, J.; Sathish, V.; Froberg, J.; Choi, Y.; Confeld, M.I.; De Fonseka, N.; Mallik, S.; Tuvin, D.M.; Feng, L. Targeted Polymeric Nanoparticles for Drug Delivery to Hypoxic, Triple-Negative Breast Tumors. *ACS Appl. Bio Mater.* **2020**, *4*, 1450–1460. [CrossRef]
17. Hama, S.; Sakai, M.; Itakura, S.; Majima, E.; Kogure, K. Rapid Modification of Antibodies on the Surface of Liposomes Composed of High-Affinity Protein A-Conjugated Phospholipids for Selective Drug Delivery. *Biochem. Biophys. Rep.* **2021**, *27*, 101067. [CrossRef]
18. Shirane, D.; Nakai, Y.; Akita, H.; Tanaka, H.; Yoshioka, H. Development of an Alcohol Dilution-Lyophilization Method for Preparing Lipid Nanoparticles Containing Encapsulated siRNA. *Biol. Pharm. Bull.* **2018**, *41*, 1291–1294. [CrossRef]
19. Nakamura, I.; Tabuchi, Y.; Itakura, S.; Hama, S.; Nishi, T.; Kogure, K.; Takasaki, I. Lipocalin2 as a Plasma Marker for Tumors with Hypoxic Regions. *Sci. Rep.* **2014**, *4*, 7235. [CrossRef] [PubMed]
20. Hama, S.; Utsumi, S.; Fukuda, Y.; Nakayama, K.; Okamura, Y.; Tsuchiya, H.; Fukuzawa, K.; Harashima, H.; Kogure, K. Development of a Novel Drug Delivery System Consisting of an Antitumor Agent, Tocopheryl Succinate. *J. Control. Release* **2012**, *161*, 843–851. [CrossRef]
21. Zhang, S.; Hosaka, M.; Negishi, K.; Iida, Y.; Tobita, S.; Yoshihara, T.; Takeuchi, T. Phosphorescent Light–Emitting Iridium Complexes Serve as a Hypoxia-Sensing Probe for Tumor Imaging in Living Animals. *Cancer Res.* **2010**, *70*, 4490–4498. [CrossRef]
22. de Jong, O.G.; van Balkom, B.W.M.; Gremmels, H.; Verhaar, M.C. Exosomes from hypoxic endothelial cells have increased collagen crosslinking activity through up-regulation of lysyl oxidase-like 2. *J. Cell. Mol. Med.* **2016**, *20*, 342–350. [CrossRef]
23. Gilkes, D.M.; Semenza, G.L.; Wirtz, D. Hypoxia and the Extracellular Matrix: Drivers of Tumour Metastasis. *Nat. Rev. Cancer* **2014**, *14*, 430–439. [CrossRef]
24. Marusak, C.; Thakur, V.; Thakur, A.; Beier, U.H.; Nussenzweig, A.; Khanna, C.; Camphausen, K.A.; Citrin, D.E.; O'Connell, M.P.; Weeraratna, A.T.; et al. Matrix Metalloproteinase-Mediated ECM Remodeling Drives Melanoma Resistance to BRAF Inhibition. *Clin. Cancer Res.* **2020**, *26*, 6039–6051. [CrossRef]

25. Shoval, H.; Kalo, E.; Benayahu, D.; Zipori, D.; Gazit, Z. Melanoma Cells Affect the Fibroblastic Microenvironment and Alter Extracellular Matrix Organization. *Sci. Rep.* **2017**, *7*, 7690.
26. Teesalu, T.; Kotamraju, V.R.; Ruoslahti, E.; Sugahara, K.N. C-End Rule Peptides Mediate Neuropilin-1-Dependent Cell, Vascular, and Tissue Penetration. *Proc. Natl. Acad. Sci. USA* **2009**, *106*, 16157–16162. [CrossRef]
27. Chen, W.; Gu, Z.; Chu, D.; Mao, G.; Paterson, D.J.; Tsao, S.C.; Xie, Z.; Sun, B.; Wang, W.; Centurion, F.; et al. Size-Dependent Penetration of Nanoparticles in Tumor Spheroids: A Multidimensional and Quantitative Study of Transcellular and Paracellular Pathways. *Small* **2023**, *20*, e2304693. [CrossRef]
28. Wang, Z.-H.; Zhang, Z.; Shi, J.; Yang, M.; Zeng, X.; Yang, Y.; Li, A.; Liu, H.; Liu, J.; Zhao, F.; et al. Bioactive Nanomotors Enable Efficient Intestinal Barrier Penetration for Colorectal Cancer Therapy. *Nat. Commun.* **2025**, *16*, 1678. [CrossRef]
29. Hama, S.; Kimura, Y.; Mikami, A.; Shiota, K.; Toyoda, M.; Tamura, A.; Nagasaki, Y.; Kanamura, K.; Kajimoto, K.; Kogure, K. Electric Stimulus Opens Intercellular Spaces in Skin. *J. Biol. Chem.* **2014**, *289*, 2450–2456. [CrossRef]
30. Yu, D.; Weber, C.R.; Raleigh, D.R.; Shen, L.; Marchiando, A.M.; Wang, Y.; Turner, J.R. MLCK-Dependent Exchange and Actin Binding Region-Dependent Anchoring of ZO-1 Regulate Tight Junction Barrier Function. *Proc. Natl. Acad. Sci. USA* **2010**, *107*, 8237–8241. [CrossRef]
31. Van Itallie, C.M.; Bridges, A.; Fanning, A.S.; Anderson, J.M. ZO-1 Stabilizes the Tight Junction Solute Barrier through Coupling to the Perijunctional Cytoskeleton. *Mol. Biol. Cell* **2009**, *20*, 3930–3940. [CrossRef]
32. Ernsting, M.J.; Murakami, M.; Roy, A.; Li, S.D. Factors controlling the pharmacokinetics, biodistribution and intratumoral penetration of nanoparticles. *J. Control. Release* **2013**, *172*, 782–794. [CrossRef]
33. Romberg, B.; Oussoren, C.; Snel, C.J.; Carstens, M.G.; Hennink, W.E.; Storm, G. Pharmacokinetics of poly(hydroxyethyl-L-asparagine)-coated liposomes is superior to PEG-liposomes at low lipid dose and upon repeated administration. *Biochim. Biophys. Acta Biomembr.* **2007**, *1768*, 737–743. [CrossRef]
34. Qi, J.; Yao, P.; He, F.; Yu, D.; Guo, Y. Blood clearance and biodistribution of cationic and anionic liposomes after intravenous injection in mice. *Colloids Surf. B Biointerfaces* **2016**, *141*, 93–100. [CrossRef]
35. Tao, W.; South, V.J.; Zhang, Y.; Davide, J.P.; Farrell, L.; Kohl, N.E.; Sepp-Lorenzino, L.; Lobell, R.B. Induction of Apoptosis by an Inhibitor of the Mitotic Kinesin KSP Requires Both Activation of the Spindle Assembly Checkpoint and Mitotic Slippage. *Cancer Cell* **2005**, *8*, 49–59. [CrossRef]
36. Tao, W.; South, V.J.; Diehl, R.E.; Davide, J.P.; Sepp-Lorenzino, L.; Fraley, M.E.; Arrington, K.L.; Lobell, R.B. An Inhibitor of Kinesin Spindle Protein Activates the Intrinsic Apoptotic Pathway Independently of P53 and de Novo Protein Synthesis. *Mol. Cell. Biol.* **2007**, *27*, 689–698. [CrossRef]
37. Zhang, Y.; Wang, F.; Li, M.; Yu, Z.; Qi, R.; Ding, J.; Chen, X. Engineering pH-Responsive Charge-Reversal Nanoparticles for Enhanced Tumor-Targeted Drug Delivery. *Biomaterials* **2021**, *276*, 120737. [CrossRef]
38. Wang, S.; Xu, H.; Li, Z.; Shen, S.; Li, J.; Ye, J.; Zhao, Z.; Guo, S.; Zhang, Y.; Liu, Y. Tumor Microenvironment-Activated Charge-Reversal Nanocarriers for Improved Cancer Therapy. *Bioact. Mater.* **2021**, *6*, 3610–3622. [CrossRef]
39. Dai, W.; Zhou, K.; Jiang, B.; Li, L.; Wang, X.; Ji, J.; Liu, Y. A comprehensive study of iRGD-modified liposomes with improved chemotherapeutic efficacy on B16 melanoma. *J. Pharm. Pharm. Sci.* **2015**, *22*, 10–20. [CrossRef]
40. Du, R.; Jiang, P.; Li, H.; Kong, F.; Kong, D.; Li, Y.; Sun, X.; Zhao, Y. Antitumor effect of iRGD-modified liposomes containing conjugated linoleic acid–paclitaxel. *Int. J. Nanomed.* **2014**, *9*, 5179–5195. [CrossRef]
41. Nikitovic, D.; Berdiaki, A.; Spyridaki, I.; Papoutsidakis, A.; Tsatsakis, A.; Tzanakakis, G.N. Enhancing Tumor Targeted Therapy: The Role of iRGD Peptide in Nanocarriers. *Cancers* **2024**, *16*, 3768. [CrossRef]
42. Zinger, A.; Koren, L.; Adir, O.; Poley, M.; Alyan, M.; Yaari, Z.; Noor, N.; Krinsky, N.; Simon, A.; Gibori, H.; et al. Collagenase Nanoparticles Enhance the Penetration of Drugs into Pancreatic Tumors. *ACS Nano* **2019**, *13*, 11008–11021. [CrossRef]
43. Sugahara, K.N.; Teesalu, T.; Karmali, P.P.; Kotamraju, V.R.; Agemy, L.; Greenwald, D.R.; Ruoslahti, E. Coadministration of a Tumor-Penetrating Peptide Enhances the Efficacy of Cancer Drugs. *Science* **2010**, *328*, 1031–1035. [CrossRef]
44. Amin, M.; Mansourian, M.; Burgers, P.C.; Amin, B.; Jaafari, M.R.; ten Hagen, T.L.M. Increased Targeting Area in Tumors by Dual-Ligand Modification of Liposomes with RGD and TAT Peptides. *Pharmaceutics* **2022**, *14*, 458. [CrossRef]

nanomaterials

MDPI

Article

Chrysin-Loaded Micelles Regulate Cell Cycle and Induce Intrinsic and Extrinsic Apoptosis in Ovarian Cancer Cells

Serife Cakir [1,2], Ummugulsum Yildiz [1,2], Turgay Yildirim [3] and Omer Aydin [1,2,4,5,*]

1 Department of Biomedical Engineering, Erciyes University, 38039 Kayseri, Turkey; genkoksrf@gmail.com (S.C.); gulsum.ugy97@gmail.com (U.Y.)
2 NanoThera Lab, Drug Application and Research Center (ERFARMA), Erciyes University, 38039 Kayseri, Turkey
3 Department of Chemical and Biomolecular Engineering, Vanderbilt University, Nashville, TN 37235, USA; turgay.yildirim@vanderbilt.edu
4 Nanotechnology Research and Application Center (ERNAM), Erciyes University, 38039 Kayseri, Turkey
5 Clinical Engineering Research and Implementation Center (ERKAM), Erciyes University, 38039 Kayseri, Turkey
* Correspondence: omeraydin@erciyes.edu.tr or biomer@umich.edu; Tel.: +90-352-207-6666 (ext. 32984)

Abstract

Effective intracellular delivery for ovarian cancer therapy remains a significant challenge. We present chrysin-loaded p(MMA-co-DMAEMA)-b-(OEGMA-co-DMA), PMOD-Chr, a nanoparticle platform precisely engineered via RAFT polymerization for advanced therapeutic delivery. This multi-functional platform features a hydrophobic p(MMA) core encapsulating chrysin (Chr), a pH-responsive p(DMAEMA) segment for endosomal escape, and a hydrophilic OEGMA (Oligo(ethylene glycol) methyl ether methacrylate) shell functionalized for enhanced cellular affinity and systemic stability. The combination of OEGMA and DMA (Dopamine methacrylamide) block facilitates passive targeting of ovarian cancer cells, enhancing internalization. Nanoparticles prepared via the nanoprecipitation method exhibited ~220 nm, demonstrating effective size modulation along with high homogeneity and spherical morphology. In A2780 and OVCAR3 ovarian cancer cells, PMOD-Chr demonstrated significantly enhanced cytotoxicity, substantially lowering the effective IC_{50} dose of Chr. Mechanistically, PMOD-Chr induced a potent G2/M cell cycle arrest, driven by the upregulation of the CDK1/Cyclin B1 complex. Furthermore, the formulation potently triggered programmed cell death by concurrently activating both the intrinsic apoptotic pathway, evidenced by the modulation of Bax, Bcl2, and caspase 9, and the extrinsic pathway involving caspase 8. These findings emphasize that precision engineering via RAFT polymerization enables the creation of sophisticated, multi-stage nanomedicines that effectively overcome key delivery barriers, offering a highly promising targeted strategy for ovarian cancer.

Keywords: raft polymerization; nanoprecipitation; chrysin; cell cycle; apoptosis; ovarian cancer

1. Introduction

Ovarian cancer is among the most fatal gynecologic malignancies worldwide, largely because the disease is typically asymptomatic until advanced stages and rapidly acquires resistance to standard therapies [1,2]. During the past two decades, clinical management has expanded beyond cytoreductive surgery and platinum-based chemotherapy to include targeted agents such as poly (ADP-ribose)-polymerase (PARP) inhibitors and anti-angiogenic drugs [3]. Nevertheless, five-year survival remains below 50%, and most patients experience recurrence driven by chemo- and PARP-inhibitor resistance [4].

With more than 300,000 new cases diagnosed globally each year, the limitations of current regimens underscore an urgent need for safer, more effective therapeutic strategies [1,2]. Standard chemotherapeutics and targeted inhibitors suffer from systemic toxicity, low selectivity, and dose-limiting adverse effects, while tumor heterogeneity continuously fuels resistance mechanisms [5,6]. Immune-checkpoint blockade has yielded only modest and inconsistent responses in ovarian cancer, often requiring complex combinatorial schedules to achieve measurable benefit [7,8]. These shortcomings collectively highlight the need to repurpose bioactive compounds with multi-target anticancer properties and to optimize their delivery.

In recent years, several natural compounds—such as curcumin, resveratrol, quercetin, and berberine—have been explored for their anticancer activities in cancers including ovarian cancer, owing to their ability to modulate key pathways [9–11]. These bioactive agents offer low systemic toxicity and multi-targeted mechanisms of action, further encouraging the search for phytochemical-based nanotherapeutics [12,13].

Among the natural compounds gaining attention, chrysin (Chr), a flavonoid found in passionflower, propolis, and honey, has emerged as a promising candidate due to its diverse pharmacological profile [14,15]. In preclinical models, chrysin has demonstrated significant anticancer activity, notably through the induction of apoptosis and inhibition of cell proliferation [16]. Mechanistically, it modulates key oncogenic signaling pathways such as NF-κB [17], MAPK [18], and PI3K/Akt [19,20] pathways intimately involved in cell growth, survival, and inflammation regulation [21]. Furthermore, chrysin's pro-apoptotic effects have been linked to the generation of intracellular reactive oxygen species (ROS), leading to oxidative stress and downstream activation of cell death mechanisms [22]. Additional studies have shown its role in disturbing calcium homeostasis and triggering mitochondrial dysfunction, including the loss of mitochondrial membrane potential, an essential step in initiating the intrinsic apoptotic cascade.

Despite its broad-spectrum anticancer potential, the clinical application of Chr is significantly limited by poor aqueous solubility, low bioavailability, and rapid metabolic inactivation through glucuronidation and sulfation [23]. These pharmacokinetic drawbacks restrict its systemic efficacy and pose major obstacles to its clinical translation as a standalone therapeutic agent [24,25].

To overcome these challenges, a wide range of drug delivery strategies—including polymeric nanoparticles, micelles, liposomes, and metal-based systems—have been developed to improve chrysin's solubility, stability, and tumor-targeting efficiency. These platforms, such as PEGylated carriers, chitosan-coated nanoparticles, and solid lipid systems, not only prolong systemic circulation and protect against premature degradation but also enable combinatorial therapeutic functions like photothermal therapy or radiosensitization [21,26,27]. Mechanistically, chrysin-loaded nanoparticles have been shown to amplify anticancer efficacy through enhanced ROS generation, mitochondrial membrane depolarization, and activation of apoptosis-related signaling pathways involving Bax, Bcl2, caspases, HIF-1α, and SIRT3 [28–30].

In this study, we developed and evaluated a well-defined, biocompatible copolymer-based nanocarrier system synthesized via Reversible Addition–Fragmentation Chain Transfer (RAFT) polymerization for the targeted delivery of chrysin in ovarian cancer models. While previous studies have primarily focused on improving chrysin's solubility and passive anticancer effects, our approach integrates precision polymer engineering with detailed mechanistic investigation of tumor cell responses. The resulting formulation (PMOD-Chr) exhibited significantly enhanced cytotoxicity in A2780 and OVCAR3 ovarian cancer cells, reducing the effective IC_{50} of free chrysin. Mechanistically, PMOD-Chr induced a robust G2/M cell cycle arrest through upregulation of the CDK1/Cyclin B1 complex and concurrently triggered both intrinsic (Bax, Bcl2, caspase 9) and extrinsic (caspase 8) apoptotic pathways. While previous work has explored various chrysin-loaded formulations, our approach provides novel molecular insight into RAFT-synthesized polymeric delivery systems for ovarian cancer treatment.

2. Materials and Methods

2.1. Synthesis

2.1.1. RAFT Polymerization

p(MMA-*co*-DMAEMA) block copolymers (P1–P3) (PMDMs) were synthesized via RAFT polymerization in 1,4-dioxane using MMA (Methyl methacrylate), DMAEMA (2-(Dimethylamino) ethyl methacrylate), CTA (4-[[[(2-Carboxyethyl)thio]thioxomethyl]thio]-4-cyanopentanoic acid; Chain transfer agent), and AIBN (Azobisisobutyronitrile) (see Table 1 for ratios). The reaction proceeded under nitrogen at 70 °C for 11.5 h (Figure 1a). Reaction aliquots (t0 and tf) were collected for ^{1}H NMR analysis. The polymers were purified by triple precipitation in n-hexane and redissolved in DCM (Dichloromethane).

Table 1. Molar ratios and masses of reagents used in the preparation of P1, P2, and P3 copolymers. P1, P2, P3—Block copolymers; MMA—Methyl methacrylate; DMAEMA—2-(Dimethylamino)ethyl methacrylate; CTA—Chain transfer agent; AIBN—Azobisisobutyronitrile.

	MMA	DMAEMA	CTA	AIBN
P1	3 g	471.63 mg	121.098 mg	6.84 mg
P2	3 g	1.179 g	151.372 mg	7.697 mg
P3	3 g	2.0122 g	0.1727 g	8.796 mg

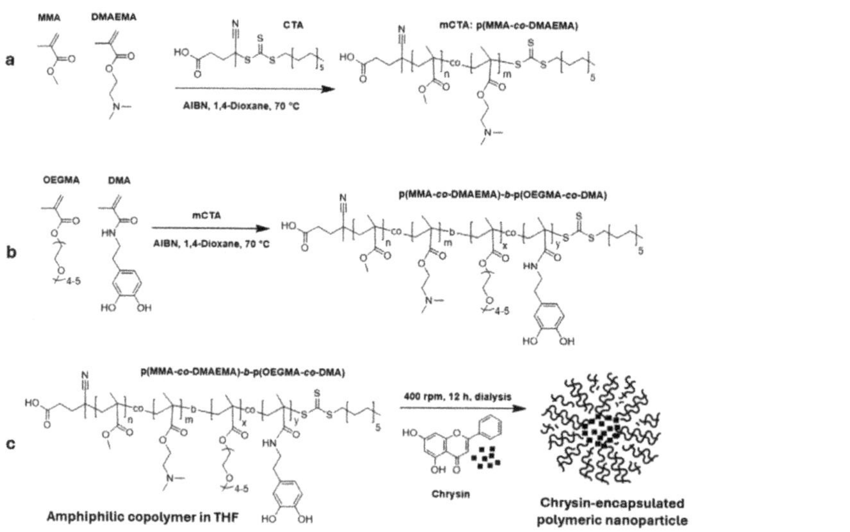

Figure 1. Schematic representation of PMOD-Chr nanoparticles synthesized via RAFT polymerization. (**a**) RAFT Polymerization of MMA and DMAEMA to synthesize p(MMA-*co*-DMAEMA) using and initiator, AIBN, and a chain transfer agent (CTA) in 1,4-dioxane at 70 °C. (**b**) Copolymerization of p(MMA-*co*-DMAEMA), DMA and OEGMA under RAFT conditions to form amphiphilic block copolymer p(MMA-*co*-DMA)-*b*-(OEGMA-*co*-DMA). (**c**) Self-assembly of the amphiphilic copolymer in aqueous medium in the presence of chrysin, forming chrysin-encapsulated polymeric nanoparticles via nanoprecipitation from THF.

p(MMA-*co*-DMAEMA)-*b*-(OEGMA-*co*-DMA) (PMOD) copolymers (BCPs) were synthesized via RAFT polymerization using OEGMA:DMA:CTA:AIBN at a molar ratio of 60:30:10:0.125 (Table 2). BCP1, BCP2, and BCP3 were synthesized using P1, P2, and P3 copolymers, respectively. Polymerization was carried out in 1,4-dioxane under nitrogen at 70 °C for 17 h with magnetic stirring (Figure 1b,c). The crude product was purified by triple precipitation in cold n-hexane, redissolved in DCM, and concentrated by rotary evaporation. Molecular weight and dispersity were determined by GPC (Gel Permeation Chromatography), and ^{1}H NMR (Proton Nuclear Magnetic Resonance) confirmed compositional control.

Table 2. Selected characterization data of block copolymers OEGMA-DMA (M2): GPC analysis and comparison with theoretical ^{1}H NMR data. M1—Monomer 1; M2—Monomer 2; CTA—Chain transfer agent; AIBN—Azobisisobutyronitrile; ^{1}H NMR—Proton nuclear magnetic resonance; M_n—Number-average molecular weight; M_w—Weight-average molecular weight; PDI—Polydispersity index (M_w/M_n); Conv.—Conversion; Theo.—Theoretical.

	M1/M2/CTA/AIBN	Conv. [%]	M_n (g mol^{-1}), Theo. ^{1}H NMR	M_w	M_n	M_w/M_n (PDI)
BCP1	90/10/1/0.125	89.4	11,017.67	1.324×10^4	1.251×10^4	1.058
BCP2	80/20/1/0.125	95.5	10,697.76	9.901×10^4	8.205×10^4	1.207
BCP3	70/30/1/0.125	95.9	10,645.75	1.636×10^5	1.444×10^5	1.113

2.1.2. Synthesis of DMA Monomer

DMA was synthesized by reacting to methacrylic anhydride with dopamine HCl in a buffered aqueous solution under nitrogen [26]. After stirring at room temperature for 12 h, the mixture was extracted with ethyl acetate to remove non-polar impurities. Following acidification, the product was re-extracted, dried over $MgSO_4$, filtered, and concentrated (Figure S1). Crude product was purified by repeated precipitation from ethyl acetate into cold n-hexane, then dried under vacuum. ^{1}H NMR spectra were obtained in DMSO-d_6 to confirm structure.

2.2. CMC Evaluation

The critical micelle concentration (CMC) of BCP3 was evaluated via Nile Red fluorescence assay [27]. A polymer stock solution (1.2 mg/mL in THF (Tetrahydrofuran)) was serially diluted (2.4 mg/mL to 0.000073 mg/mL), and micelles were formed by dropwise addition of 2 mL deionized water to each dilution, followed by THF evaporation over 48 h. Nile Red (10 μL of 0.1 mM in THF) was added to each sample and incubated for 2 h. Fluorescence intensity was measured at 550 ± 2 nm excitation and 620 ± 2 nm emission.

2.3. Nanoprecipitation of PMOD-Chr

Chrysin-loaded p(MMA-co-DMAEMA)-b-(OEGMA-co-DMA), PMOD-Chr nanoparticles were synthesized via a nanoprecipitation approach, based on solvent displacement and self-assembly principles [28,29]. Briefly, BCP3 (70 mg) and chrysin (7 mg) were dissolved in THF (7 mL), yielding 11 mg/mL stock. The nanoprecipitation method was applied: dropwise addition of the copolymer-Chr solution into 6 mL distilled water, followed by overnight stirring at 400 rpm (Figure 1c). Empty nanoparticles were prepared similarly. Solutions were dried by rotary evaporation (2 h), then diluted and analyzed by DLS (Dynamic Light Scattering). Final characterization included NTA (Nanoparticle Tracking Analysis) (after 100-fold dilution of 1 mg/mL samples) and SEM (Scanning Electron Microscope) (samples diluted to 0.01 mg/mL and further 100×).

A 1 mg/mL of Chr stock solution was prepared in DMSO. To determine the Encapsulation Efficacy (EE%) of PMOD-Chr nanoparticles, the nanoparticle solution was centrifuged at 15,000 rpm for 5 min to separate the PMOD from the supernatant containing unencapsulated Chr. The concentration of free Chr in the supernatant was then quantified by absorbance measurements at 351 nm using Synergy™ HTX Multi-Mode Microplate Reader (Agilent BioTek Instruments Inc., Winooski, VT, USA).

The amount of encapsulated Chr was calculated by subtracting the amount of free Chr in the supernatant from the initial amount of Chr used. The EE% was then determined using the following formula:

$$EE\% = \frac{\text{weight of loaded Chr}}{\text{weight of initial Chr}} \times 100$$

The Drug Loading Content (DLC) was calculated as the ratio of the weight of Chr encapsulated in the nanoparticles to the total weight of the nanoparticles, expressed as a percentage, using the following equation:

$$DLC\% = \frac{\text{weight of encapsulated Chr}}{\text{total weight of polymer} + \text{Chr}} \times 100$$

2.4. Cell Culture

A2780 and OVCAR3, human ovarian cancer cell lines and HACAT human immortalized keratinocyte cell lines, were cultured in RPMI-1640 or DMEM supplemented with 10% FBS (Fetal Bovine Serum), 1% L-glutamine, 1% penicillin/streptomycin, and 1% amphotericin B, and incubated at 37 °C in a humidified atmosphere containing 5% CO_2. The cells were seeded in flasks and monitored daily for viability and confluency.

2.5. Cell Proliferation Assay

IC_{50} (half maximal inhibitory concentration) values of Chr, PMOD, and PMOD-Chr in A2780 and OVCAR3 cancer cells and HACAT cells were determined. Five thousand cells were seeded in 96 wells and incubated for 24 h. A2780 and OVCAR3 cells were treated with treatment agents for 24, 48, and 72 h. Cell viability was determined using the MTT (3-(4,5-dimethylthiazol-2-yl)-2,5-diphenyltetrazolium bromide) assay. Cell viability was normalized by considering the untreated (control) well. IC_{50} values were calculated using GraphPad Prism 8 (Graphpad Software Inc., San Diego, CA, USA).

2.6. Colony Formation Assay

A2780 and OVCAR3 cells were seeded in 6-well plates at 1000 cells per well. The cells were incubated for 48–72 h to form colonies. Then, the cells were applied at 12 µM and 24 µM Chr, 12 µM and 24 µM PMOD, and 12 µM and 24 µM PMOD-Chr concentrations. After 7–10 days, when colonies of sufficient size were formed in the untreated wells, the cells were stained with 1% crystal violet in 10% methanol at room temperature for 30 min. The dye was removed, and the wells were washed three times with PBS. The number of colonies was counted. Then, the dye in the cells was dissolved with a 10% acetic acid solution and measured at 595 nm in a Synergy™ HTX Multi-Mode Microplate Reader (Agilent BioTek Instruments Inc., Winooski, VT, USA, ABD).

2.7. Apoptosis and Cell Cycle Assay

Cell cycle and apoptosis analyses of A2780 and OVCAR3 cells were quantified using BD FlowJoTM Software v10.10, Boston, MA, USA. Cells were seeded in 6 wells as 10×10^5 cells/well and treated with Chr, PMOD, and PMOD-Chr at 12 µM and 24 µM concentrations for 48 h. Cells were collected after drug incubations. Cells were directly stained using the Biolegend 640914 Annexin V Apoptosis kit for apoptosis analysis. For the cell cycle assay, cells were fixed overnight using 70% ethanol at 4 °C. Fixed cells were incubated with RNase A solution at 37 °C for 30 min and then stained with PI (Propidium Iodide). Analysis was performed using a BD FAcs Canto II device. Graphs were obtained using FlowJo 7 (BD FlowJoTM Software v10.10, Boston, MA, USA) and GraphPad Prism 8 software (Graphpad Software Inc., San Diego, CA, USA, ABD).

2.8. Gene Expression Analysis

A2780 and OVCAR3 cells were grown in 25 cm^2 flasks, and cells were treated with selected concentrations of Chr, PMOD, or PMOD-Chr for 48 h. After treatment, total RNA was isolated using Trizol to examine the effects on the mRNA expression levels of apoptosis-associated Bax, Bcl2, caspase 3, 8, and 9, and cell cycle-associated CDK1, CDK4, CDK6 (Cyclin-Dependent Kinases), Cyclin B1, and Cyclin D. Then, total RNA was obtained by Biorad cDNA kit (iScript™ cDNA Synthesis Kit, Biorad, CA, USA) for reverse transcription. The mRNA expression levels of target genes were analyzed by ABI StepOnePlus QPCR (Quantitative Polymerase Chain Reaction) device (Biorad Itaq, SYBR Green PCR Master Mix, Biorad, CA, USA) using cDNA Gene amplification. QPCR reaction conditions were initially set up at 95 °C for 10 min, followed by 45 cycles, each cycle consisting of 15 s denaturation at 95 °C and 1 min annealing/extension at 60 °C. Quantitative RT-PCR (Reverse Transcription Polymerase Chain Reaction) data were analyzed by the Comparative threshold (Ct) method, and fold inductions of samples were compared with untreated samples. Relative gene expression levels were assessed and normalized to β-Actin and GAPDH (Glyceraldehyde 3-phosphate dehydrogenase) housekeeping genes.

2.9. Statistical Analysis

All experiments were performed in triplicate, and data are presented as mean ± standard error of mean (SEM). Statistical significance was defined as a p value < 0.05. One-way ANOVA was used to determine the statistical significance of a variable in two or more samples, and two-way ANOVA was used to determine the statistical significance of two variables. Comparisons were made using Tukey's post hoc multiple comparisons test and Dunnett's post hoc multiple comparisons test. GraphPad Prism 8 (Graphpad Software Inc., San Diego, CA, USA) was used for graphing and statistical analysis.

3. Results

3.1. Characterizations of First Block Copolymers (Ps) as macroCTA Agent (mCTA)

Details of the chemical reaction for the synthesis of copolymers P1, P2, and P3 by RAFT are illustrated in Figure S2. The transformation of the monomer within P1 and P2 is supported by key alterations in the ^{1}H NMR spectra when comparing the *tf* to the initial state, t0. Specifically, the disappearance or significant attenuation of vinylic region signals (δ 4.5–7.0 ppm), characteristic of the monomer's C=C double bond protons, along with the emergence of new, broader signals in the aliphatic region (δ 0.5–2.5 ppm), indicates the conversion of monomeric vinyl groups into a saturated polymer backbone. In the t0 spectrum (Figure S3a,c), peaks labelled t0-'c' (δ~5.6 ppm) and t0-'d' (δ~6.1 ppm) are attributed to vinyl protons, while t0-'f' (δ~2.0 ppm) indicates a methyl group. Peak t0-'a' (δ~3.5 ppm) is primarily assigned to the solvent. In the *tf* spectrum of P1 and P2 (Figure S3b,d), the notable persistence of peaks tf-'c' and tf-'d' (δ~5.5–5.9 ppm) suggests incomplete monomer conversion, despite the appearance of broad signals tf-'e' and tf-'f' (δ~0.8–2.5 ppm) corresponding to aliphatic protons and tf-'b' (δ~3.9 ppm) characteristic of side chains. In contrast, the ^{1}H NMR analysis of the P3 copolymer provided compelling evidence of successful synthesis (Figure 2a). The characteristic monomer signals 'c' and 'd', present in the t0 spectrum (Figure 2b), were completely absent in the tf spectrum, signifying full monomer consumption and quantitative conversion into the P3 copolymer. Changes in peak morphology further supported copolymerization, consistent with the presence of macromolecules.

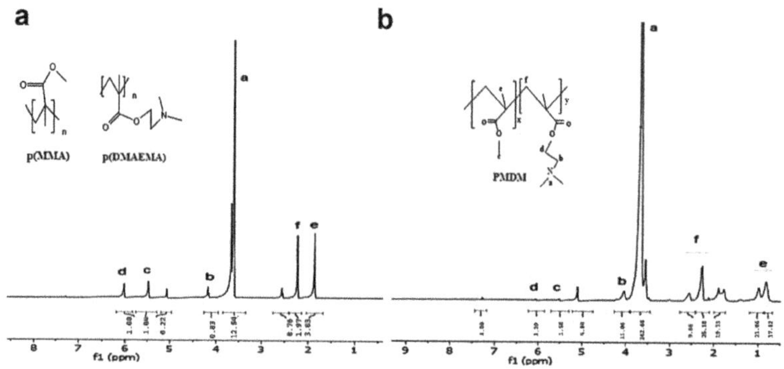

Figure 2. Representative ^{1}H NMR spectra (recorded in DMSO-d_6) of the reaction mixture at the initial stage (t0) and final state (tf) demonstrate the progress of the RAFT polymerization of the PMDM copolymer. (**a**) The MMA monomers exhibit distinctive vinyl proton signals at $\delta \approx 5.5$–6.2 ppm, attributed to the terminal -C=CH_2 groups. Upon completion of the reaction, these signals are no longer detectable, indicating the full consumption of monomeric C=C bonds. (**b**) Broad new peaks emerge in the $\delta \approx 0.8$–2.2 ppm region, corresponding to the saturated polymethacrylate backbone ('e' and 'f'). The continued presence of side-chain resonances, including protons 'a' [$N(CH_3)_2$] and 'b' [O–CH_2–N–] from the DMAEMA unit, 'c' (OCH_3 from MMA ester side chain), and 'd' (CH_2 adjacent to ether or tertiary amine groups from OEGMA or DMAEMA units), confirms successful incorporation of the functional monomers into the polymer chain.

3.2. Characterizations of Second Block Copolymers (BCPs)

The block copolymers BCP1, BCP2, and BCP3 were synthesized via RAFT and exhibited high monomer conversions. GPC analysis revealed low Polydispersity Indices (PDIs) for all copolymers (Table 2), indicating good polymerization control and narrow molecular weight distributions. For BCP1, the theoretical (^{1}H NMR) and measured (GPC) number-average molecular weights (M_n) showed reasonable agreement (Table 2). However, BCP2 and BCP3 displayed significantly higher GPC-determined molecular weights compared to theoretical ^{1}H NMR values, warranting further investigation. Nevertheless, BCP3, despite its highest M2 monomer content (30%), maintained high conversion and low PDI, suggesting successful synthesis of a well-defined copolymer with the targeted higher hydrophilic block content. The GPC-determined M_n values consistently increased from BCP1 to BCP3 with increasing M2 monomer ratio.

Analysis of the GPC chromatograms (Figure S4) revealed distinct differences among the BCPs. Based on elution times, BCP2 eluted earliest, followed by BCP1, and finally BCP3. Peak widths indicated that BCP2 exhibited the narrowest peak, BCP1 showed an intermediate PDI, and BCP3 displayed the broadest peak. The largely monomodal main polymer peaks suggested successful block copolymer formation, with minor low molecular weight species potentially present in BCP1 and BCP3. BCP3 (70/30/1/0.125) was chosen for nanoparticle preparation due to its advantageous polymerization characteristics. It exhibited the highest monomer conversion rate (95.97%), which reflects efficient chain growth during RAFT polymerization. Furthermore, its molecular weight profile was optimal for forming uniform and stable nanoparticles: GPC analysis revealed a number-average molecular weight (M_n) of 1.444 × 10^5 g/mol and a weight-average molecular weight (M_w) of 1.636 × 10^5

g/mol, with a narrow dispersity index (M_w/M_n = 1.113) (Table 2). These features are critical for ensuring controlled drug loading and release behavior, making BCP3 the most suitable candidate for chrysin encapsulation and subsequent formulation development.

The CMC of BCP3 was determined to be 0.01618 mg/mL (Figure 3a). This relatively low CMC value signifies a high propensity for self-assembly and indicates considerable micelle stability against dilution, which is advantageous for biological applications.

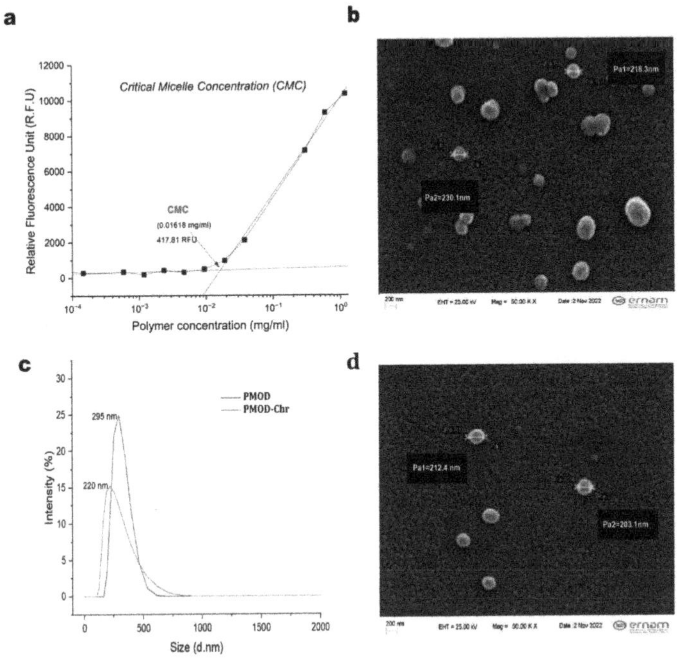

Figure 3. Chemical synthesis and characterization of self-assembling PMOD copolymer and PMOD-Chr nanoparticles. (**a**) Relative fluorescence units (RFUs) were measured across a range of polymer concentrations (mg mL^{-1}). The CMC was determined to be 0.07416 mg/mL, corresponding to an RFU value of 214.89, as indicated in red on the graph. (**b**) PMOD-Chr surface morphology of the sample obtained via SEM. The image was acquired at an accelerating voltage (EHT) of 25.00 kV and a magnification (Mag) of 50,000×. The scale bar represents 200 nm. The measurements on the image indicate particle size (Pa). (**c**) Particle size distribution profiles from DLS measurements: PMOD nanoparticles (empty) and PMOD-Chr nanoparticles. (**d**) PMOD surface morphology of the sample obtained via SEM. The image was acquired at an accelerating voltage (EHT) of 25.00 kV and a magnification (Mag) of 50,000×. The scale bar represents 200 nm. The measurements on the image indicate particle size (Pa).

The successful incorporation of DMA into the PMDM-OEGMA copolymer architecture was confirmed by ^{1}H NMR spectroscopy (Figure S5). Characteristic resonance signals corresponding to the N-methyl protons of DMA were clearly observed at approximately δ 2.8–3.1 ppm. Integration of these signals, relative to specific proton resonances from the PMMA, DMAEMA, and OEGMA segments, quantified DMA incorporation, substantiating its covalent linkage within the copolymer structure.

3.3. *Physicochemical Characterization of PMOD-Chr Nanoparticles*

FTIR (Fourier Transform Infrared) spectroscopy confirmed Chr incorporation. Pure Chr (Figure S6) exhibited characteristic bands at ~1600 and 1580 cm^{-1} (aromatic C=C), ~1685 cm^{-1}

(conjugated C=O), and above 3000 cm^{-1} (aromatic C-H). In the PMOD-Chr nanoparticles (Figure 4), while the matrix showed a strong C=O stretch around 1720–1725 cm^{-1}, key Chr features persisted: aromatic C-H stretches above 3000 cm^{-1} and a shoulder at ~1600 cm^{-1}. The shifted and less intense Chr carbonyl band, along with masked minor peaks, indicated Chr–matrix interactions and successful encapsulation.

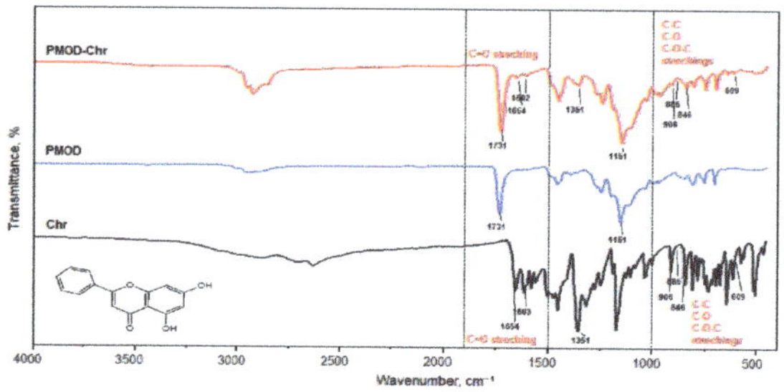

Figure 4. FTIR spectroscopy analysis to confirm the loading of Chr into PMOD. The spectra shown are for pure Chr (black), PMOD (blue), and PMOD-Chr (red). The Chr spectrum exhibits characteristic peaks for the C=O stretching vibration at 1654 cm^{-1} and aromatic ring vibrations at 1603 cm^{-1}. The PMOD nanoparticle spectrum is dominated by a strong ester carbonyl (C=O) peak at 1731 cm^{-1} from the polymer backbone. The spectrum for the PMOD-Chr is a clear superposition of both components, retaining the polymer peak at 1731 cm^{-1} while also displaying the signature peaks of encapsulated Chr.

DLS analysis revealed that the applied preparation protocol markedly influenced nanoparticle size (Figure 3c). Drug-loaded PMOD nanoparticles exhibited a reduced hydrodynamic diameter of approximately 220 nm, likely due to drug-induced polymer chain folding. This optimum small size is particularly advantageous for drug delivery applications due to EPR (Enhanced Permeability and Retention Effect), suggesting the protocol's robustness in consistently producing compact nanoparticles.

NTA analysis corroborated the DLS findings, confirming the size distribution of the nanoformulation. PMOD-Chr demonstrated a more concentrated and stable distribution profile than empty nanoparticles. Post-processing analyses revealed an increased particle density within the same size range, highlighting the efficiency and reproducibility of the production process (Figure S7).

SEM showed generally spherical PMOD-Chr (Figure 3b) and PMOD (Figure 3d). Chr presence could affect morphology and aggregation, but the formulations often maintained spherical shape and homogeneity despite loading. These observations confirm the impact of formulation on nanoparticle characteristics.

3.4. Determination of EE (%) and DLC (%) of PMOD-Chr

Based on the drug amount used in the formulation (Figure S8) and subsequent calculations, the EE% of Chr into the nanoparticles was determined to be 57.96%. The total amount of drug loaded within the nanoparticles was found to be 5.79% of Chr, representing the DLC% for this preparation.

3.5. PMOD-Chr Inhibits Cell Proliferation and Clonogenic Survival in Human Ovarian Cancer Cells

The effects of Chr and PMOD-Chr on cell viability in the dose range of 1–128 μM in A2780 and OVCAR3 ovarian cancer cells were observed for 24, 48, and 72 h. The doses used in our study were calculated based on the Chr content of the nanomaterial. In two different ovarian cancer cells, doses of approximately 8 μM and above decreased viability at all time durations. In the comparison of Chr and PMOD-Chr, PMOD-Chr provided cell viability at lower doses, and this effect was more efficient in OVCAR3 cells (Figures 5a–h and S9). It was seen that the designed PMOD without Chr did not affect cell viability (Figure S10). This showed that PMOD did not have a toxic effect in cells and that the designed nanoparticle could be an ideal cellular therapeutic carrier agent. On healthy human keratinocyte cells (HACATs), Chr showed toxic effects at doses of 32 μM and above, and PMOD-Chr only at a dose of 128 μM (Figure S11). PMOD-Chr has an effect on cancer cells; it only showed toxic effects on healthy cells at high doses. There was no significant change in colony numbers between the untreated and PMOD groups in A2780 and OVCAR3 cells. However, in both cell lines, the number of colonies decreased compared to the untreated cells with 6 μM of Chr and 6 μM of PMOD-Chr. In A2780 cells, a decrease of approximately 15% was observed when 6 μM Chr or PMOD-Chr groups were compared, and a decrease of approximately 25% was observed in OVCAR3 cells. Results showed that administration of PMOD-Chr resulted in increased inhibition of clonogenic survival of ovarian cancer cells (Figure 5i,j).

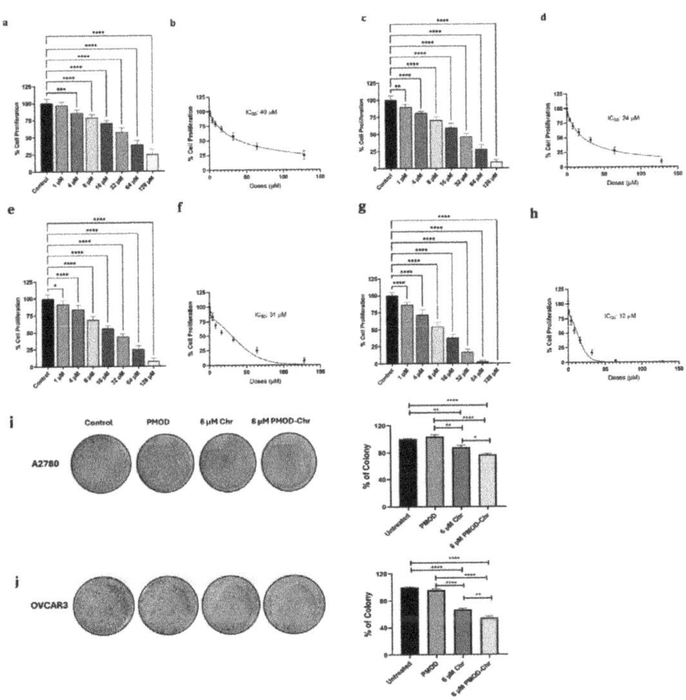

Figure 5. Showing the cell proliferation test results of Chr and PMOD-Chr in ovarian cancer cells. (**a**) Cell viability result of Chr in A2780 cells at 72 h and (**b**) IC_{50} dose at 72 h, (**c**) cell viability result of Chr in OVCAR3 cells at 72 h and (**d**) IC_{50} dose at 72 h, (**e**) cell viability result of PMOD-Chr in A2780 cells at 72 h and (**f**) IC_{50} dose at 72 h, (**g**) cell viability result of PMOD-Chr in OVCAR3 cells at 72 h and (**h**) IC_{50} dose at 72 h. Clonogenic survival of ovarian cancer cells was decreased by PMOD-Chr and Chr. (**i**) Colony formation of A2780 ovarian cancer cells and percent of colonies formed at each treatment. (**j**) Colony formation of OVCAR3 ovarian cancer cells and percent of colonies formed at each treatment. Each treatment was repeated 3 times, and results were displayed as the average ± SEM (standard error of the mean) of the three repeats and normalized to untreated groups. p value represents, *: $p \leq 0.05$, **: $p \leq 0.01$, ***: $p \leq 0.001$, ****: $p \leq 0.0001$.

3.6. PMOD-Chr Increased the Rate of Apoptosis Compared to Free Chr

Cells that do not undergo apoptosis are constantly dividing, and cancer occurs. Since the most basic feature of cancer is to escape apoptosis, anticancer drugs are required to lead the cells to apoptosis. Both ovarian cancer cells were stained with Annexin V-FITC (Fluorescein Isothiocyanate) and PI (Propidium Iodide) in order to examine the apoptotic feature with Chr and PMOD-Chr at 12 μM and 24 μM doses, and were examined by flow cytometry. In both cell lines, an increase in the rate of cells going to apoptosis was observed depending on the dose of Chr and PMOD-Chr.

As a result of the comparison of A2780 and OVCAR3 cells, Chr and PMOD-Chr were more effective in OVCAR3 cells at both 12 μM and 24 μM doses, and a high rate of apoptotic cell death was observed (Figures 6a,b and S12). In both cell lines, the gene expression levels of Bax, Bcl2, caspase 3, 8, and 9 related to apoptosis of Chr and PMOD-Chr were examined by QPCR. In PMOD-Chr, A2780 and OVCAR3 ovarian cancer cells, proapoptotic or apoptotic Bax, caspase 3 and 9 gene expressions significantly increased and anti-apoptotic Bcl2 decreased (Figure 6c,d). In OVCAR3 ovarian cancer cells, it was observed that the intrinsic apoptotic pathway was activated with caspase 8 (Figure 6d). According to both ovarian cancer cells, PMOD-Chr most effectively leads to apoptosis in OVCAR3 cells, confirming the results obtained by flow cytometry.

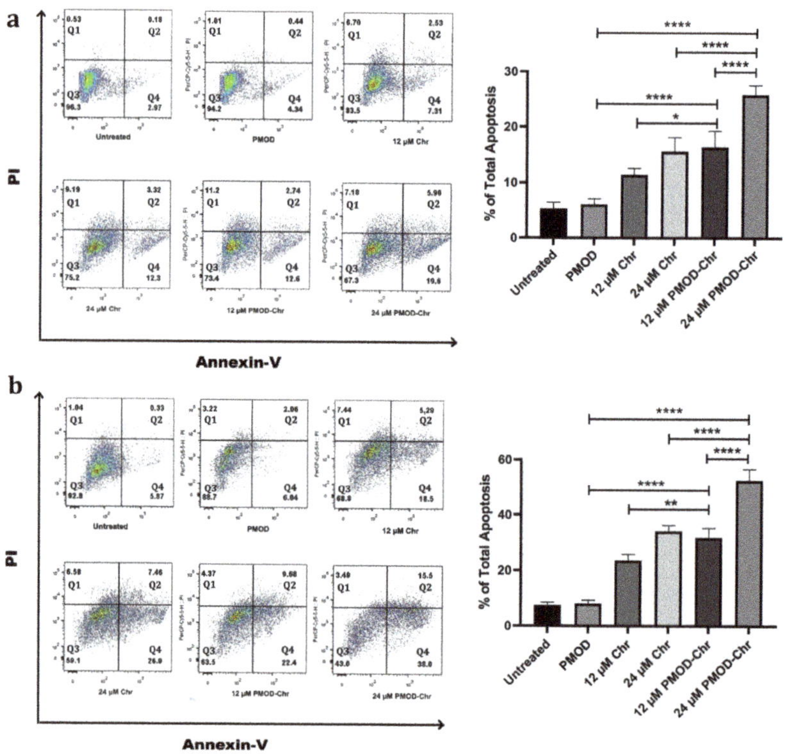

Figure 6. *Cont.*

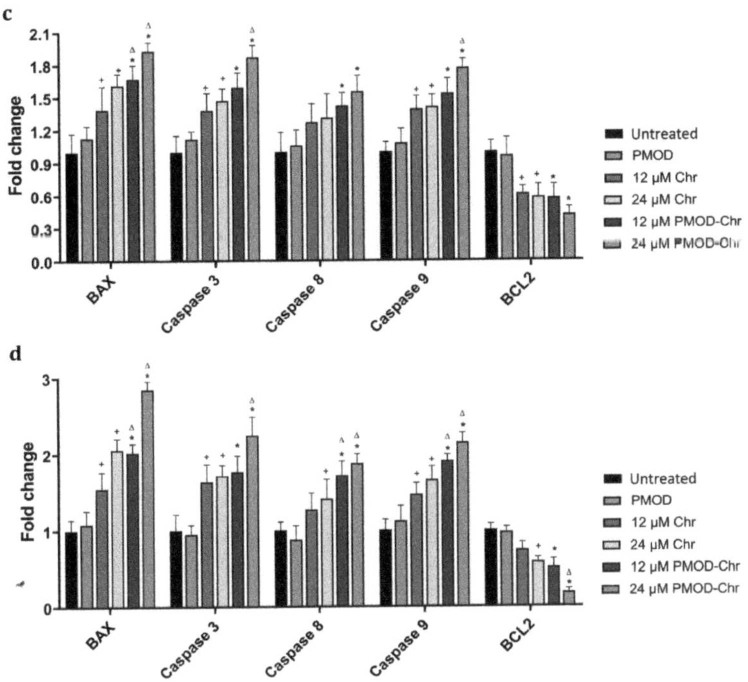

Figure 6. Analysis of apoptotic status of A2780 and OVCAR3 cells by flow cytometry and QPCR. (**a**,**b**): Flow cytometry analysis of A2780 (**a**) and OVCAR3 (**b**) cells treated with 12 μM and 24 μM free Chr and PMOD-Chr for 48 h, followed by staining with Annexin V-FITC and PI, and then flow cytometry analysis. PMOD-Chr and Chr significantly increased the apoptotic rate compared to untreated cells. The percentage of early and late apoptotic cells and necrotic cells is shown in the flow cytometry chart. The quandra in flow cytometry represent necrosis Q1 (Annexin$^+$/PI$^+$), late apoptosis Q2 (Annexin$^+$/PI$^+$), live Q3 (Annexin$^-$/PI$^-$), and early apoptosis Q4 (Annexin$^+$/PI$^-$) cells. Total apoptosis refers to all cells undergoing early (Q2) and late apoptosis (Q4). (**c**,**d**): RNA isolation and cDNA were obtained from A2780 (**c**) and OVCAR3 (**d**) cells treated with 12 μM and 24 μM Chr or PMOD-Chr for 48 h and apoptosis-related gene expression was examined. Normalization was performed with β-Actin and GAPDH housekeeping genes. Gene expression fold changes were obtained compared to control group cells.. Each treatment was repeated 3 times, and results were displayed as the average ± SEM of the three repeats and normalized to untreated groups. p-value represents in (**a**,**b**), *: $p \leq 0.05$, **: $p \leq 0.01$, ****: $p \leq 0.0001$. p value represents in (**c**,**d**), +: $p \leq 0.05$ compared to untreated group, *: $p \leq 0.05$ compared to PMOD group, Δ: $p \leq 0.05$ compared to Chr group.

3.7. Cell Cycle Is Arrested by PMOD-Chr in the Sub-G1 and G2/M Phase

Cancer cells constantly divide and proliferate uncontrollably. Drug candidates to be used in cancer are expected to slow down or stop cell division. In our study, we examined the effect of PMOD-Chr on the cell cycle in two different ovarian cancers by staining the DNA content with PI using flow cytometry. In A2780 cells, PMOD-Chr allowed cells to remain in the Sub-G1 phase more than Chr. In OVCAR3 cells, PMOD-Chr was observed to increase cell accumulation in the G2/M phase as well as Sub-G1 (Figure 7a,b). This shows that PMOD-Chr not only leads to apoptosis with Sub-G1 in OVCAR3 cells but also stops cell proliferation at the mitotic stage by accumulating in the G2/M phase. Gene expression studies conducted with both cell lines provided regulation of CDK1, CDK4, CDK6, Cyclin B, and Cyclin D genes, which have important roles in the progression of the cell cycle. In

A2780 cells, PMOD-Chr increased the expression of genes effective in G1 and G2/M phases. In OVCAR3 cells, when PMOD-Chr and Chr were compared, it upregulated the expression level of CDK1, and Cyclin B associated with the G2/M phase quite effectively (Figure 7c,d). When the results obtained by both flow cytometry and QPCR were examined, PMOD-Chr arrested the cell cycle in the Sub-G1 and G2/M phases more effectively in OVCAR3 cells.

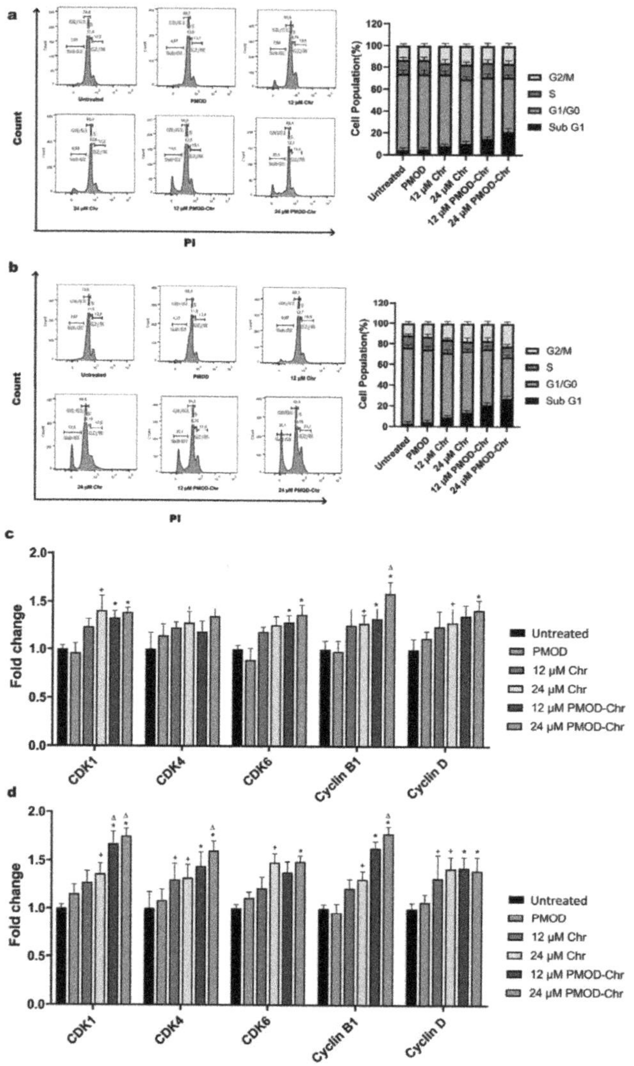

Figure 7. Effect of PMOD-Chr and Chr on DNA content in A2780 (**a**) and OVCAR3 (**b**) ovarian cancer cells. (**a**,**b**): Cells collected after 48 h of treatment with 12 and 24 μM Chr and PMOD-Chr were stained with PI. Cell cycle distribution was determined by flow cytometry analysis. Statistical charts of the cell population in G0/G1, S, and G2/M phases. (**c**,**d**): RNA isolation and cDNA were obtained from A2780 (**c**) and OVCAR3 (**d**) cells treated with 12 μM and 24 μM Chr or PMOD-Chr for 48 h, and cell cycle-related gene expression was examined. Normalization was performed with β-Actin and GAPDH housekeeping genes.. Each treatment was repeated 3 times, and results were displayed as the average ± SEM of the three repeats and normalized to untreated groups. p value represents; +: $p \leq 0.05$ compared to untreated group, *: $p \leq 0.05$ compared to PMOD group, Δ: $p \leq 0.05$ compared to Chr group.

4. Discussion

Ovarian cancer is the leading cause of gynecological cancer-related deaths [31,32]. One of the most important difficulties in the treatment of this disease is that drug use and options are limited due to side effects that occur as a result of drug use, toxic effects due to dose, systemic effects, and drug resistance [33,34]. In order to reduce this limitation and provide effective treatment options, targeted drug delivery strategies are used in treatments, and new systems to carry these drugs are needed [35]. The most current practical approach to overcome the limitations in the use of molecules that can be used in cancer treatment due to their hydrophobic properties and to protect the drug against corrosive reactions, destruction, and to increase its durability and effectiveness is to encapsulate these compounds in drug delivery systems [36].

Our rational nanoparticle design, precisely engineered via RAFT polymerization, centered on PMOD-Chr, leverages the pH-responsive behavior [37] of p(DMAEMA) [38] to overcome critical barriers in intracellular drug and gene delivery, particularly for challenges like ovarian cancer therapy. DMAEMA's unique "switchable" nature, with a pKa around ~7.4 [39], dictates its hydrophilicity in acidic conditions and hydrophobicity in basic environments. This property allows precise control over nanoparticle morphology and size by simply adjusting ambient pH.

This intelligent design employs a dual mechanism to enhance therapeutic delivery [40]. Firstly, the OEGMA shell ensures high biocompatibility and prolongs systemic circulation by minimizing protein adsorption and immune recognition, promoting effective tumor accumulation [41]. Crucially, upon endocytosis into acidic endosomes, the PMOD block actively triggers the proton sponge effect [42]. This mechanism involves the dynamic protonation of the tertiary amine groups, leading to an osmotic influx of ions and water that swells and ultimately ruptures the endosomal membrane. This ensures the rapid and protected cytosolic release of encapsulated cargo, such as Chr in our ovarian cancer model, thereby preventing lysosomal degradation and maximizing therapeutic efficacy.

In the synthesis of novel polymeric micelles, RAFT polymerization proved instrumental in the successful creation of the P3 copolymer. Evidence of complete monomer transformation was compellingly derived from significant changes observed in the ^{1}H NMR spectra. In the initial state's (t0) spectrum, characteristic vinylic region signals at δ~5.6 ppm (t0-'c') and δ~6.1 ppm (t0-'d'), attributed to the monomer's C=C double bond protons, were clearly present. Crucially, in the final (tf) spectrum, these defining monomer signals were entirely absent, unequivocally indicating full monomer conversion and quantitative incorporation into the P3 copolymer. Concurrently, the emergence of new, broader signals in the aliphatic region (tf-'e' and *tf*-'f' at δ~0.8–2.5 ppm), corresponding to the saturated polymer backbone, along with a distinct signal at tf-'b' (δ~3.9 ppm), characteristic of side chains, further supported successful macromolecule formation. These findings collectively suggest that P3 was synthesized with a notably higher conversion efficiency, potentially leading to enhanced performance for its intended applications.

The CMC of our PMOD-Chr nanoparticles, determined by plotting polymer concentration against Nile Red fluorescence intensity, was found to be 0.01618 mg/mL (corresponding to a relative fluorescence unit of 417.81). This remarkably low CMC is a highly advantageous characteristic for drug delivery applications [43,44]. It signifies a strong propensity for the polymer to self-assemble into stable micellar structures and indicates that these nanoparticles can maintain their integrity and encapsulated cargo even under significant dilution in physiological environments, such as the bloodstream. This inherent stability directly enhances their potential as robust and effective drug delivery systems, preventing premature drug release and promoting targeted accumulation.

Characteristic resonance signals corresponding to the N-methyl protons of DMA were clearly observed at approximately δ 2.8–3.1 ppm. Integration of these signals, relative to specific proton resonances from the PMMA, DMAEMA, and OEGMA segments, quantified DMA incorporation, substantiating its covalent linkage within the copolymer structure [40]. In our PMOD-Chr nanoparticles, the DMA block functions as a crucial passive targeting agent for delivering Chr to ovarian cancer cells. Rather than active receptor binding, DMA exploits the unique characteristics of the tumor microenvironment; its inherently cationic nature enhances electrostatic interactions with the often negatively charged surfaces of ovarian cancer cells, promoting increased cellular affinity and internalization. This augmented cellular uptake, combined with the EPR effect [45] that passively accumulates nanoparticles in tumors, ensures more effective delivery of the therapeutic payload, Chr, to the target cells within the ovarian cancer tissue.

FTIR spectroscopy further confirmed the successful encapsulation of Chr within the PMOD nanoparticles. Pure Chr exhibited characteristic bands at ~1600 and 1580 cm^{-1} (aromatic C=C), ~1685 cm^{-1} (conjugated C=O), and above 3000 cm^{-1} (aromatic C-H). In the Chr-loaded nanoparticles, while the polymer matrix showed a strong C=O stretch around 1720–1725 cm^{-1}, key Chr features persisted, including aromatic C-H stretches above 3000 cm^{-1} and a shoulder at ~1600 cm^{-1}. The shifted and less intense Chr carbonyl band, alongside masked minor peaks, indicated interactions between Chr and the matrix, confirming successful encapsulation. This suggests that Chr is effectively retained within the nanoparticle matrix, potentially enhancing its bioactive benefits.

DLS analysis revealed that the nanoprecipitation method offers an optimized approach for nanocarrier design. Beyond in situ synthesis parameters, high-pressure homogenization further refines the physicochemical properties of these nanoparticles, critically influencing their therapeutic potential. While these specific cases sometimes exhibited slightly broader distributions, the overall size reduction is highly beneficial for improving in vivo pharmacokinetics by reducing premature clearance by the reticuloendothelial system and enhancing tumor accumulation through the EPR effect. While our nanoparticles slightly exceed the sub-200 nm threshold often cited for optimal tumor penetration, their hydrophilic OEGMA-rich corona and size profile still support efficient passive targeting. Ongoing efforts to modulate polymer architecture and formulation conditions aim to further enhance their therapeutic potential.

SEM consistently exhibits a spherical morphology; a common characteristic observed in nanoparticle systems and one that aligns perfectly with our production methodology. This spherical shape arises from the inherent tendency of the system to minimize its surface tension and achieve a thermodynamically more stable configuration. Synthesized via RAFT polymerization, our PMOD copolymers spontaneously self-assemble in an aqueous environment. During this self-assembly process, the hydrophobic PMMA core segregates from the aqueous phase to concentrate internally, while the hydrophilic OEGMA and pH-responsive DMAEMA blocks orient towards the particle's outer surface. As this core–shell structure forms, the spherical geometry naturally represents the lowest free energy state at the molecular level. This observed spherical morphology offers significant advantages for both drug delivery and cellular uptake mechanisms. Spherical nanoparticles tend to exhibit a more uniform distribution within the body and can be more efficiently internalized by target cells via endocytosis.

Given the well-established proton sponge effect of p(DMAEMA), its inclusion in our nanoparticle design is expected to facilitate endosomal escape, thereby enhancing intracellular delivery efficiency. Although this mechanism was not directly quantified in our study, its contribution is supported by prior work demonstrating successful cytoplasmic release and bioactivity of delivered agents.

Chr-treated ES2 and OV90 ovarian cancer cells were observed to reduce cell proliferation. Chr has been studied in many different ovarian cancer cell lines, but there is no study on OVCAR3 cells [26,46–48]. A selenium-Chr polyurea dendrimer nanoformulation has been included in a study on the efficacy of OVCAR3 ovarian cancer cells, but only Chr was not examined [47]. Chr-loaded chitosan-folic acid-coated solid lipid nanoparticles were effective in pancreatic cancer, but Chr did not show much effect in ovarian cancer cells [26]. The IC_{50} value of Chr in A2780 cells at 72 h was seen to be 50 μM and above. In our study, it was found to be 40 μM. However, although the IC_{50} value was partially effective in PMOD-Chr, it was quite effective in OVCAR3 cells and reduced the IC_{50} value to 12 μM. According to these results, it is seen that PMOD-Chr inhibits cell proliferation more effectively than Chr alone. PMOD-Chr is non-toxic and can be a more effective targeted transport system.

Cell cycle checkpoints aim to ensure that the cell cycle progresses properly and healthily, and if the cell is not repaired, the cell will either stop itself or undergo a controlled death [49]. The Sub-G1 portion is the phase of the cell cycle where cells shrink in size and are more commonly identified as apoptotic cells [50]. In breast and prostate cancers, the Chr caused the cell cycle to remain in the Sub-G1 phase [51,52]. It has been reported in previous studies that Chr arrests A2780 cells in Sub-G1 of the cell cycle [48]. As a result of our cell cycle analyses performed with propidium iodide staining of A2780 and OVCAR3 cells, a greater accumulation in the Sub-G1 phase was observed in both PMOD-Chr groups compared to the other groups. This result shows that PMOD-Chr effectively drives the cells to the apoptotic pathway. Lim et al. revealed that Chr in urinary bladder cancer cells occurred in the G2/M phase of the cell cycle [50]. It gradually increased the percentage of cells in the Sub-G1 phase of the cell cycle in two different ovarian cancers. An increase in the percentage of G2/M phase cells was observed in ES2 ovarian cancer cells, but no change was observed in OV90 cells [22]. In our study with A2780 and OVCAR3 cells, an increase in the Sub-G1 phase was observed in the Chr and PMOD-Chr groups depending on the dose. The most dramatic increase was seen in the PMOD-Chr group in OVCAR3 cells at a dose of 24 μM. The G2/M checkpoint of the cell cycle prevents the cell from entering mitosis to repair the damaged genome and is involved in several pro-survival signaling pathways [53]. In the cell cycle study conducted with A2780 and OVCAR3 cells, an accumulation in the G2/M cell cycle was observed in the PMOD-Chr group. PMOD-Chr keeps cells in the Sub-G1 and G2/M phases in ovarian cancer, preventing the cells from completing mitosis or causing apoptosis, causing the cells to die via the apoptotic pathway.

The control mechanisms of cells, such as apoptosis, are one of the most important mechanisms in preventing normal cells from becoming cancerous. Apoptosis, which is the main mechanism of cancer cell death, is a programmed cell death process used by normal and cancer cells [54]. Chr initiates apoptosis, or cell death, in many cancer cells by regulating apoptotic cell pathways [55,56]. Lim et al. Chr has been shown to regulate the induction of apoptosis in ES2 and OV90 ovarian cancer cells via mitochondria-mediated intracellular Ca^{2+} accumulation. Chr triggered early and late apoptosis depending on the dose [55]. As reported in the literature, Chr has been reported to cause apoptosis in different ovarian cancer cells. The study has shown the effectiveness of Chr with two different cell types. In particular, it is observed that A2780 and OVCAR3 cell groups treated with PMOD-Chr, our nanocarrier produced by RAFT polymerization encapsulated in Chr, lead to cell death by apoptosis.

CDKs and cyclins serve as important control points in the cell cycle [49]. CDK4/6 and the cyclin complex are activated, especially in the G1 period of cells, and enable them to pass from G1 to S [57]. CDK1 and the cyclin B complex are activated towards the end of G2 and throughout the M phases of the cell cycle [58]. The complex of cyclin B-CDK1 migrates to the nucleus to activate several downstream pathways to allow cells to enter mitosis [56]. According to the results of cell cycle-related gene expressions, it is seen that PMOD-Chr increases gene activity related to the G2/M phase.

Among the master regulator genes involved in the activation or inhibition of apoptosis, the most important genes are proapoptotic BAX, antiapoptotic Bcl2, caspase 3/9 involved in intrinsic apoptosis, and caspase 3/8 involved in the extrinsic pathway [54]. In MDA MB-231 cells, Chr induced apoptosis by changing caspases 3 and 8 gene expression. Chr analog 8-bromo-7-methoxy Chr has been shown to induce apoptosis in A2780 cells [48]. According to the previous literature, Chr dose-dependently leads A2780 cells to apoptosis via Bax, Bcl, caspases 3, 8, and 9 [48]. In cells treated with Chr and Chr Niosome complex, decreased gene expression of apoptosis-related Bcl2 and increased gene expression of Bax and caspase 3 were observed [59]. As a result of gene expression studies conducted in A2780 and OVCAR3 ovarian cancer cells, it was understood that it regulates apoptosis-related genes and leads cells to apoptosis. PMOD-Chr, unlike Chr, activates intrinsic and extrinsic apoptosis and leads cells to apoptosis.

5. Conclusions

In this study, we developed a RAFT polymerization-based diblock copolymer nanocarrier (PMOD-Chr) for the targeted delivery of Chr in ovarian cancer cells. While free Chr exhibited moderate antiproliferative effects, PMOD-Chr significantly enhanced therapeutic efficacy at substantially lower doses. The formulation induced potent G2/M and Sub-G1 cell cycle arrest, particularly in OVCAR3 cells, and activated both intrinsic and extrinsic apoptotic pathways, highlighting its dual-action cytotoxic mechanism. These results demonstrate that precise molecular encapsulation via RAFT polymerization not only improves Chr's bioavailability and intracellular delivery but also amplifies its antitumor activity. Our findings suggest that the PMOD-Chr platform offers a robust and versatile nanotherapeutic strategy with strong potential for further application across diverse cancer models and in vivo systems.

Supplementary Materials: The following supporting information can be downloaded at https://www.mdpi.com/article/10.3390/nano15171362/s1, Figure S1. Chemical reaction scheme illustrating the synthesis of dopamine methacrylamide (DMA) from methacrylic anhydride and dopamine hydrochloride., Figure S2. Chemical Reaction Scheme for the Synthesis of p(MMA-co-DMAEMA) (PMDM) First Block Copolymers (Ps). Schematic illustration of the RAFT polymerization process employed for the synthesis of the initial block copolymers P1, P2, and P3., Figure S3. 1H NMR Spectroscopy Analysis of First Block Copolymers P1 and P2. Representative 1H NMR spectra of copolymers P1 and P2 comparing the initial monomer mixture (t0, A and C, respectively) with the final polymer product (tf, B and D, respectively), indicating incomplete monomer conversion as evidenced by the persistence of vinylic proton signals. Figure S4. Gel Permeation Chromatography (GPC) Chromatograms of Block Copolymers (BCPs), BCP1, BCP2, and BCP3. RI (A) and LS (B) Overlays. Figure S5. 1H NMR spectrum confirming Dopamine Methacrylamide (DMA) incorporation into PMOD copolymer, Figure S6. Fourier Transform Infrared (FTIR) Spectrum of Pure Chr., Figure S7. Nanoparticle Tracking Analysis (NTA). Particle size distribution and intensity plots obtained from NTA for PMOD and PMOD-Chr nanoparticles, Figure S8. Calibration curve of Chr., Figure S9. Effects of Chr and PMOD-Chr on ovarian cancer cell viability, Figure S10. Effects of empty (PMOD) nanoparticle on ovarian cancer cell viability, Figure S11. Investigation of the toxic effects of Chr and PMOD-Chr on healthy human keratinocyte cells, Figure S12. Analysis of the effects of PMOD, Chr, and PMOD-Chr on early and late apoptosis.

Author Contributions: S.C.: Conceptualization, Data Curation, Investigation, Methodology, Writing—Original Draft, Visualization, and Writing—Review and Editing; U.Y.: Conceptualization, Methodology, Writing—Review and Editing; T.Y.: Conceptualization, Methodology, Writing—Review and Editing; O.A.: Funding Acquisition, Supervision, Writing—Review and Editing. All authors have read and agreed to the published version of the manuscript.

Funding: This study has been supported by Erciyes University Scientific Research Project Coordination Unit under grant number FDK-2022-11852. Serife Cakir is supported by the TUBITAK-2211C.

Data Availability Statement: All data supporting the findings of this study are included in this published article and its Supplementary Materials.

Acknowledgments: The authors would like to express their sincere gratitude to Mustafa Cakir for his support in cell culture evaluations and his significant contributions to the subsequent data analysis.

Conflicts of Interest: The authors declare no conflicts of interest.

Abbreviations

The following abbreviations are used in this manuscript:

Chr	Chrysin
PMMA	Poly (Methyl Methacrylate)
DMAEMA	2-(Dimethyl Amino) Ethyl Methacrylate
OEGMA	Oligo (Ethylene Glycol) Methacrylate
PMDM	p(MMA-co-DMAEMA)
DMA	Dopamine Metacrylamide
PMOD	p(MMA-co-DMAEMA)-b-(OEGMA-co-DMA)
BCP	Block Copolymer
RAFT	Reversible Addition–Fragmentation Chain Transfer
EPR	Enhanced Permeability and Retention Effect

References

1. Maleki, Z.; Vali, M.; Nikbakht, H.A.; Hassanipour, S.; Kouhi, A.; Sedighi, S.; Farokhi, R.; Ghaem, H. Survival Rate of Ovarian Cancer in Asian Countries: A Systematic Review and Meta-Analysis. *BMC Cancer* **2023**, *23*, 558. [CrossRef]
2. González-Martín, A.; Harter, P.; Leary, A.; Lorusso, D.; Miller, R.E.; Pothuri, B.; Ray-Coquard, I.; Tan, D.S.P.; Bellet, E.; Oaknin, A.; et al. Newly Diagnosed and Relapsed Epithelial Ovarian Cancer: ESMO Clinical Practice Guideline for Diagnosis, Treatment and Follow-up. *Ann. Oncol.* **2023**, *34*, 833–848. [CrossRef]
3. Baradács, I.; Teutsch, B.; Vincze, Á.; Hegyi, P.; Szabó, B.; Nyirády, P.; Ács, N.; Melczer, Z.; Bánhidy, F.; Lintner, B. Efficacy and Safety of Combination Therapy with PARP Inhibitors and Anti-Angiogenic Agents in Ovarian Cancer: A Systematic Review and Meta-Analysis. *J. Clin. Med.* **2025**, *14*, 1776. [CrossRef]
4. Satora, M.; Kułak, K.; Zaremba, B.; Grunwald, A.; Świechowska-Starek, P.; Tarkowski, R. New Hopes and Promises in the Treatment of Ovarian Cancer Focusing on Targeted Treatment—A Narrative Review. *Front. Pharmacol.* **2024**, *15*, 1416555. [CrossRef]
5. Meneses Do Rêgo, A.C.; Araújo-Filho, I. Decoding PARP Inhibitor Resistance in Ovarian Cancer: Molecular Insights and Emerging Therapeutic Strategies. *Int. J. Innov. Res. Med. Sci.* **2024**, *9*, 652–664. [CrossRef]
6. McMullen, M.; Karakasis, K.; Madariaga, A.; Oza, A.M. Overcoming Platinum and Parp-Inhibitor Resistance in Ovarian Cancer. *Cancers* **2020**, *12*, 1607. [CrossRef]
7. Colombo, I.; Karakasis, K.; Suku, S.; Oza, A.M. Chasing Immune Checkpoint Inhibitors in Ovarian Cancer: Novel Combinations and Biomarker Discovery. *Cancers* **2023**, *15*, 3220. [CrossRef]
8. Dumitru, A.; Dobrica, E.C.; Croitoru, A.; Cretoiu, S.M.; Gaspar, B.S. Focus on PD-1/PD-L1 as a Therapeutic Target in Ovarian Cancer. *Int. J. Mol. Sci.* **2022**, *23*, 12067. [CrossRef]
9. Sun, S.; Fang, H. Curcumin Inhibits Ovarian Cancer Progression by Regulating Circ-PLEKHM3/MiR-320a/SMG1 Axis. *J. Ovarian Res.* **2021**, *14*, 158. [CrossRef]
10. Ribeiro, E.; Vale, N. The Role of Resveratrol in Cancer Management: From Monotherapy to Combination Regimens. *Targets* **2024**, *2*, 307–326. [CrossRef]

11. Maleki Dana, P.; Jahanshahi, M.; Badehnoosh, B.; Shafabakhsh, R.; Asemi, Z.; Hallajzadeh, J. Inhibitory Effects of Berberine on Ovarian Cancer: Beyond Apoptosis. *Med. Chem. Res.* **2021**, *30*, 1605–1613. [CrossRef]
12. Eksi, O.B.; Guler, A.; Akdeniz, M.; Atalay, P.; Hamurcu, Z.; Aydin, O. Development of Silver-Based Hybrid Nanoparticles Loaded with EEF2 K-SiRNA and Quercetin against Triple-Negative Breast Cancer. *Drug Deliv. Transl. Res.* **2025**, 1–23. [CrossRef]
13. Fu, Y.; Ge, Y.; Yi, S.; Peng, Q.; Jiang, H.; Zhou, J. A Review of Synergistic Strategies in Cancer Therapy: Resveratrol-Loaded Hydrogels for Targeted and Multimodal Treatment. *Discov. Oncol.* **2025**, *16*, 1382. [CrossRef]
14. Garg, A.; Chaturvedi, S. A Comprehensive Review on Chrysin: Emphasis on Molecular Targets, Pharmacological Actions and Bio-Pharmaceutical Aspects. *Curr. Drug Targets* **2021**, *23*, 420–436. [CrossRef]
15. Mani, R.; Natesan, V. Chrysin: Sources, Beneficial Pharmacological Activities, and Molecular Mechanism of Action. *Phytochemistry* **2018**, *145*, 187–196. [CrossRef]
16. Mehdi, S.H.; Nafees, S.; Zafaryab, M.; Khan, M.A.; Alam Rizvi, M.M. Chrysin: A Promising Anticancer Agent Its Current Trends and Future Perspectives. *Eur. J. Exp. Biol.* **2018**, *8*, 16. [CrossRef]
17. Li, H.; Chen, A.; Yuan, Q.; Chen, W.; Zhong, H.; Teng, M.; Xu, C.; Qiu, Y.; Cao, J. NF-KB/Twist Axis Is Involved in Chysin Inhibition of Ovarian Cancer Stem Cell Features Induced by Co-Treatment of TNF-α and TGF-β. *Int. J. Clin. Exp. Pathol.* **2019**, *12*, 101–112. [PubMed]
18. Jung, G.H.; Lee, J.H.; Han, S.H.; Woo, J.S.; Choi, E.Y.; Jeon, S.J.; Han, E.J.; Jung, S.H.; Park, Y.S.; Park, B.K.; et al. Chrysin Induces Apoptosis via the MAPK Pathway and Regulates ERK/MTOR-Mediated Autophagy in MC-3 Cells. *Int. J. Mol. Sci.* **2022**, *23*, 15747. [CrossRef] [PubMed]
19. Woo, K.J.; Jeong, Y.J.; Park, J.W.; Kwon, T.K. Chrysin-Induced Apoptosis Is Mediated through Caspase Activation and Akt Inactivation in U937 Leukemia Cells. *Biochem. Biophys. Res. Commun.* **2004**, *325*, 1215–1222. [CrossRef]
20. Cai, S.; Li, Q.; Zhou, H.; Xu, Y.; Song, J.; Gan, C.; Qi, Z.; Qi, S. Mechanism of PI3K/AKT/MTOR Signaling Pathway for Mediating Anti-Inflammatory and Anti-Oxidant Effects of Chrysin: A Protein Microarray-Based Study. *J. South. Med. Univ.* **2021**, *41*, 1554–1561.
21. Dabiri, S.; Jafari, S.; Molavi, O. Advances in Nanocarrier-Mediated Delivery of Chrysin: Enhancing Solubility, Bioavailability, and Anticancer Efficacy. *Bioimpacts* **2024**, *15*, 30269. [CrossRef]
22. Lim, W.; Ryu, S.; Bazer, F.W.; Kim, S.; Song, G. Chrysin Attenuates Progression of Ovarian Cancer Cells by Regulating Signaling Cascades and Mitochondrial Dysfunction. *J. Cell Physiol.* **2018**, *233*, 3129–3140. [CrossRef]
23. Dong, X.; Cao, Y.; Wang, N.; Wang, P.; Li, M. Systematic Study on Solubility of Chrysin in Different Organic Solvents: The Synergistic Effect of Multiple Intermolecular Interactions on the Dissolution Process. *J. Mol. Liq.* **2021**, *325*, 115180. [CrossRef]
24. Jasim, A.J.; Sulaiman, G.M.; Ay, H.; Mohammed, S.A.A.; Mohammed, H.A.; Jabir, M.S.; Khan, R.A. Preliminary Trials of the Gold Nanoparticles Conjugated Chrysin: An Assessment of Anti-Oxidant, Anti-Microbial, and in Vitro Cytotoxic Activities of a Nanoformulated Flavonoid. *Nanotechnol. Rev.* **2022**, *11*, 2726–2741. [CrossRef]
25. Nosrati, H.; Rakhshbahar, A.; Salehiabar, M.; Afroogh, S.; Kheiri Manjili, H.; Danafar, H.; Davaran, S. Bovine Serum Albumin: An Efficient Biomacromolecule Nanocarrier for Improving the Therapeutic Efficacy of Chrysin. *J. Mol. Liq.* **2018**, *271*, 639–646. [CrossRef]
26. Farhadi, A.; Homayouni Tabrizi, M.; Sadeghi, S.; Vala, D.; Khosravi, T. Targeted Delivery and Anticancer Effects of Chrysin-Loaded Chitosan-Folic Acid Coated Solid Lipid Nanoparticles in Pancreatic Malignant Cells. *J. Biomater. Sci. Polym. Ed.* **2023**, *34*, 315–333. [CrossRef] [PubMed]
27. Jangid, A.K.; Solanki, R.; Patel, S.; Medicherla, K.; Pooja, D.; Kulhari, H. Improving Anticancer Activity of Chrysin Using Tumor Microenvironment PH-Responsive and Self-Assembled Nanoparticles. *ACS Omega* **2022**, *7*, 15919–15928. [CrossRef]
28. Ragab, E.M.; El Gamal, D.M.; Mohamed, T.M.; Khamis, A.A. Impairment of Electron Transport Chain and Induction of Apoptosis by Chrysin Nanoparticles Targeting Succinate-Ubiquinone Oxidoreductase in Pancreatic and Lung Cancer Cells. *Genes Nutr.* **2023**, *18*, 4. [CrossRef]
29. Alanazi, S.T.; Salama, S.A.; El-ebiary, A.M.; Altowairqi, A.K.; Alharthi, A.T.; Alzahrani, S.M.; Althagafi, S.H.; Alotaibi, R.A.; Tammam, A.A.E. Targeting SIRT1, NLRP3 Inflammasome, and Nrf2 Signaling with Chrysin Alleviates the Iron-Triggered Hepatotoxicity in Rats. *Toxicology* **2024**, *504*, 153766. [CrossRef]
30. Kim, K.M.; Jung, J. Upregulation of G Protein-Coupled Estrogen Receptor by Chrysin-Nanoparticles Inhibits Tumor Proliferation and Metastasis in Triple Negative Breast Cancer Xenograft Model. *Front. Endocrinol.* **2020**, *11*, 560605. [CrossRef]
31. Jayson, G.C.; Kohn, E.C.; Kitchener, H.C.; Ledermann, J.A. Ovarian Cancer. *Lancet* **2014**, *384*, 1376–1388. [CrossRef] [PubMed]
32. Matulonis, U.A.; Sood, A.K.; Fallowfield, L.; Howitt, B.E.; Sehouli, J.; Karlan, B.Y. Ovarian Cancer. *Nat. Rev. Dis. Primers* **2016**, *2*, 16061. [CrossRef]
33. Guan, L.; Lu, Y. New Developments in Molecular Targeted Therapy of Ovarian Cancer. *Discov. Med.* **2018**, *26*, 219–229.

34. Radu, M.R.; Prădatu, A.; Duică, F.; Micu, R.; Creţoiu, S.M.; Suciu, N.; Creţoiu, D.; Varlas, V.N.; Rădoi, V.E. Ovarian Cancer: Biomarkers and Targeted Therapy. *Biomedicines* **2021**, *9*, 693. [CrossRef]
35. Raj, S.; Khurana, S.; Choudhari, R.; Kesari, K.K.; Kamal, M.A.; Garg, N.; Ruokolainen, J.; Das, B.C.; Kumar, D. Specific Targeting Cancer Cells with Nanoparticles and Drug Delivery in Cancer Therapy. In *Seminars in Cancer Biology*; Elsevier: Amsterdam, The Netherlands, 2021; Volume 69, pp. 166–177.
36. Sánchez-López, E.; Guerra, M.; Dias-Ferreira, J.; Lopez-Machado, A.; Ettcheto, M.; Cano, A.; Espina, M.; Camins, A.; Garcia, M.L.; Souto, E.B. Current Applications of Nanoemulsions in Cancer Therapeutics. *Nanomaterials* **2019**, *9*, 821. [CrossRef] [PubMed]
37. Mu, Y.; Gong, L.; Peng, T.; Yao, J.; Lin, Z. Advances in PH-Responsive Drug Delivery Systems. *OpenNano* **2021**, *5*, 100031. [CrossRef]
38. Ramezanian, S.; Moghaddas, J.; Roghani-Mamaqani, H.; Rezamand, A. Dual PH- and Temperature-Responsive Poly(Dimethylaminoethyl Methacrylate)-Coated Mesoporous Silica Nanoparticles as a Smart Drug Delivery System. *Sci. Rep.* **2023**, *13*, 20194. [CrossRef]
39. Singh, J.; Nayak, P. PH-Responsive Polymers for Drug Delivery: Trends and Opportunities. *J. Polym. Sci.* **2023**, *61*, 2828–2850. [CrossRef]
40. Ahmed, A.; Sarwar, S.; Hu, Y.; Munir, M.U.; Nisar, M.F.; Ikram, F.; Asif, A.; Rahman, S.U.; Chaudhry, A.A.; Rehman, I.U. Surface-Modified Polymeric Nanoparticles for Drug Delivery to Cancer Cells. *Expert. Opin. Drug Deliv.* **2021**, *18*, 1–24. [CrossRef]
41. Hussain, Z.; Khan, S.; Imran, M.; Sohail, M.; Shah, S.W.A.; de Matas, M. PEGylation: A Promising Strategy to Overcome Challenges to Cancer-Targeted Nanomedicines: A Review of Challenges to Clinical Transition and Promising Resolution. *Drug Deliv. Transl. Res.* **2019**, *9*, 721–734. [CrossRef]
42. Qiu, C.; Xia, F.; Zhang, J.; Shi, Q.; Meng, Y.; Wang, C.; Pang, H.; Gu, L.; Xu, C.; Guo, Q.; et al. Advanced Strategies for Overcoming Endosomal/Lysosomal Barrier in Nanodrug Delivery. *Research* **2023**, *6*, 0148. [CrossRef]
43. Aydin, O.; Youssef, I.; Yuksel Durmaz, Y.; Tiruchinapally, G.; Elsayed, M.E.H. Formulation of Acid-Sensitive Micelles for Delivery of Cabazitaxel into Prostate Cancer Cells. *Mol. Pharm.* **2016**, *13*, 1413–1429. [CrossRef]
44. Memet, B.C.; Yildiz, U.; Eksi, O.B.; Aydin, O. DOX-Couped Polymeric Micelles as a State-of-the-Art Strategy Against Triple Negative Breast Cancer. In Proceedings of the 9th World Congress on Recent Advances in Nanotechnology, RAN 2024, London, UK, 8–10 April 2024. [CrossRef]
45. Shinde, V.R.; Revi, N.; Murugappan, S.; Singh, S.P.; Rengan, A.K. Enhanced Permeability and Retention Effect: A Key Facilitator for Solid Tumor Targeting by Nanoparticles. *Photodiagnosis Photodyn. Ther.* **2022**, *39*, 102915. [CrossRef]
46. Bose, S.; Saha, P.; Alam, M.T.; Chatterjee, B.; Sarkar, M.; Dixit, A.K.; Kumar, D.; Pathak, R.K.; Tripathi, P.P.; Srivastava, A.K. Inhibition of DNA Polymerase Eta-mediated Translesion DNA Synthesis with Small Molecule Sensitises Ovarian Cancer Stem-like Cells to Chemotherapy. *Br. J. Pharmacol.* **2025**, *182*, 3594–3611. [CrossRef] [PubMed]
47. Santos, I.; Ramos, C.; Mendes, C.; Sequeira, C.O.; Tomé, C.S.; Fernandes, D.G.H.; Mota, P.; Pires, R.F.; Urso, D.; Hipólito, A. Targeting Glutathione and Cystathionine β-Synthase in Ovarian Cancer Treatment by Selenium–Chrysin Polyurea Dendrimer Nanoformulation. *Nutrients* **2019**, *11*, 2523. [CrossRef] [PubMed]
48. Ding, Q.; Chen, Y.; Zhang, Q.; Guo, Y.; Huang, Z.; Dai, L.; Cao, S. 8-Bromo-7-Methoxychrysin Induces Apoptosis by Regulating Akt/FOXO3a Pathway in Cisplatin-Sensitive and Resistant Ovarian Cancer Cells. *Mol. Med. Rep.* **2015**, *12*, 5100–5108. [CrossRef] [PubMed]
49. Matthews, H.K.; Bertoli, C.; de Bruin, R.A.M. Cell Cycle Control in Cancer. *Nat. Rev. Mol. Cell Biol.* **2022**, *23*, 74–88. [CrossRef]
50. Lima, A.P.B.; Melo, A.S.; Ferreira, G.M.; da Silva, G.N. Chrysin Inhibits the Cell Viability, Induces Apoptosis and Modulates Expression of Genes Related to Epigenetic Events in Bladder Cancer Cells. *Nat. Prod. Res.* **2023**, *37*, 1877–1881. [CrossRef]
51. Han, H.; Lee, S.-O.; Xu, Y.; Kim, J.-E.; Lee, H.-J. SPHK/HIF-1α Signaling Pathway Has a Critical Role in Chrysin-Induced Anticancer Activity in Hypoxia-Induced PC-3 Cells. *Cells* **2022**, *11*, 2787. [CrossRef]
52. Ramos, P.S.; Ferreira, C.; Passos, C.L.A.; Silva, J.L.; Fialho, E. Effect of Quercetin and Chrysin and Its Association on Viability and Cell Cycle Progression in MDA-MB-231 and MCF-7 Human Breast Cancer Cells. *Biomed. Pharmacother.* **2024**, *179*, 117276. [CrossRef]
53. Jung, Y.; Kraikivski, P.; Shafiekhani, S.; Terhune, S.S.; Dash, R.K. Crosstalk between Plk1, P53, Cell Cycle, and G2/M DNA Damage Checkpoint Regulation in Cancer: Computational Modeling and Analysis. *NPJ Syst. Biol. Appl.* **2021**, *7*, 46. [CrossRef] [PubMed]
54. Singh, P.; Lim, B. Targeting Apoptosis in Cancer. *Curr. Oncol. Rep.* **2022**, *24*, 273–284. [CrossRef]
55. Talebi, M.; Talebi, M.; Farkhondeh, T.; Simal-Gandara, J.; Kopustinskiene, D.M.; Bernatoniene, J.; Samarghandian, S. Emerging Cellular and Molecular Mechanisms Underlying Anticancer Indications of Chrysin. *Cancer Cell Int.* **2021**, *21*, 214. [CrossRef] [PubMed]

56. Salari, N.; Faraji, F.; Jafarpour, S.; Faraji, F.; Rasoulpoor, S.; Dokaneheifard, S.; Mohammadi, M. Anti-Cancer Activity of Chrysin in Cancer Therapy: A Systematic Review. *Indian. J. Surg. Oncol.* **2022**, *13*, 681–690. [CrossRef]
57. Topacio, B.R.; Zatulovskiy, E.; Cristea, S.; Xie, S.; Tambo, C.S.; Rubin, S.M.; Sage, J.; Kõivomägi, M.; Skotheim, J.M. Cyclin D-Cdk4, 6 Drives Cell-Cycle Progression via the Retinoblastoma Protein's C-Terminal Helix. *Mol. Cell* **2019**, *74*, 758–770. [CrossRef] [PubMed]
58. Brown, N.R.; Korolchuk, S.; Martin, M.P.; Stanley, W.A.; Moukhametzianov, R.; Noble, M.E.M.; Endicott, J.A. CDK1 Structures Reveal Conserved and Unique Features of the Essential Cell Cycle CDK. *Nat. Commun.* **2015**, *6*, 6769. [CrossRef]
59. Tarahomi, M.; Firouzi Amandi, A.; Eslami, M.; Yazdani, Y.; Salek Farrokhi, A.; Ghorbani, F.; Taherian, M.; Yousefi, B. Niosomes Nanoparticles as a Novel Approach in Drug Delivery Enhances Anticancer Properties of Chrysin in Human Ovarian Carcinoma Cells (SKOV3): An in Vitro Study. *Med. Oncol.* **2023**, *40*, 87. [CrossRef]

nanomaterials

MDPI

Article

Biogenic ZnO Nanoparticles Synthesized by *B. licheniformis*: A Selective Cytotoxicity Against NG-108 Glioblastoma Cells

Alberto Bacilio Quispe Cohaila [1,2,*], Gabriela de Lourdes Fora Quispe [1], César Julio Cáceda Quiroz [3], Roxana Mamani Anccasi [4], Telmo Agustin Mejia García [5], Rocío María Tamayo Calderón [6], Francisco Gamarra Gómez [1,7] and Elisban Juani Sacari Sacari [1,8,*]

1 Grupo de Investigación GIMAECC, Facultad de Ingeniería, Universidad Nacional Jorge Basadre Grohmann, Avenida Miraflores S/N, Ciudad Universitaria, Tacna 23003, Peru
2 Escuela de Metalurgia y Materiales, Facultad de Ingeniería, Universidad Nacional Jorge Basadre Grohmann, Avenida Miraflores S/N, Ciudad Universitaria, Tacna 23003, Peru
3 Laboratorio de Biorremediación, Facultad de Ciencias, Universidad Nacional Jorge Basadre Grohmann, Avenida Miraflores S/N, Ciudad Universitaria, Tacna 23003, Peru
4 Departamento de Biología y Microbiología, Facultad de Ciencias, Universidad Nacional Jorge Basadre Grohmann, Avenida Miraflores S/N, Ciudad Universitaria, Tacna 23003, Peru
5 Centro de Investigación E-Health, Departamento Ciencias de la Salud, Universidad de Ciencias y Humanidades, Av. Universitaria Norte 5175, Los Olivos Lima 150101, Peru
6 Laboratorio de Compuestos, Departamento de Ingeniería de Materiales, Facultad de Ingeniería de Procesos, Universidad Nacional de San Agustín de Arequipa, Arequipa 04001, Peru
7 Laboratorio de Nanotecnología, Facultad de Ingeniería, Universidad Nacional Jorge Basadre Grohmann, Avenida Miraflores S/N, Ciudad Universitaria, Tacna 23003, Peru
8 Facultad de Ciencias, Universidad Nacional de Ingeniería, Av. Túpac Amaru 210, Lima 15333, Peru
* Correspondence: aquispec@unjbg.edu.pe (A.B.Q.C.); esacaris@unjbg.edu.pe (E.J.S.S.); Tel.: +51-925573684 (E.J.S.S.)

Abstract

Glioblastoma multiforme (GBM) remains the most aggressive primary brain tumor with median survival of 14.6 months, necessitating novel therapeutic approaches. Here, we report the biogenic synthesis of zinc oxide nanoparticles (ZnO NPs) using *Bacillus licheniformis* strain TT14s isolated from mining environments and demonstrate their selective anti-glioma efficacy. ZnO NPs exhibited hexagonal wurtzite structure (crystallite size: 15.48 nm) with spherical morphology (19.37 ± 5.28 nm diameter) as confirmed by XRD, HRTEM, and comprehensive physicochemical characterization. Colloidal stability analysis revealed an isoelectric point at pH 7.46, ensuring optimal dispersion in biological media. Cytotoxicity evaluation revealed remarkable selectivity: at 100 μg/mL, ZnO NPs reduced NG-108 glioblastoma cell viability to 36.07 ± 1.89% within 1 h while maintaining 78.9 ± 0.94% viability in primary retinal cells. The selective cytotoxicity was attributed to the interplay of convergent mechanisms acting under dark conditions, including defect-mediated ROS generation supported by photoluminescence analysis revealing a characteristic oxygen vacancy emission at 550 nm, pH-dependent dissolution enhanced in the acidic tumor microenvironment, and preferential cellular uptake by rapidly proliferating cancer cells with compromised antioxidant defenses. Time-course analysis demonstrated concentration-dependent effects with therapeutic windows favoring normal cell preservation. The intrinsic cytotoxic activity under dark laboratory conditions eliminates the need for external activation, providing practical advantages for therapeutic applications. These findings establish ZnO NPs as promising candidates for targeted glioblastoma therapy, warranting further in vivo validation and mechanistic elucidation for clinical translation.

Keywords: biogenic synthesis; zinc oxide nanoparticles; glioblastoma; selective cytotoxicity; apoptosis; *Bacillus licheniformis*; green synthesis; cancer therapy

1. Introduction

Glioblastoma multiforme (GBM) represents the most aggressive and prevalent primary brain malignancy, accounting for approximately 51% of all malignant brain tumors and exhibiting an incidence rate of 3.19–3.26 per 100,000 individuals annually [1,2]. Despite significant advances in neurosurgical techniques, radiotherapy protocols, and chemotherapeutic regimens, GBM remains among the most challenging oncological conditions to treat, with a median overall survival of 14.6 months and a devastating 5-year survival rate of approximately 5–7% [3,4].

The dismal prognosis of GBM stems from several intrinsic biological characteristics that render conventional therapeutic approaches inadequate. Recent breakthrough studies have revealed that glioblastoma cells hijack neuronal mechanisms for brain invasion, exploiting synaptic input and migration pathways that facilitate organized collective invasion through "oncostreams" [5]. This infiltrative behavior, guided by tumor-associated microglia through plexin-B2-mediated alignment, renders complete surgical resection virtually impossible even with maximal safe resection techniques [6]. Furthermore, the heterogeneous tumor microenvironment harbors glioma stem cells (GSCs) that exhibit intrinsic resistance to radiotherapy and chemotherapy, contributing to inevitable tumor recurrence [7,8].

Current standard-of-care treatment, established by the landmark Stupp protocol, combines maximal safe surgical resection followed by concurrent radiotherapy with temozolomide chemotherapy [3]. However, primary resistance to temozolomide affects approximately 50% of patients due to O6-methylguanine-DNA methyltransferase (MGMT) overexpression, while acquired resistance develops through multiple mechanisms including enrichment of treatment-resistant stem cell populations [9,10]. These GSCs, characterized by expression of markers such as CD133, CD44, and SOX2, demonstrate enhanced DNA repair capabilities and altered metabolic profiles that enable survival under therapeutic stress [11].

The blood–brain barrier (BBB) presents an additional formidable obstacle to effective drug delivery, limiting the penetration of therapeutic agents to tumor sites while creating sanctuary sites for residual cancer cells [12,13]. Conventional chemotherapeutic agents often fail to achieve therapeutically relevant concentrations within brain parenchyma, necessitating higher systemic doses that increase toxicity to normal tissues [14].

Active targeting strategies utilizing specific ligand conjugation enhance nanomedicine selectivity beyond passive targeting mechanisms [15]. Peptide-based approaches have demonstrated improved therapeutic indices through receptor-mediated cellular uptake, offering promising strategies for optimizing ZnO nanoparticle selectivity in glioblastoma therapy via brain tumor-specific ligand conjugation.

In recent years, nanotechnology has emerged as a transformative approach in cancer therapeutics, offering unprecedented opportunities to overcome the limitations of conventional treatments [16,17]. Among various nanomaterials, ZnO NPs have garnered significant attention due to their unique physicochemical properties, including biocompatibility, selective cytotoxicity toward cancer cells, and the ability to generate reactive oxygen species (ROS) [18,19]. ZnO NPs exhibit a wide band gap energy of approximately 3.3 eV, enabling photocatalytic activation that enhances their therapeutic efficacy through photodynamic mechanisms [20].

The anticancer activity of ZnO NPs operates through multiple convergent mechanisms, including pH-responsive dissolution in the acidic tumor microenvironment (pH 5.5–6.5), mitochondrial targeting with subsequent apoptosis induction, and selective accumulation in rapidly proliferating cells [18,21]. Studies have demonstrated preferential cytotoxicity against various glioma cell lines (A172, U87, LNZ308) while showing minimal effects on normal astrocytes, suggesting a favorable therapeutic window [22]. The mechanism of selective toxicity involves differential cellular uptake, enhanced ROS generation in cancer cells with compromised antioxidant systems, and zinc ion-mediated disruption of mitochondrial membrane potential [23].

Traditional chemical synthesis methods for ZnO NPs often involve high-temperature processes (500–1000 °C), hazardous chemicals, and significant environmental impact [24]. In contrast, green synthesis approaches utilizing biological organisms offer environmentally sustainable alternatives with enhanced biocompatibility and potentially superior therapeutic properties [25,26]. Bacterial-mediated synthesis, particularly using non-pathogenic species such as *Bacillus licheniformis*, has emerged as a promising approach for producing ZnO NPs with controlled morphology, enhanced stability, and reduced cytotoxicity to normal cells [27,28].

The biogenic synthesis process typically involves extracellular enzyme-mediated reduction of zinc salts through NADH-dependent reductases, with bacterial exopolysaccharides providing natural capping and stabilization [29]. This approach offers several advantages over chemical methods, including ambient temperature and pressure conditions, renewable biological resources, minimal waste generation, and the production of nanoparticles with enhanced colloidal stability and biocompatibility [30,31].

Bacillus licheniformis, a Gram-positive, non-pathogenic bacterium widely distributed in natural environments, has demonstrated particular efficacy in nanoparticle biosynthesis due to its robust enzyme systems and ability to secrete stabilizing biomolecules [32]. Previous studies have shown that *B. licheniformis*-mediated synthesis produces ZnO NPs with hexagonal wurtzite crystal structure, optimal size distribution (10–50 nm), and enhanced antimicrobial and anticancer activities compared to chemically synthesized counterparts [33].

The potential of biogenically synthesized ZnO NPs in glioblastoma therapy is particularly noteworthy given their ability to address multiple therapeutic challenges simultaneously. Their nanoscale dimensions may facilitate BBB penetration through transcytosis mechanisms, while their pH-responsive behavior enables selective activation within the acidic tumor microenvironment [34]. Additionally, their capacity to generate multiple ROS species upon UV or visible light activation presents opportunities for photodynamic therapy applications in conjunction with surgical interventions [18].

Recent advances in nanomedicine have demonstrated clinical viability, with over 50 FDA-approved nanotherapeutics currently in clinical use and more than 200 nanomedicine formulations in various stages of clinical trials [35,36]. The translation of ZnO NPs to clinical applications is supported by zinc's essential biological role (recommended daily intake of 8–11 mg) and its Generally Recognized as Safe (GRAS) status for topical applications [37].

In this study, we aim to evaluate the cytotoxic effects of biogenically synthesized ZnO NPs using *B. licheniformis* on primary cultures of chicken embryonic retinal cells and the NG-108 glioblastoma cell line. By comparing the responses of these two cell types to various concentrations and exposure times of ZnO NPs, we seek to determine the potential for selective cytotoxicity toward glioma cells while preserving normal neural tissue viability.

Our research contributes to the growing body of evidence supporting the development of environmentally sustainable nanotherapeutics for glioblastoma treatment. By elucidating the complex interactions between biogenic ZnO NPs and glioma cells, we aim to establish a foundation for the development of more effective, targeted, and environmentally responsible treatments for this devastating disease, offering new hope for patients suffering from this aggressive form of brain cancer while advancing the principles of green nanotechnology in biomedical applications.

2. Materials and Methods

2.1. Sample Collection and Bacterial Isolation

The bacterial strain TT14s, employed for ZnO biosynthesis, was originally isolated from soil samples collected at the Tutupaca mining environmental liability site in the Candarave province, Tacna, Peru (coordinates: east 0358602; north 8113940; altitude 4687 m.a.s.l.). The environmental samples underwent systematic homogenization and granulometric separation followed by aqueous extraction under sterile conditions to obtain a cell suspension. The isolation protocol specifically targeted alkaliphilic microorganisms through a selective enrichment strategy, employing sequential subcultures in nutrient broth medium adjusted to pH 10.5 with 0.1 N NaOH. Subsequently, discrete bacterial colonies exhibiting alkaline-tolerance were isolated through streak-plate technique on nutrient agar under identical pH conditions. The isolate was preserved in the Bioremediation Laboratory culture collection at Jorge Basadre Grohmann National University, where it was assigned the designation TT14s for taxonomic and experimental purposes.

2.2. Molecular Identification and Sequencing

DNA extraction from bacterial isolate TT14s was accomplished employing the InnuPREP Bacteria DNA Kit according to the protocol provided by the manufacturer. Quantitative and qualitative assessment of the extracted DNA was performed through fluorescence-based measurement using a Qubit 4 fluorometer (Life Technologies, Carlsbad, CA, USA).

Sequencing library construction was carried out using the Illumina DNA Prep protocol (Illumina, Granta Park, Cambridge, UK), with incorporation of distinctive dual barcodes from the Nextera DNA CD Indexes system (Illumina, Granta Park, Cambridge, UK) to facilitate sample multiplexing during sequencing runs. Whole-genome sequencing was performed on the Illumina MiSeq Control platform (Version 2.6.0) employing a V3 600-cycle chemistry kit, which produced paired-end sequence reads of 2×151 nucleotides.

Species-level identification was achieved through phylogenetic analysis of the 16S ribosomal RNA gene sequence recovered from the complete genome assembly. Taxonomic assignment was determined by sequence homology searching against the National Center for Biotechnology Information (NCBI) reference database using the Basic Local Alignment Search Tool (BLAST+, Version 2.15.0) algorithm. This genomic approach for bacterial identification, consistent with methodologies established by Cáceda et al. (2023) [38], enabled precise taxonomic placement of isolate TT14s.

2.3. Biogenic ZnO Synthesis by Bacillus licheniformis

The green synthesis of ZnO nanoparticles was conducted using *Bacillus licheniformis* strain TT14s, previously isolated and molecularly characterized from our laboratory's microbial repository. The synthetic approach was adapted from established protocols reported by Tripathi and colleagues (2021) [39], incorporating specific modifications to optimize production yield and scalability. The experimental procedure initiated with bacterial reactivation in Brain Heart Infusion (BHI) medium at 35 °C for 24 h. Following reactivation, cultures were transferred onto nutrient agar plates and incubated at 35 °C for an additional 24 h period to obtain viable bacterial colonies. Morphological verification was conducted through Gram staining techniques to confirm the characteristic features of *B. licheniformis*: Gram-positive, rod-shaped bacterial cells with terminal rounding and endospore-forming capability.

Active bacterial cultures were grown in nutrient broth at 35 °C with continuous agitation at 180 rpm over a 48 h period. Biomass collection was achieved through centrifugation at 8000 rpm for 10 min. The biosynthetic reaction was initiated by combining 5 g of harvested bacterial biomass with a precursor solution containing 375 mL of 0.2 M zinc acetate dihydrate and 375 mL of 0.5 M sodium bicarbonate, both dissolved in sterile deionized water. The reaction vessel was maintained at 35 °C under continuous agitation at 170 rpm for 72 h to promote bacterial-mediated nanoparticle formation.

Product recovery involved centrifugal separation at 8000 rpm for 10 min, followed by six consecutive washing steps using sterile deionized water to eliminate impurities and residual bacterial byproducts. The recovered precipitate underwent thermal dehydration at 60 °C for 18 h. Thermal analysis and crystallization enhancement were conducted using thermogravimetric and differential scanning calorimetric techniques. Final crystallization was accomplished through mechanical homogenization using an agate pestle and mortar, followed by calcination at 350 °C for 2 h in a ceramic vessel. A final grinding step was performed to deaggregate the calcined ZnO nanoparticles, producing a uniform powder suitable for comprehensive characterization and biological evaluation.

2.4. Characterization Techniques

Thermal characterization of the biosynthesized ZnO nanoparticles was conducted using multiple complementary analytical approaches. Weight loss behavior and thermal transitions were evaluated through thermogravimetric analysis (TGA) coupled with differential scanning calorimetry (DSC) on samples prior to calcination, employing an SDT650 thermal analyzer (TA Instruments, New Castle, DE, USA) operating under nitrogen atmosphere (100 mL/min flow rate) with a controlled heating ramp of 20 °C/min. Crystallographic structure determination was accomplished via X-ray powder diffraction (XRD) analysis of calcined specimens using a PANalytical Aeris Research diffractometer operating with Cu Kα radiation at 40 kV and 15 mA (Malvern Panalytical Ltd., Almelo, The Netherlands). Vibrational spectroscopic characterization was performed using attenuated total reflectance Fourier-transform infrared (ATR-FTIR) analysis across the 200–4000 cm^{-1} spectral range with a Bruker Invenio R spectrometer (Ettlingen, Germany). Morphological assessment was conducted through transmission electron microscopy (TEM) employing a Thermo Scientific Talos 200i instrument operating at 200 kV (Thermo Scientific Co., Eindhoven, The Netherlands). Surface topography investigation was carried out using scanning electron microscopy with a Thermo Scientific Quattro S microscope at 30 kV under high vacuum conditions (Thermo Scientific Co., Eindhoven, The Netherlands). Photoluminescence (PL) measurements were performed using a fluorescence spectrometer (Fluorormax Plus, Horiba Scientific, Irvine, CA, USA) at room temperature with multiple excitation wavelengths (325, 365, and 405 nm) to comprehensively analyze defect-related emission characteristics. The defect-related emission intensity was correlated with potential ROS generation capacity based on established literature relationships between oxygen vacancy defects and reactive species formation.

2.5. Zeta Potential and Colloidal Stability Assessment

Zeta potential measurements were conducted using the salt addition method to determine the isoelectric point and assess colloidal stability. Solutions containing 50 mg ZnO nanoparticles in 50 mL of 0.1 M NaCl were prepared at pH values ranging from 3 to 11, adjusted using 0.1 M HCl and NaOH solutions. Samples were maintained under constant stirring for 24 h at 25 °C and subsequently filtered through 0.45 µm membrane filters, and the final pH was measured using a calibrated pH meter. The isoelectric point was determined from the intersection of initial vs. final pH plots. Chemical stability was verified by X-ray diffraction analysis of samples recovered after pH stability studies.

2.6. Primary Cultures of Embryonic Chicken Retinal Neuronal Cells

Fertilized chicken eggs (*Gallus gallus*, White Leghorn breed) at embryonic development stage 7 (E7) were sourced from a local commercial supplier (Arequipa, Peru) and maintained in controlled incubation at 37.4 °C with 5% moisture content. Embryonic tissue extraction was conducted under aseptic conditions within a laminar airflow workstation. After rapid enucleation of ocular structures, eyes were transferred to 40 mm glass culture vessels containing calcium and magnesium-free saline buffer (CMF). Retinal tissues were carefully extracted and subjected to enzymatic disaggregation using 0.2% trypsin solution (Sigma-Aldrich, St. Louis, MO, USA) for 15 min at 37 °C. Post-dissociation, cellular material was cultured in Dulbecco's Modified Eagle Medium (DMEM, Invitrogen, Carlsbad, CA, USA) supplemented with 10% fetal bovine serum (FBS, Sigma-Aldrich, St. Louis, MO, USA) and antimicrobial agents comprising 100 U/mL penicillin and 100 µg/mL streptomycin (Invitrogen, USA). Cellular suspensions were distributed onto 35 mm culture vessels (Corning, Corning, NY, USA) at a plating density of 2.5×10^6 cells per dish. Incubation conditions were maintained at 37 °C in a humidified environment containing 5% CO_2 and 95% atmospheric air.

2.7. NG-108 Glioblastoma Cell Line Culture

The NG-108 cellular model, representing a fusion hybrid of mouse N18TG-2 neuroblastoma and rat C6 BV-1 glioma lineages, was utilized in this investigation. Cryopreserved NG-108 stocks (comprising 40% DMEM, 50% FBS, 10% DMSO) were rapidly thawed at 37 °C in a water bath. Cellular material was gently resuspended and transferred to 10 mL culture tubes containing 1 mL DMEM medium. Cell lines were established in DMEM medium supplemented with 2 mM L-glutamine (Sigma-Aldrich, USA) and 10% FBS. Following 24 h of initial culture, the growth medium was refreshed to eliminate residual DMSO. Cellular maintenance was performed in 25 cm^2 culture flasks (Corning, USA) at 37 °C with 5% CO_2 atmosphere for extended periods. Medium renewal was conducted every 48 h. Subculturing procedures were executed at 70% confluency using 0.25% trypsin-EDTA solution (Gibco, USA). Cell suspensions underwent centrifugation at 3500 rpm for 5 min, and cellular quantification was accomplished using a Neubauer hemocytometer.

2.8. Zinc Oxide NP Concentrations and Treatment Times

Primary retinal cell cultures (E7C2) were maintained in DMEM medium supplemented with 10% FBS, 100 U/mL penicillin, and 100 µg/mL streptomycin at 37 °C in a 5% CO_2 humidified environment. ZnO nanoparticles underwent sterilization through autoclave treatment at 121 °C for 10 min at 15 psi pressure. A master stock solution of 1 mg/mL ZnO nanoparticles was prepared in sterile ultrapure water. Before cellular treatment, nanoparticles were subjected to ultrasonic dispersion (Branson Ultrasonics, Danbury, CT, USA) at ambient temperature for 10 min to ensure uniform distribution. Embryonic retinal and NG-108 cell cultures were exposed to ZnO nanoparticles at final working concentrations of 10, 50, and 100 µg/mL for exposure periods of 1, 6, 12, and 24 h. Control cultures were maintained without nanoparticle exposure for each experimental condition.

2.9. MTT Cell Viability Assays

Cytotoxic effects of ZnO nanoparticles were evaluated using the MTT (3-(4,5-Dimethylthiazol-2-yl)-2,5-diphenyltetrazolium bromide) colorimetric viability assay. Embryonic retinal cellular preparations (2.5×10^6 cells/dish) and NG-108 glioblastoma cultures (1×10^6 cells/20 mm dish) underwent treatment with ZnO nanoparticles as previously described. Post-treatment, MTT reagent (5 mg/mL in PBS, Sigma-Aldrich, USA) was introduced to each culture well and maintained for 4 h at 37 °C. Subsequently, cultures underwent washing with Hank's Balanced Salt Solution (HBSS composition: NaCl 128 mM, KCl 4 mM, Na_2HPO_4 1 mM, KH_2PO_4 0.5 mM, $MgCl_2$ 1 mM, $CaCl_2$ 3 mM, HEPES 20 mM, glucose 12 mM, pH 7.4). Formazan crystal dissolution was accomplished using an isopropanol-HCl mixture (1:0.6 ratio). Optical density measurements were performed at 570 nm using a spectrophotometer (Ultrospec 1000, Pharmacia Biotech, Amersham, UK).

2.10. DAPI Nuclear Staining in NG-108 Glioma Cultures

NG-108 cellular cultures exposed to ZnO nanoparticles (10, 50, and 100 μg/mL) for 1 h underwent nuclear morphology assessment. Culture preparations were washed twice with phosphate-buffered saline (0.16 M, pH 7.6) and fixed using 4% paraformaldehyde in PBS for 15 min at ambient temperature. Nuclear visualization was performed using DAPI (4′,6-diamidino-2-phenylindole, Invitrogen, USA) following the manufacturer's protocols. Cellular specimens were observed and documented using an Olympus IX73 Inverted Fluorescence Microscope (Olympus, Tokyo, Japan). Quantification of cells displaying normal versus fragmented nuclear morphology was conducted across 6 randomly selected fields at 37.8× magnification.

2.11. Statistical Analysis

Data analysis was performed using GraphPad Prism 8.4 statistical software (GraphPad Software Inc., San Diego, CA, USA). All experimental results are reported as mean values ± standard error of the mean (SEM) calculated from a minimum of three independent experimental replicates. Statistical significance evaluation was conducted through one-way analysis of variance (ANOVA), with post hoc analysis utilizing Tukey's honestly significant difference test and Dunnett's multiple comparison methodology. Statistical significance threshold was established at $p < 0.05$.

3. Results

3.1. Results of Molecular Identification

The taxonomic identification of bacterial strain TT14s was conducted through comprehensive molecular analysis of the 16S rRNA gene sequence using high-resolution comparative genomics approaches. The amplified sequence, comprising 1242 base pairs, was subjected to nucleotide BLAST analysis against the curated NCBI GenBank database to establish phylogenetic relationships. Sequence similarity analysis revealed 100% sequence identity with reference strains of *Bacillus licheniformis* (GenBank accession numbers NR_118996.1, NR_116023.1), indicating precise taxonomic placement within this species. To establish evolutionary relationships and confirm the taxonomic position, phylogenetic reconstruction was performed using the Neighbor Joining algorithm, incorporating appropriate nucleotide substitution models and statistical parameters. The resulting phylogenetic tree topology unambiguously positioned strain TT14s within the *B. licheniformis* clade, providing robust support for its taxonomic classification. This molecular systematic approach, based on 16S rRNA gene sequence analysis, represents a reliable method for bacterial identification, particularly within the genus Bacillus, where this genetic marker exhibits sufficient discriminatory power for species-level resolution. For reference purposes and to facilitate future comparative analyses, the 16S rRNA gene sequence of strain TT14s has been deposited in the GenBank nucleotide database and assigned the accession number PV034281.

3.2. Thermogravimetric and Scanning Differential Calorimetry

Figure 1 presents the thermal behavior of biosynthesized ZnO NPs through combined TGA-DSC analysis, revealing critical insights into material composition and transformation processes. The TGA curve demonstrates gradual mass reduction from ambient temperature, with pronounced weight loss between 200 °C and 300 °C, accounting for approximately 33% of initial sample weight, confirming substantial organic content from bacterial synthesis.

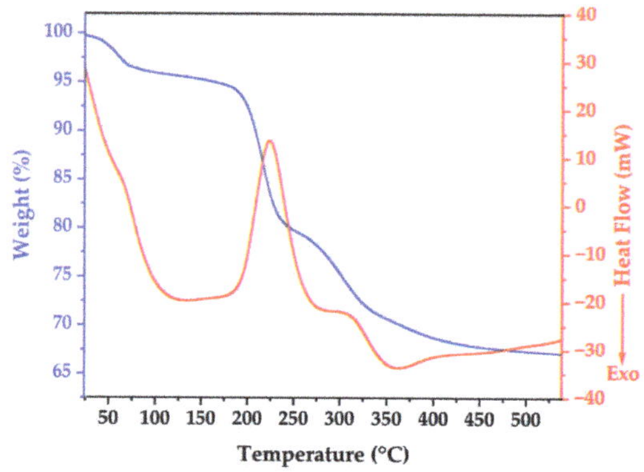

Figure 1. Thermogravimetric and differential scanning calorimetry of as prepared sample.

The DSC profile exhibits five distinct thermal events with specific mechanistic significance. An initial endothermic peak at 139.58 °C is attributed to desorption of surface-bound water molecules and volatile organic compounds from bacterial cell walls. A subsequent endothermic transition at 220 °C corresponds to zinc acetate complex decomposition and bacterial protein degradation. The major endothermic event at 285 °C represents primary bacterial biomass and organic capping agent decomposition, coinciding with maximum TGA mass loss. An exothermic peak at 365 °C indicates ZnO crystallization and annealing processes, while thermal events beyond 419 °C signify final organic matter elimination and crystal structure optimization.

The thermal profile stabilization beyond 350 °C validates the chosen calcination parameters (350 °C for 2 h), confirming effective removal of bacterial remnants while facilitating high-quality crystalline ZnO formation. This systematic thermal analysis elucidates the crucial role of controlled thermal treatment in transforming biogenic precursors into pure, crystalline ZnO nanoparticles with optimized structural properties [40,41].

3.3. X-Ray Diffraction Results

The X-ray diffraction pattern of the biogenically synthesized ZnO NPs is presented in Figure 2. The diffractogram reveals well-defined diffraction peaks that can be indexed to the hexagonal wurtzite crystal structure of zinc oxide, consistent with the Crystallography Open Database (COD) reference 96-210-7060. All observed diffraction peaks at 2θ values of 32.03°, 34.67°, 36.51°, 47.78°, 56.83°, 63.07°, 68.16°, 69.30°, 72.76°, and 77.16° correspond to the (100), (002), (101), (102), (110), (103), (112), (201), (004), and (202) crystallographic planes, respectively, confirming the formation of phase-pure ZnO without detectable impurities [42].

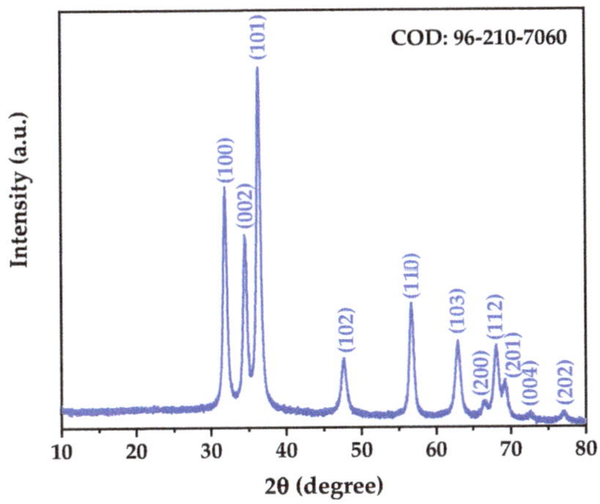

Figure 2. X-ray diffraction of obtained material.

Rietveld structure refinement was performed to obtain precise crystallographic parameters and structural information. The refinement confirmed that the synthesized ZnO NPs crystallize in the hexagonal crystal system with space group P 63 m c (No. 186), characteristic of the wurtzite structure [43]. The refined lattice parameters were determined to be a = b = 3.25018 Å and c = 5.2087 Å, yielding a c/a ratio of 1.603, which is consistent with the ideal wurtzite structure. These values are in excellent agreement with the COD Card 96-210-7060 standard reference for synthetic wurtzite.

The quality of the Rietveld refinement was assessed through statistical reliability factors. The weighted profile R-factor (R_{wp}) of 4.61% and profile R-factor (R_p) of 3.36% indicate an excellent fit between the experimental and calculated diffraction patterns [44]. The goodness of fit (GOF) value of 0.94 confirms the reliability of the structural model, as values close to unity indicate optimal refinement quality [45].

The average crystallite size was determined directly from the Rietveld refinement through the analysis of peak broadening parameters, yielding a value of 15.48 nm. This microstructural parameter was extracted from the refined profile function that accounts for both instrumental and sample-dependent broadening effects, providing a more accurate determination compared to single-peak analysis methods [46]. The obtained crystallite size falls within the typical range reported for biogenically synthesized ZnO NPs, confirming the nanocrystalline nature of the synthesized material [47].

Microstrain analysis revealed a low microstrain value of 0.016%, indicating minimal lattice distortions in the crystal structure. This low microstrain suggests well-ordered crystalline domains with minimal defects, which is characteristic of high-quality ZnO NPs synthesized under controlled conditions [48]. The calculated theoretical density of 5.67 g/cm^3 matches the standard bulk density of ZnO, confirming the formation of dense crystalline nanoparticles without significant porosity or structural defects.

The sharp and well-resolved diffraction peaks, combined with the absence of additional peaks, confirm the formation of single-phase hexagonal ZnO with high crystallinity. The narrow peak widths indicate good structural ordering and uniform crystallite size distribution, which is advantageous for potential biomedical applications where consistent physicochemical properties are crucial [49].

3.4. FTIR Spectroscopy Results

The FTIR spectrum of the calcined ZnO NPs (Figure 3) confirms successful biogenic synthesis and thermal processing. The spectrum exhibits four distinct absorption bands at 2346, 1533, 1377, and 381 cm^{-1}, which validate both ZnO formation and controlled surface chemistry following the 350 °C calcination treatment.

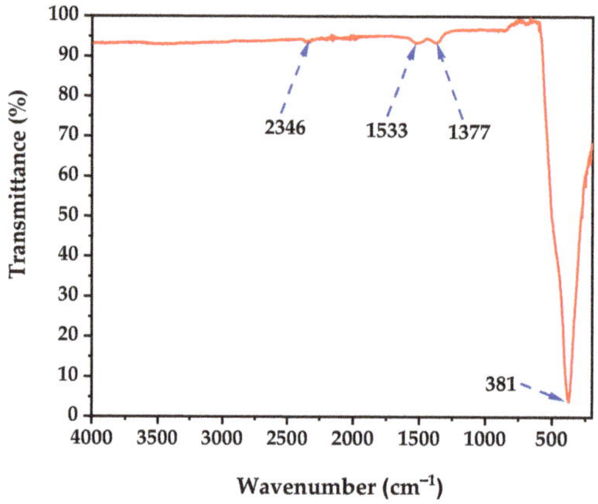

Figure 3. FTIR spectra of ZnO nanoparticles.

The broad transmission baseline above 3000 cm^{-1} indicates minimal O-H stretching vibrations from surface hydroxyl groups, confirming effective removal of bacterial biomass and adsorbed water during calcination [50]. This reduced hydration is consistent with the 33% mass loss observed in TGA analysis and demonstrates successful thermal processing.

The sharp absorption band at 2346 cm^{-1} corresponds to atmospheric CO_2 contamination during spectroscopic measurement [51]. However, minor contributions may arise from residual carbonate species formed during the synthesis process involving sodium bicarbonate as a co-precursor.

The absorption band at 1533 cm^{-1} is assigned to N-H bending vibrations and C-N stretching modes from thermally stable nitrogen-containing compounds, likely originating from bacterial cell wall components that survived the calcination process, indicating selective preservation of specific biomolecular functionalities that may contribute to enhanced biocompatibility. Additionally, the band at 1377 cm^{-1} corresponds to symmetric stretching vibrations of carboxylate groups (COO^-), formed during the interaction between zinc acetate precursors and bacterial exudates during the 72-h incubation period [52,53].

The most diagnostic absorption band appears at 381 cm^{-1}, corresponding to the characteristic Zn-O stretching vibrations of the hexagonal wurtzite crystal structure [54]. The sharp, intense nature of this fundamental vibration confirms formation of highly crystalline ZnO NPs, validating the effectiveness of the thermal treatment in promoting complete crystallization while maintaining the desired wurtzite structure.

The FTIR spectrum demonstrates that the biogenic synthesis protocol using *B. licheniformis* followed by 350 °C calcination produces ZnO NPs with minimal organic surface modification. The thermal treatment effectively removed the majority of bacterial biomass while preserving only essential surface functionalities (carboxylate and amine groups) that enhance biocompatibility without compromising the intrinsic crystalline properties of ZnO [25]. This controlled surface chemistry represents an optimal balance for biomedical applications, maintaining the crystalline integrity necessary for anticancer activity while preserving selective organic modifications derived from the bacterial synthesis process.

3.5. Scanning and Transmission Electron Microscopy

Scanning electron microscopy analysis (Figure 4a,b) revealed that biogenic ZnO NPs exhibit a complex morphological architecture characterized by primary structures ranging from spherical to plate-like configurations. These secondary structures comprise aggregates of individual nanoparticles, as clearly evidenced in transmission electron microscopy (TEM) images (Figure 4c), where apparently compact structures are observed to consist of discrete nanoparticles with an average size of 19.37 ± 5.28 nm. This hierarchical organization is characteristic of biogenic synthesis, wherein biomolecules function as stabilizing and morphology-directing agents, promoting the formation of controlled aggregates rather than isolated particles [55,56].

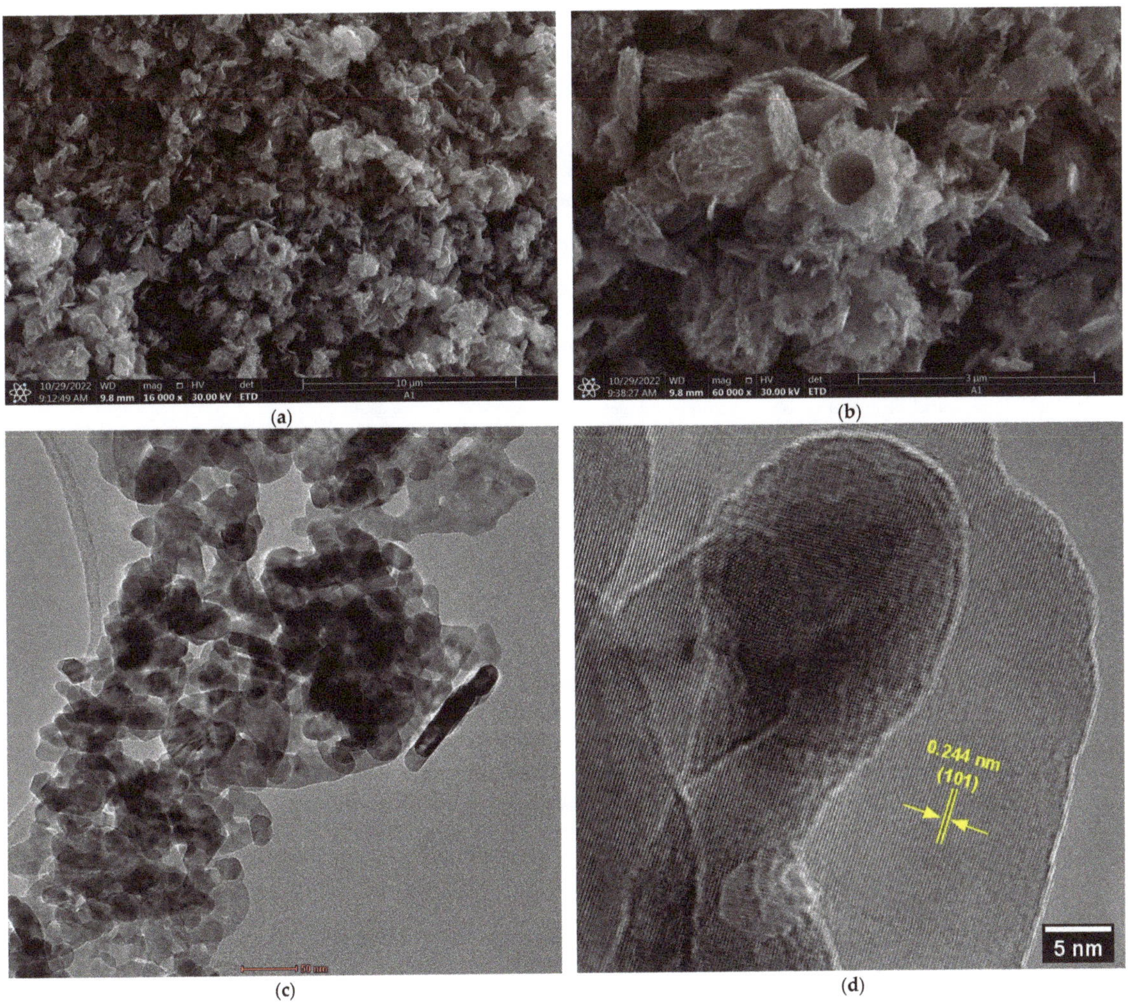

Figure 4. Electron microscopy characterization of biogenic ZnO NPs. (**a**,**b**) Scanning Electron Microscopy images at different magnifications showing spherical and plate-like ZnO structures. (**c**) TEM image revealing individual nanoparticles with average size of 19.37 ± 5.28 nm forming the observed structures. (**d**) HRTEM image showing lattice fringes with d-spacing of 0.244 nm corresponding to the (101) plane of wurtzite ZnO.

High-resolution transmission electron microscopy (HRTEM) (Figure 4d) analysis confirmed the crystalline nature of the synthesized material, revealing an interplanar spacing of 0.244 nm corresponding to the (101) crystallographic plane of the wurtzite ZnO structure [57]. This measurement demonstrates excellent agreement with reported values for high-quality crystalline ZnO and confirms that the biogenic synthesis process does not compromise the structural integrity of the material, maintaining the crystallographic properties necessary for advanced applications [58].

3.6. Photoluminescence Analysis and Defect Characterization

Photoluminescence spectroscopy at multiple excitation wavelengths provides critical insights into the electronic defect structure correlating with biological activity (Figure 5), where the PL spectra at excitation wavelengths of 254, 325, and 355 nm reveal distinct emission characteristics that depend on the excitation energy and penetration depth, with 254 nm excitation (4.88 eV) showing a dominant sharp emission peak centered at approximately 380 nm attributed to near-band-edge (NBE) recombination of free excitons consistent with established ZnO nanostructure while 325 nm excitation [59] (3.82 eV) produces NBE emission at 380 nm accompanied by broad defect-related emission extending from 500–600 nm with moderate intensity, and most significantly, 355 nm excitation (3.49 eV) generates the most intense broad emission band centered at approximately 550 nm in the green-yellow region accompanied by a secondary defect band around 500 nm. The broad emission at 550 nm observed under 355 nm excitation is attributed to oxygen vacancy (V_o) and zinc interstitial (Zn_i) defects within the ZnO lattice formed during the biogenic synthesis process under reducing bacterial metabolite conditions, where the excitation wavelength dependence indicates different defect levels being accessed, with 254 nm primarily activating band-edge states with minimal defect activation, 325 nm accessing intermediate defect levels with moderate oxygen vacancy contribution, and 355 nm achieving maximum defect state population particularly oxygen vacancies [60]. According to established literature correlations, the intensity of defect emission under 355 nm excitation directly correlates with ROS generation capacity under dark conditions through defect-mediated mechanisms where oxygen vacancies interact with water molecules to form hydroxyl radicals and subsequently generate superoxide anions and hydrogen peroxide in aqueous cellular environments through the reaction sequence [61] $V_O + H_2O \rightarrow V_O^+ + OH^- + e^-$, followed by $e^- + O_2 \rightarrow O_2^{-\bullet}$ and $2O_2^{-\bullet} + 2H^+ \rightarrow H_2O_2 + O_2$. The high defect-to-band-edge emission ratio indicates substantial oxygen vacancy concentration explaining the observed cytotoxic activity through defect-mediated ROS generation in aqueous cellular environments without external photoactivation, where this intrinsic ROS generation capacity maximally activated under near-UV conditions (355 nm) provides the mechanistic foundation for the selective anticancer activity observed in biological assays, as demonstrated by studies showing direct correlation between photoluminescence defect emission intensity and enhanced reactive oxygen species production capacity in ZnO nanoparticles with formation energies of oxygen vacancies calculated as 1.3–1.52 eV using thermodynamic approaches [21,62].

3.7. Colloidal Stability and Surface Charge Characteristics

Zeta potential analysis reveals critical information about colloidal stability and surface interactions relevant to biological applications (Figure 6), where the zeta potential varies systematically with pH exhibiting an isoelectric point (IEP) at pH 7.46 that is particularly significant as it closely matches physiological pH conditions (7.4), resulting in near-neutral surface charge (±2.5 mV) during biological assays, which differs from typical literature values of pH 9.7–10.3 for pristine ZnO nanoparticles indicating substantial surface modification effects from the biogenic synthesis process involving bacterial metabolites and extracellular polymeric substances [63]. At the experimental pH conditions used for cytotoxicity studies (pH 7.4–7.6), the minimal surface charge indicates reduced electrostatic interactions with cell membranes enabling controlled cellular uptake, minimal protein corona formation preserving nanoparticle bioactivity, and optimal dispersion stability without excessive aggregation or repulsion, as established by studies demonstrating that surface neutrality achieved through natural capping agents significantly limits binding sites for undesired serum proteins thereby reducing immune system activation and improving bioavailability in biological systems [64].

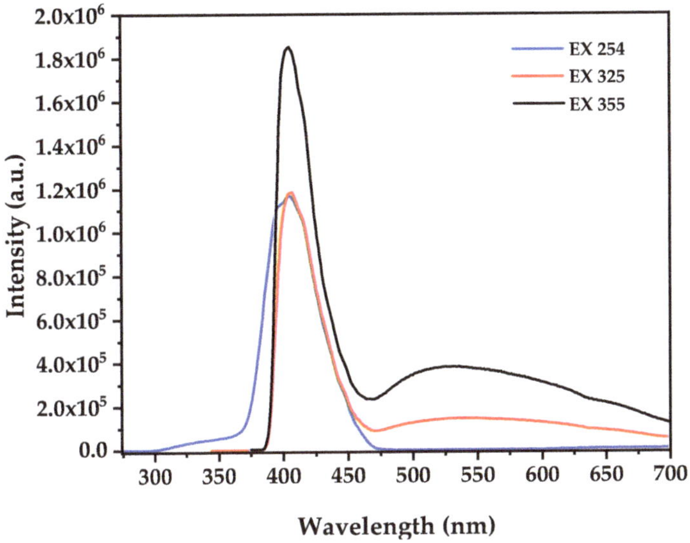

Figure 5. PL Spectra of biogenic ZnO NPs at distinct excitation wavelengths.

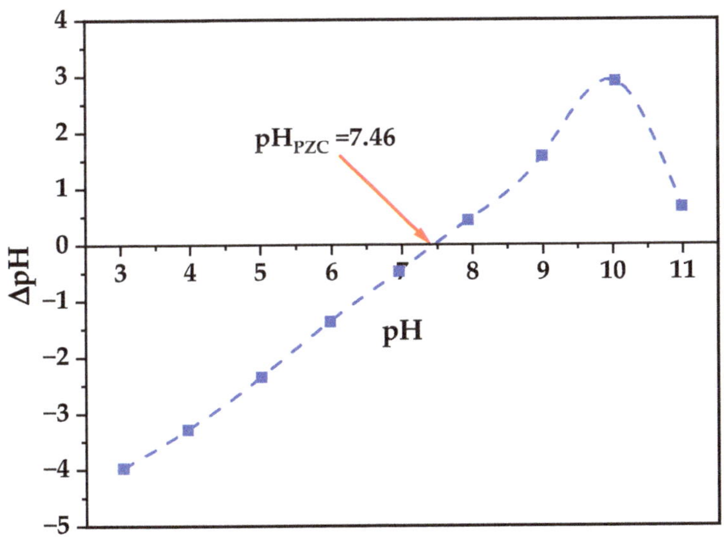

Figure 6. Zeta Potential of biogenic ZnO.

The pH stability study demonstrates remarkable chemical robustness across the tested pH range (3–11) where XRD analysis of samples recovered after 24 h exposure to extreme pH conditions (Figure 7) confirms retention of the pure hexagonal wurtzite structure without phase transitions or decomposition products, validating the suitability of biogenic ZnO nanoparticles for diverse biological applications spanning different physiological pH environments, which is consistent with reports showing ZnO nanoparticles maintain structural integrity despite size-dependent dissolution at circumneutral pH where organic chelators can enhance dissolution for smaller particles while larger aggregates remain stable [65]. This stability profile combined with the near-neutral zeta potential at physiological pH optimizes the balance between cellular uptake efficiency and biocompatibility contributing to the observed selective cytotoxicity profile against glioblastoma cells, where the enhanced stability compared to chemically synthesized ZnO is attributed to the protective biomolecular capping layer formed during bacterial synthesis providing steric stabilization and preventing aggregation-induced precipitation [66–68].

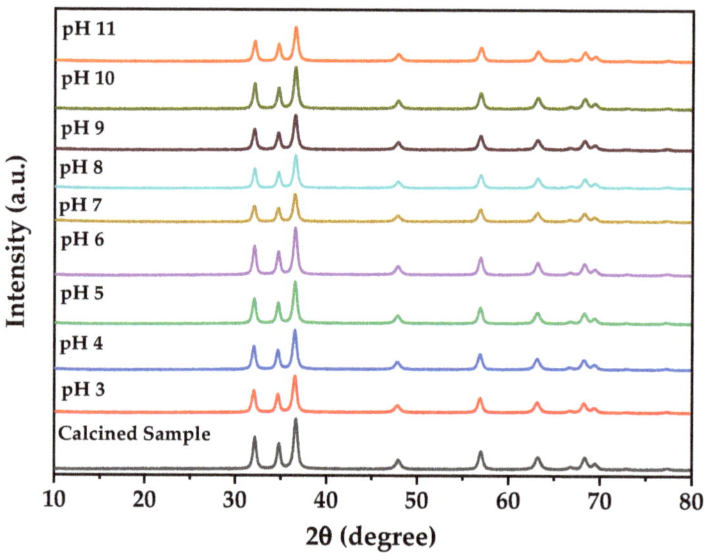

Figure 7. XRD analysis after different pH exposure of biogenic ZnO.

3.8. Effect of ZnO NPs on Primary Retinal Cells

The cytotoxic effect of ZnO NPs on primary cultures of embryonic chicken retinal cells was evaluated using the MTT assay. Our results demonstrated a concentration- and time-dependent cytotoxicity (Table 1, Figure 8). At lower concentrations (10 and 50 μg/mL), no significant cytotoxicity was observed after 1 h of incubation (96.53 ± 0.22% and 89.68 ± 1.70% viability, respectively; $p > 0.05$). However, significant cell death was observed at these concentrations after 6, 12, and 24 h of exposure ($p < 0.001$).

At the highest concentration tested (100 μg/mL), significant cytotoxicity was observed at all time points. Cell viability decreased to 78.90 ± 0.94% after 1 h and further declined to 43.48 ± 2.34% after 24 h of exposure ($p < 0.0001$). These results indicate that primary retinal cells are relatively resistant to short-term exposure to ZnO NPs at lower concentrations but become increasingly susceptible over time and at higher concentrations.

Table 1. Dose- and time-dependent cytotoxicity of zinc oxide nanoparticles (ZnO NPs) in primary retinal cells: cell viability percentages (mean ± SEM) from n = 3–4 independent experiments with N = 3–4 replicates per condition. Values with letter (a) show no significant differences (Tukey, $p > 0.05$).

Concentration	10 µg/mL			50 µg/mL			100 µg/mL		
Time	**Mean %**	**SEM**	**N**	**Mean %**	**SEM**	**N**	**Mean %**	**SEM**	**N**
Control	100.00 [a]	0.12	3	100.00 [a]	3.52	4	100	0.74	4
1 h	96.53 [a]	0.22	3	89.68 [a]	1.7	4	78.9	0.94	4
6 h	74.7	1.57	3	58.21	1.8	4	48.31	3.73	4
12 h	79.39	2.64	3	62.76	0.09	4	46.61	2.34	4
24 h	72.36	1.09	3	54.77	2.24	4	43.48	2.34	4

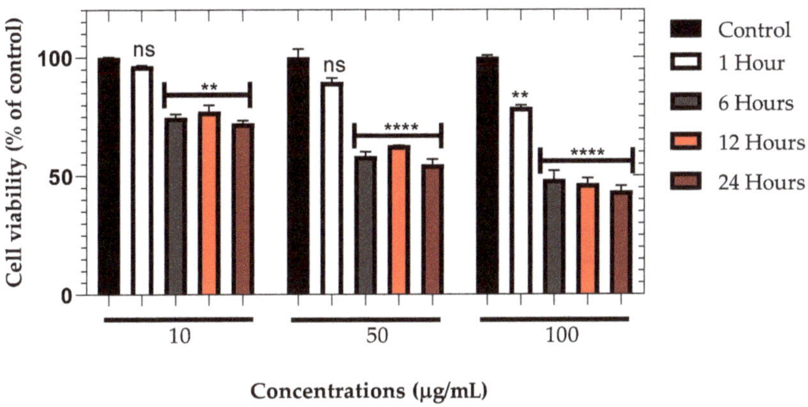

Figure 8. Time- and concentration-dependent cytotoxicity of ZnO NPs on primary retinal cells. Data are presented as mean ± SEM. ** $p < 0.01$, **** $p < 0.0001$ compared to control (one-way ANOVA followed by Dunnett's multiple comparisons test).

The observed time-dependent increase in cytotoxicity suggests a cumulative effect of ZnO NPs on retinal cells. This could be due to gradual nanoparticle uptake, sustained generation of reactive oxygen species (ROS), or progressive disruption of cellular processes. The concentration-dependent effect indicates a dose-threshold for significant cytotoxicity, which could be important for determining safe exposure levels in potential therapeutic applications [69–71].

3.9. Effect of ZnO NPs on NG-108 Glioblastoma Cells

The cytotoxic effect of ZnO NPs on NG-108 glioblastoma cells was more pronounced compared to primary retinal cells (Table 2, Figure 9). After just 1 h of incubation, significant reductions in cell viability were observed across all tested concentrations. Cell viability decreased to 70.41 ± 4.29%, 58.23 ± 3.15%, and 36.07 ± 1.89% at 10, 50, and 100 µg/mL, respectively ($p < 0.001$ for 10 µg/mL and $p < 0.0001$ for 50 and 100 µg/mL).

These results demonstrate a higher sensitivity of NG-108 glioblastoma cells to ZnO NPs compared to primary retinal cells, particularly at shorter exposure times. This differential cytotoxicity suggests a potential therapeutic window for targeting glioblastoma cells while minimizing damage to healthy neural tissue. The rapid onset of cytotoxicity in glioblastoma cells could be attributed to their higher metabolic rate, increased proliferation, or altered membrane permeability compared to normal cells [72–74].

Table 2. Effect of ZnO NPs on NG-108 glioblastoma cell viability after 1 h of exposure. Data are presented as mean percentage of viable cells ± SEM (n = 3 independent experiments).

Concentrations (μg/mL)	Zinc Oxide NPs Incubated for 1 h		
	Mean %	**SEM**	**N**
Control	100	1.22	3
10	70.41	4.29	3
50	58.23	3.15	3
100	36.07	1.89	3

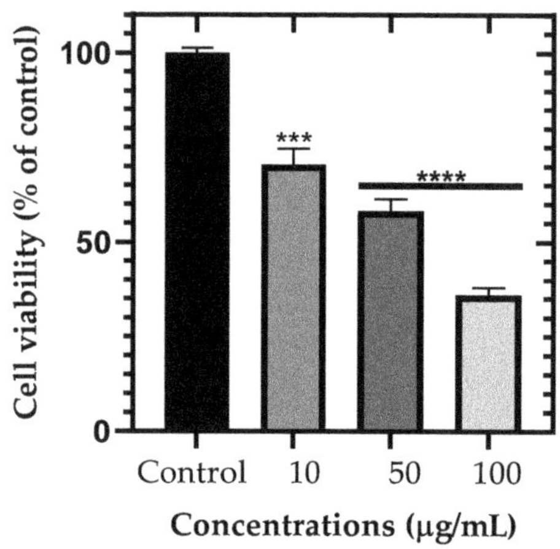

Figure 9. Dose-dependent cytotoxicity of ZnO NPs on NG-108 glioblastoma cells after 1 h of exposure. Data are presented as mean ± SEM. *** $p < 0.001$, **** $p < 0.0001$ compared to control (one-way ANOVA followed by Dunnett's multiple comparisons test).

3.10. Comparative Analysis of ZnO NPs' Sensitivity

To quantitatively assess the differential cytotoxic response and establish the therapeutic window of ZnO NPs, we performed a comprehensive comparative analysis be-tween primary retinal cells and NG-108 glioblastoma cells after 1 h of exposure (Figure 10). Two-way ANOVA revealed highly significant main effects of both cell type ($F_{1,14} = 89.42$, $p < 0.0001$) and ZnO concentration ($F_{3,14} = 11.00$, $p < 0.001$), with no significant interaction effect ($F_{3,14} = 1.25$, $p = 0.326$), indicating that the differential sensitivity pattern remains consistent across all tested concentrations.

Post hoc analysis using Sidak's multiple comparisons test revealed a concentration-dependent pattern of selectivity: no significant difference between cell types at 10 μg/mL (ns), but highly significant differences at 50 μg/mL ($p < 0.0001$, mean difference: 40%) and 100 μg/mL ($p < 0.0001$, mean difference: 45%). These results demonstrate that selective cytotoxicity emerges at concentrations $\geq$ 50 μg/mL, establishing a clear threshold for therapeutic selectivity.

The concentration-dependent selectivity analysis revealed minimal selectivity at 10 μg/mL (1.4-fold, ns), but substantial therapeutic windows at higher concentrations: 1.7-fold at 50 μg/mL and 2.3-fold at 100 μg/mL (both $p < 0.0001$). Notably, NG-108 glioblastoma cells exhibited significant cytotoxicity at all concentrations compared to un-treated controls ($p < 0.001$ for all concentrations), whereas primary retinal cells showed significant viability reduction only at 100 μg/mL ($p < 0.05$), confirming the safety profile for normal neural tissue at therapeutically relevant concentrations.

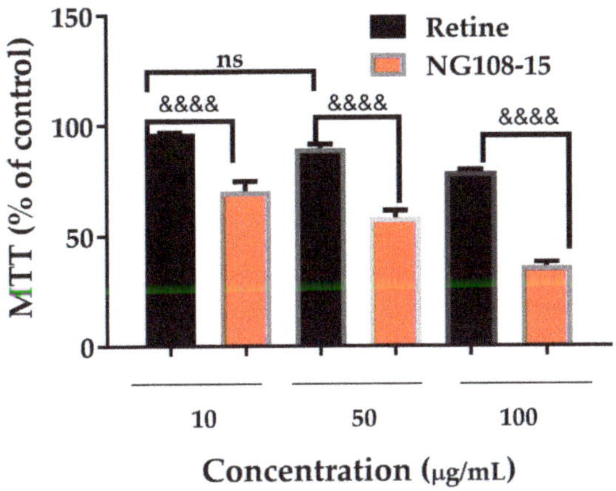

Figure 10. Comparative cell viability of retinal cells and NG-108 glioblastoma cells exposed to ZnO NPs for 1 h. Data represent mean ± SEM (n = 3–4). Two-way ANOVA with Sidak's post hoc test. &&&& $p < 0.0001$ between cell types; ns = not significant.

The rapid onset of selective cytotoxicity within 1 h suggests efficient cellular up-take and immediate biological activity in cancer cells, likely mediated by enhanced membrane permeability and compromised antioxidant systems characteristic of malignant phenotypes. This temporal pattern, combined with the dose-dependent selectivity profile, positions ZnO NPs as promising candidates for targeted glioblastoma therapy with minimal off-target effects on healthy neural tissue [19,24,69].

3.11. Nuclear Morphology Analysis of NG-108 Cells

To investigate the mechanism of cell death induced by ZnO NPs, we performed DAPI staining to assess nuclear morphology in NG-108 cells after 1 h of exposure (Figure 11). Quantitative analysis revealed a dose-dependent increase in the percentage of cells exhibiting nuclear fragmentation, a hallmark of apoptosis (Table 3, Figure 11).

Table 3. Percentage of NG-108 cells with normal nuclei after 1 h of exposure to ZnO NPs. Data are presented as mean percentage ± SEM (n = 3 independent experiments).

Zinc Oxide Concentration	Mean (%)	SEM	n
Control	100	0.441	3
10 μg/mL	74.6	1.076	3
50 μg/mL	63.6	1.417	3
100 μg/mL	54.54	0.782	3

The percentage of cells with normal nuclei decreased significantly with increasing ZnO NPs concentration. At 10 μg/mL, approximately 25.40% of cells showed nuclear fragmentation, increasing to 36.40% at 50 μg/mL and 45.46% at 100 μg/mL. These results suggest that ZnO NPs induce apoptosis in NG-108 glioblastoma cells in a dose-dependent manner.

The observed nuclear fragmentation correlates well with the decreased cell viability measured by the MTT assay, supporting apoptosis as a primary mechanism of ZnO NPs -induced cell death in glioblastoma cells. This finding is consistent with previous studies suggesting that ZnO NPs can trigger apoptotic pathways through ROS generation and DNA damage [67,75–77].

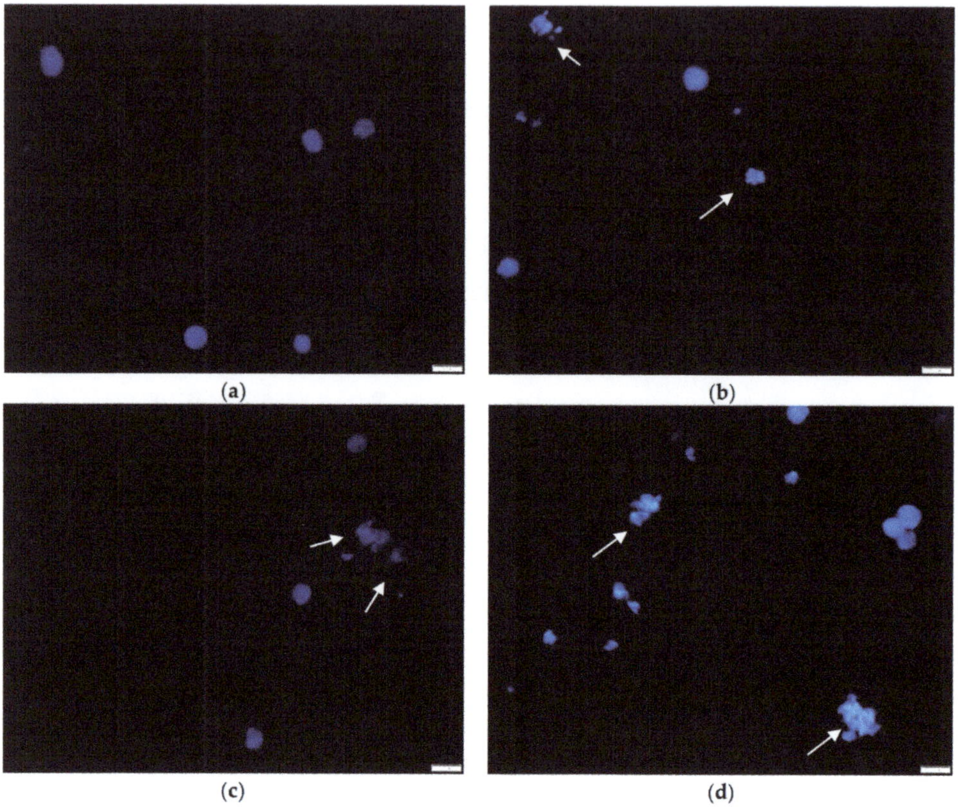

Figure 11. Representative fluorescence microscopy images of DAPI-stained NG-108 cells after 1 h of exposure to ZnO NPs. (a) Control, (b) 10 μg/mL, (c) 50 μg/mL, and (d) 100 μg/mL. White arrows indicate fragmented nuclei (Scale bar, 20 μm).

3.12. Mechanistic Insights into Selective Cytotoxicity

The observed selectivity results from three convergent mechanisms operating under dark laboratory conditions. First, defect-mediated ROS generation occurs through oxygen vacancy defects (characteristic 550 nm photoluminescence emission) that generate reactive species via surface reactions: $V_O + H_2O \rightarrow V_O^+ + OH^- + e^-$, followed by O_2 reduction to superoxide and hydrogen peroxide, explaining cytotoxic activity without external activation. Second, pH-responsive dissolution enabled by zeta potential characteristics (pHPZC = 7.46) promotes preferential zinc ion release in acidic tumor microenvironments (pH 5.5–6.5) versus physiological pH (7.4), while maintaining optimal dispersion stability (±2.5 mV) for cellular uptake. Third, differential cellular vulnerability manifests as rapid cytotoxicity in glioblastoma cells (1 h) versus delayed effects in retinal cells, reflecting enhanced nanoparticle uptake in cancer cells and their compromised antioxidant capacity compared to normal cells. This multi-factorial mechanism explains the 2.3-fold selectivity index, demonstrating intrinsic anticancer activity without external activation.

4. Discussion

The present study investigated the cytotoxic effects of biogenically synthesized ZnO NPs on primary retinal cells and NG-108 glioblastoma cells, aiming to evaluate their potential as a targeted therapy for glioblastoma. Our findings demonstrate a significant difference in sensitivity between normal neural cells and glioblastoma cells, suggesting a potential therapeutic window for ZnO NP-based treatments. The observed differential cytotoxicity is a key finding, with primary retinal cells showing resistance to ZnO NPs at lower concentrations (10 and 50 μg/mL) during short-term exposure (1 h), while glioblastoma cells exhibited significant cytotoxicity under the same conditions. This selective toxicity aligns with previous studies on other cancer cell types [78,79] and could be attributed to several factors, including differences in metabolic rates, cell membrane properties, and antioxidant capacities between normal and cancer cells [80].

Importantly, our study demonstrates that biogenic ZnO NPs achieve selective anticancer activity under standard dark laboratory conditions, eliminating the need for external photoactivation. The mechanistic foundation was elucidated through photoluminescence analysis revealing substantial oxygen vacancy defects (evidenced by intense 550 nm emission), which enable intrinsic ROS generation through defect-mediated surface reactions with cellular environments. This dark cytotoxicity, combined with pH-responsive dissolution behavior (pHPZC = 7.46), provides practical advantages over photodynamic approaches requiring controlled irradiation.

The time-dependent effects observed in primary retinal cells, where prolonged exposure resulted in increased cytotoxicity, suggest a cumulative impact of ZnO NPs on cellular function. This phenomenon could be explained by gradual nanoparticle uptake, zinc ion release from dissolving NPs, or the accumulation of cellular damage over time [66]. These findings emphasize the importance of considering exposure time when developing ZnO NP-based therapies and highlight the need for careful dosing strategies to minimize damage to healthy tissue [71].

Our nuclear morphology analysis of NG-108 cells exposed to ZnO NPs revealed a dose-dependent increase in nuclear fragmentation, a hallmark of apoptosis. This observation is consistent with previous studies suggesting that ZnO NPs induce apoptosis in various cancer cell types. The apoptotic mechanism involves defect-mediated ROS generation operating continuously under dark conditions, leading to sustained oxidative stress, DNA damage, and mitochondrial dysfunction. This intrinsic activity pattern differs from burst-type ROS generation in photodynamic therapy, potentially offering advantages in terms of sustained therapeutic pressure on cancer cells while allowing normal cells time for repair mechanisms. The predominance of apoptosis as the cell death mechanism is favorable for potential therapeutic applications, as it is less likely to induce inflammation compared to necrotic cell death [71,81,82].

The selective cytotoxicity of ZnO NPs towards glioblastoma cells observed in this study has significant implications for the development of novel glioblastoma therapies. The differential sensitivity between normal and cancer cells suggests that ZnO NPs could potentially target glioblastoma cells while sparing healthy neural tissue, addressing a major challenge in current glioblastoma treatments [83–85]. Furthermore, the unique mechanism of action of ZnO NPs, involving ROS generation and direct cellular damage, may help overcome the resistance mechanisms that limit the efficacy of conventional chemotherapies [71,86]. The nanoscale size of ZnO particles may also facilitate their passage through the blood–brain barrier, a significant obstacle in glioblastoma treatment. The dark cytotoxicity eliminates complications associated with light penetration in brain tissue, a significant advantage for glioblastoma treatment where deep tissue access is challenging.

Additionally, the apoptosis-inducing properties of ZnO NPs could potentially synergize with existing treatments, such as temozolomide or radiation therapy [87].

While our results are promising, several limitations should be addressed in future studies. The use of cell cultures, while informative, does not fully recapitulate the complexities of the tumor microenvironment or the blood–brain barrier. In vivo studies using appropriate animal models are necessary to validate these findings. Further investigation into the long-term effects of ZnO NPs on both normal and cancer cells is needed to assess their safety profile for potential clinical applications. More detailed mechanistic studies are required to fully understand the basis for the differential sensitivity between normal and cancer cells. The influence of nanoparticle size, shape, and surface modifications on their biological effects should be explored to optimize their therapeutic potential. Studies on the in vivo distribution, accumulation, and clearance of ZnO NPs are crucial for assessing their potential for clinical translation.

5. Conclusions

This study demonstrates the successful biogenic synthesis of ZnO NPs using *Bacillus licheniformis* strain TT14s, producing crystalline nanoparticles with a mean diameter of 19.37 ± 5.28 nm and a hexagonal wurtzite structure confirmed by XRD analysis. The synthesis protocol yielded high-purity ZnO with minimal organic residue after 350 °C calcination, as evidenced by the characteristic Zn-O stretching vibration at 381 cm^{-1} in FTIR analysis and lattice spacing of 0.244 nm corresponding to the (101) plane in HRTEM imaging.

The primary finding is the intrinsic selective cytotoxicity of ZnO NPs against NG-108 glioblastoma cells versus primary retinal cells under dark laboratory conditions. This selectivity is mechanistically attributed to: (1) defect-mediated ROS generation confirmed by characteristic 550 nm photoluminescence emission from oxygen vacancy defects, (2) pH-responsive dissolution behavior (pHPZC = 7.46) favoring zinc ion release in acidic tumor microenvironments, and (3) differential cellular uptake in rapidly proliferating cancer cells with compromised antioxidant systems. At 100 μg/mL, glioblastoma cell viability dropped to 36.07% within 1 h, while retinal cells maintained 78.9% viability under identical conditions, representing a practical 2.3-fold higher selectivity index.

DAPI nuclear staining quantitatively confirmed apoptotic cell death, with nuclear fragmentation increasing from 25.4% at 10 μg/mL to 45.46% at 100 μg/mL in glioblastoma cells. This dose-dependent response validates apoptosis as the primary death mechanism rather than necrosis. The 1 h exposure time required for significant cytotoxicity suggests rapid cellular uptake and immediate biological activity. The intrinsic cytotoxic activity under dark laboratory conditions eliminates the need for external activation, providing practical advantages for therapeutic applications.

ZnO NPs demonstrate promise as selective anti-glioma agents with intrinsic dark cytotoxicity through validated mechanisms: defect-mediated ROS generation, pH-responsive dissolution, and enhanced cancer cell uptake. The elimination of external activation requirements provides significant practical advantages for therapeutic development. The green synthesis approach using *B. licheniformis* offers a scalable, environmentally sustainable platform with superior biocompatibility compared to chemical methods, suitable for clinical translation as a dark-active therapeutic agent for glioblastoma treatment.

Author Contributions: Conceptualization, A.B.Q.C., G.d.L.F.Q. and E.J.S.S.; methodology, A.B.Q.C., G.d.L.F.Q., C.J.C.Q. and E.J.S.S.; software, G.d.L.F.Q. and E.J.S.S.; validation, A.B.Q.C., G.d.L.F.Q. and E.J.S.S.; formal analysis, A.B.Q.C.; investigation, A.B.Q.C., G.d.L.F.Q., R.M.A., T.A.M.G., C.J.C.Q., R.M.T.C., F.G.G. and E.J.S.S.; resources, A.B.Q.C., F.G.G. and C.J.C.Q.; data curation, R.M.A., T.A.M.G., G.d.L.F.Q. and E.J.S.S.; writing—original draft preparation, A.B.Q.C., G.d.L.F.Q. and E.J.S.S.; writing—review and editing, A.B.Q.C., G.d.L.F.Q. and E.J.S.S.; visualization, A.B.Q.C., G.d.L.F.Q. and

E.J.S.S.; supervision, A.B.Q.C.; project administration, A.B.Q.C.; funding acquisition, A.B.Q.C. All authors have read and agreed to the published version of the manuscript.

Funding: This research was funded by the Universidad Nacional Jorge Basadre Grohmann through "Fondos del canon sobrecanon y regalias mineras" with the project "Nanotecnologia microbiana: Síntesis verde de nanomateriales" approved with the resolution N° 10086-2022-UNJBG.

Data Availability Statement: The authors verify that all data obtained in this study are presented in this published article.

Acknowledgments: The gratitude to Universidad Nacional Jorge Basadre Grohmann for institutional support and access to research facilities essential for this investigation. We acknowledge the equipment support provided through multiple research projects that enabled this work: "Study of the application of nanotechnology for the purification of water with arsenic in the Tacna region" (Rectoral Resolution No. 3780-2014-UN/JBG), "Determination of the optical fingerprints of solid, liquid and organic materials using visible and infrared spectroscopy" (Rectoral Resolution No. 5854-2019-UN/JBG), "Estudio de materiales ferroeléctricos ($BiFeO_3$ y Bi_2FeCrO_6) y su aplicación en celdas solares" (Rectoral Resolution No. 4516-2018-UN/JBG), and "Análisis genómico de microorganismos degradadores de cianuro para la remediación, en fase de biorreactor de pasivos ambientales mineros de la región de Tacna" (Rectoral Resolution No. 4723-2015-UN/JBG). These projects provided crucial access to material characterization facilities and the bioremediation laboratory infrastructure necessary for nanoparticle biosynthesis and comprehensive analysis. The author Elisban Juani Sacari Sacari gratefully acknowledges the financial support provided by CONCYTEC through the PROCIENCIA program under the "Becas en programas de doctorado en alianzas interinstitucionales" competition, according to contracts N°PE501088673-2024-PROCIENCIA-BM and N°PE501084296-2023-PROCIENCIA-BM for undertaking a Doctoral program in Physics at the Universidad Nacional de Ingeniería, Peru.

Conflicts of Interest: The authors declare no conflicts of interest.

References

1. Miller, K.D.; Ostrom, Q.T.; Kruchko, C.; Patil, N.; Tihan, T.; Cioffi, G.; Fuchs, H.E.; Waite, K.A.; Jemal, A.; Siegel, R.L.; et al. Brain and other central nervous system tumor statistics, 2021. *CA Cancer J. Clin.* **2021**, *71*, 381–406. [CrossRef] [PubMed]
2. Price, M.; Ballard, C.; Benedetti, J.; Neff, C.; Cioffi, G.; Waite, K.A.; Kruchko, C.; Barnholtz-Sloan, J.S.; Ostrom, Q.T. CBTRUS Statistical Report: Primary Brain and Other Central Nervous System Tumors Diagnosed in the United States in 2017–2021. *Neuro Oncol.* **2024**, *26*, vi1–vi85. [CrossRef]
3. Stupp, R.; Mason, W.P.; van den Bent, M.J.; Weller, M.; Fisher, B.; Taphoorn, M.J.B.; Belanger, K.; Brandes, A.A.; Marosi, C.; Bogdahn, U.; et al. Radiotherapy plus concomitant and adjuvant temozolomide for glioblastoma. *N. Engl. J. Med.* **2005**, *352*, 987–996. [CrossRef]
4. Tamimi, A.F.; Juweid, M. *Glioblastoma: Epidemiology and Outcome of Glioblastoma*; Exon Publications: Brisbane, Australia, 2017; ISBN 9780994438126.
5. Venkataramani, V.; Yang, Y.; Schubert, M.C.; Reyhan, E.; Tetzlaff, S.K.; Wißmann, N.; Botz, M.; Soyka, S.J.; Beretta, C.A.; Pramatarov, R.L.; et al. Glioblastoma hijacks neuronal mechanisms for brain invasion. *Cell* **2022**, *185*, 2899–2917.e31. [CrossRef]
6. Kang, S.; Ughetta, M.E.; Zhang, J.Y.; Marallano, V.J.; Sattiraju, A.; Hannah, T.; Wahane, S.; Ramakrishnan, A.; Estill, M.; Tsankova, N.M.; et al. Glioblastoma shift from bulk to infiltrative growth is guided by plexin-B2-mediated microglia alignment in invasive niches. *Nat. Cancer* **2025**, 1–19. [CrossRef]
7. Bao, S.; Wu, Q.; McLendon, R.E.; Hao, Y.; Shi, Q.; Hjelmeland, A.B.; Dewhirst, M.W.; Bigner, D.D.; Rich, J.N. Glioma stem cells promote radioresistance by preferential activation of the DNA damage response. *Nature* **2006**, *444*, 756–760. [CrossRef]
8. Lathia, J.D.; Mack, S.C.; Mulkearns-Hubert, E.E.; Valentim, C.L.L.; Rich, J.N. Cancer stem cells in glioblastoma. *Genes Dev.* **2015**, *29*, 1203–1217. [CrossRef] [PubMed]
9. Hegi, M.E.; Diserens, A.-C.; Gorlia, T.; Hamou, M.-F.; de Tribolet, N.; Weller, M.; Kros, J.M.; Hainfellner, J.A.; Mason, W.; Mariani, L.; et al. MGMT gene silencing and benefit from temozolomide in glioblastoma. *N. Engl. J. Med.* **2005**, *352*, 997–1003. [CrossRef] [PubMed]
10. Lee, S.Y. Temozolomide resistance in glioblastoma multiforme. *Genes Dis.* **2016**, *3*, 198–210. [CrossRef]

11. Wu, B.; Sun, C.; Feng, F.; Ge, M.; Xia, L. Do relevant markers of cancer stem cells CD133 and Nestin indicate a poor prognosis in glioma patients? A systematic review and meta-analysis. *J. Exp. Clin. Cancer Res.* **2015**, *34*, 44. [CrossRef]
12. Abbott, N.J.; Patabendige, A.A.K.; Dolman, D.E.M.; Yusof, S.R.; Begley, D.J. Structure and function of the blood-brain barrier. *Neurobiol. Dis.* **2010**, *37*, 13–25. [CrossRef]
13. van Tellingen, O.; Yetkin-Arik, B.; de Gooijer, M.C.; Wesseling, P.; Wurdinger, T.; de Vries, H.E. Overcoming the blood-brain tumor barrier for effective glioblastoma treatment. *Drug Resist. Updat.* **2015**, *19*, 1–12. [CrossRef]
14. Pardridge, W.M. The blood-brain barrier: Bottleneck in brain drug development. *NeuroRx* **2005**, *2*, 3–14. [CrossRef]
15. Cavallaro, P.A.; de Santo, M.; Belsito, E.L.; Longobucco, C.; Curcio, M.; Morelli, C.; Pasqua, L.; Leggio, A. Peptides Targeting HER2-Positive Breast Cancer Cells and Applications in Tumor Imaging and Delivery of Chemotherapeutics. *Nanomaterials* **2023**, *13*, 2476. [CrossRef]
16. Shi, J.; Kantoff, P.W.; Wooster, R.; Farokhzad, O.C. Cancer nanomedicine: Progress, challenges and opportunities. *Nat. Rev. Cancer* **2017**, *17*, 20–37. [CrossRef]
17. Mitchell, M.J.; Billingsley, M.M.; Haley, R.M.; Wechsler, M.E.; Peppas, N.A.; Langer, R. Engineering precision nanoparticles for drug delivery. *Nat. Rev. Drug Discov.* **2021**, *20*, 101–124. [CrossRef] [PubMed]
18. Jiang, J.; Pi, J.; Cai, J. The Advancing of Zinc Oxide Nanoparticles for Biomedical Applications. *Bioinorg. Chem. Appl.* **2018**, *2018*, 1062562. [CrossRef] [PubMed]
19. Bisht, G.; Rayamajhi, S. ZnO Nanoparticles: A Promising Anticancer Agent. *Nanobiomedicine (Rij)* **2016**, *3*, 9. [CrossRef]
20. Djurišić, A.B.; Ng, A.; Chen, X.Y. ZnO nanostructures for optoelectronics: Material properties and device applications. *Prog. Quantum Electron.* **2010**, *34*, 191–259. [CrossRef]
21. Wang, J.; Lee, J.S.; Kim, D.; Zhu, L. Exploration of Zinc Oxide Nanoparticles as a Multitarget and Multifunctional Anticancer Nanomedicine. *ACS Appl. Mater. Interfaces* **2017**, *9*, 39971–39984. [CrossRef]
22. Ostrovsky, S.; Kazimirsky, G.; Gedanken, A.; Brodie, C. Selective cytotoxic effect of ZnO nanoparticles on glioma cells. *Nano Res.* **2009**, *2*, 882–890. [CrossRef]
23. Wang, H.; Joseph, J.A. Quantifying cellular oxidative stress by dichlorofluorescein assay using microplate reader. *Free Radic. Biol. Med.* **1999**, *27*, 612–616. [CrossRef] [PubMed]
24. Kalpana, V.N.; Devi Rajeswari, V. A Review on Green Synthesis, Biomedical Applications, and Toxicity Studies of ZnO NPs. *Bioinorg. Chem. Appl.* **2018**, *2018*, 3569758. [CrossRef] [PubMed]
25. Siddiqi, K.S.; Husen, A.; Rao, R.A.K. A review on biosynthesis of silver nanoparticles and their biocidal properties. *J. Nanobiotechnol.* **2018**, *16*, 14. [CrossRef]
26. Kuppusamy, P.; Yusoff, M.M.; Maniam, G.P.; Govindan, N. Biosynthesis of metallic nanoparticles using plant derivatives and their new avenues in pharmacological applications—An updated report. *Saudi Pharm. J.* **2016**, *24*, 473–484. [CrossRef] [PubMed]
27. Agarwal, H.; Venkat Kumar, S.; Rajeshkumar, S. A review on green synthesis of zinc oxide nanoparticles—An eco-friendly approach. *Resour. Effic. Technol.* **2017**, *3*, 406–413. [CrossRef]
28. Jameel, M.S.; Aziz, A.A.; Dheyab, M.A. Green synthesis: Proposed mechanism and factors influencing the synthesis of platinum nanoparticles. *Green Process. Synth.* **2020**, *9*, 386–398. [CrossRef]
29. Das, R.K.; Pachapur, V.L.; Lonappan, L.; Naghdi, M.; Pulicharla, R.; Maiti, S.; Cledon, M.; Dalila, L.M.A.; Sarma, S.J.; Brar, S.K. Biological synthesis of metallic nanoparticles: Plants, animals and microbial aspects. *Nanotechnol. Environ. Eng.* **2017**, *2*, 18. [CrossRef]
30. Singh, P.; Kim, Y.-J.; Zhang, D.; Yang, D.-C. Biological Synthesis of Nanoparticles from Plants and Microorganisms. *Trends Biotechnol.* **2016**, *34*, 588–599. [CrossRef]
31. Kalimuthu, K.; Suresh Babu, R.; Venkataraman, D.; Bilal, M.; Gurunathan, S. Biosynthesis of silver nanocrystals by Bacillus licheniformis. *Colloids Surf. B Biointerfaces* **2008**, *65*, 150–153. [CrossRef]
32. Schallmey, M.; Singh, A.; Ward, O.P. Developments in the use of Bacillus species for industrial production. *Can. J. Microbiol.* **2004**, *50*, 1–17. [CrossRef]
33. Salem, S.S.; Fouda, A. Green Synthesis of Metallic Nanoparticles and Their Prospective Biotechnological Applications: An Overview. *Biol. Trace Elem. Res.* **2021**, *199*, 344–370. [CrossRef] [PubMed]
34. Hanahan, D.; Weinberg, R.A. Hallmarks of cancer: The next generation. *Cell* **2011**, *144*, 646–674. [CrossRef] [PubMed]
35. Rodríguez, F.; Caruana, P.; de La Fuente, N.; Español, P.; Gámez, M.; Balart, J.; Llurba, E.; Rovira, R.; Ruiz, R.; Martín-Lorente, C.; et al. Nano-Based Approved Pharmaceuticals for Cancer Treatment: Present and Future Challenges. *Biomolecules* **2022**, *12*, 784. [CrossRef]
36. Wicki, A.; Witzigmann, D.; Balasubramanian, V.; Huwyler, J. Nanomedicine in cancer therapy: Challenges, opportunities, and clinical applications. *J. Control. Release* **2015**, *200*, 138–157. [CrossRef]

37. National Academies Press (US). *Dietary Reference Intakes for Vitamin A, Vitamin K, Arsenic, Boron, Chromium, Copper, Iodine, Iron, Manganese, Molybdenum, Nickel, Silicon, Vanadium, and Zinc*; National Academies Press: Washington, DC, USA, 2001; ISBN 0309072794.
38. Cáceda Quiroz, C.J.; Fora Quispe, G.d.L.; Carpio Mamani, M.; Maraza Choque, G.J.; Sacari Sacari, E.J. Cyanide Bioremediation by Bacillus subtilis under Alkaline Conditions. *Water* **2023**, *15*, 3645. [CrossRef]
39. Tripathi, R.M.; Bhadwal, A.S.; Gupta, R.K.; Singh, P.; Shrivastav, A.; Shrivastav, B.R. ZnO nanoflowers: Novel biogenic synthesis and enhanced photocatalytic activity. *J. Photochem. Photobiol. B* **2014**, *141*, 288–295. [CrossRef] [PubMed]
40. Jebali, M.; Colangelo, G.; Gómez-Merino, A.I. Green Synthesis, Characterization, and Empirical Thermal Conductivity Assessment of ZnO Nanofluids for High-Efficiency Heat-Transfer Applications. *Materials* **2023**, *16*, 1542. [CrossRef] [PubMed]
41. Ebadi, M.; Zolfaghari, M.R.; Aghaei, S.S.; Zargar, M.; Shafiei, M.; Zahiri, H.S.; Noghabi, K.A. A bio-inspired strategy for the synthesis of zinc oxide nanoparticles (ZnO NPs) using the cell extract of cyanobacterium Nostoc sp. EA03: From biological function to toxicity evaluation. *RSC Adv.* **2019**, *9*, 23508–23525. [CrossRef]
42. Abrahams, S.C.; Bernstein, J.L. Remeasurement of the structure of hexagonal ZnO. *Struct. Sci.* **1969**, *25*, 1233–1236. [CrossRef]
43. Özgür, Ü.; Alivov, Y.I.; Liu, C.; Teke, A.; Reshchikov, M.A.; Doğan, S.; Avrutin, V.; Cho, S.-J.; Morkoç, H. A comprehensive review of ZnO materials and devices. *J. Appl. Phys.* **2005**, *98*, 041301. [CrossRef]
44. McCusker, L.B.; von Dreele, R.B.; Cox, D.E.; Louër, D.; Scardi, P. Rietveld refinement guidelines. *J. Appl. Crystallogr.* **1999**, *32*, 36–50. [CrossRef]
45. Toby, B.H. R factors in Rietveld analysis: How good is good enough? *Powder Diffr.* **2006**, *21*, 67–70. [CrossRef]
46. Langford, J.I.; Wilson, A.J.C. Scherrer after sixty years: A survey and some new results in the determination of crystallite size. *J. Appl. Crystallogr.* **1978**, *11*, 102–113. [CrossRef]
47. Jan, H.; Shah, M.; Usman, H.; Khan, M.A.; Zia, M.; Hano, C.; Abbasi, B.H. Biogenic Synthesis and Characterization of Antimicrobial and Antiparasitic Zinc Oxide (ZnO) Nanoparticles Using Aqueous Extracts of the Himalayan Columbine (Aquilegia pubiflora). *Front. Mater.* **2020**, *7*, 249. [CrossRef]
48. Prabhu, Y.T.; Rao, K.V.; Kumar, V.S.S.; Kumari, B.S. X-Ray Analysis by Williamson-Hall and Size-Strain Plot Methods of ZnO Nanoparticles with Fuel Variation. *WJNSE* **2014**, *04*, 21–28. [CrossRef]
49. Meulenkamp, E.A. Synthesis and Growth of ZnO Nanoparticles. *J. Phys. Chem. B* **1998**, *102*, 5566–5572. [CrossRef]
50. Faisal, S.; Jan, H.; Shah, S.A.; Shah, S.; Khan, A.; Akbar, M.T.; Rizwan, M.; Jan, F.; Wajidullah; Akhtar, N.; et al. Green Synthesis of Zinc Oxide (ZnO) Nanoparticles Using Aqueous Fruit Extracts of Myristica fragrans: Their Characterizations and Biological and Environmental Applications. *ACS Omega* **2021**, *6*, 9709–9722. [CrossRef]
51. Stuart, B. *Infrared Spectroscopy: Fundamentals and Applications*; John Wiley and Sons Ltd.: Chichester, UK; Hoboken, NJ, USA, 2004; ISBN 978-0-470-85428-0.
52. Pavithra, S.; Mani, M.; Mohana, B.; Jayavel, R.; Kumaresan, S. Precursor Dependent Tailoring of Morphology and Crystallite Size of Biogenic ZnO Nanostructures with Enhanced Antimicrobial Activity—A Novel Green Chemistry Approach. *BioNanoScience* **2021**, *11*, 44–52. [CrossRef]
53. Quispe Cohaila, A.B.; Fora Quispe, G.d.L.; Lanchipa Ramos, W.O.; Cáceda Quiroz, C.J.; Tamayo Calderón, R.M.; Medina Salas, J.P.; Rajendran, S.; Sacari Sacari, E.J. Synthesis of ZnO Nanoparticles by Bacillus subtilis for Efficient Photocatalytic Degradation of Cyanide. *Nanomaterials* **2025**, *15*, 501. [CrossRef] [PubMed]
54. Matei, A.; Cernica, I.; Cadar, O.; Roman, C.; Schiopu, V. Synthesis and characterization of ZnO–polymer nanocomposites. *Int. J. Mater. Form.* **2008**, *1*, 767–770. [CrossRef]
55. Xia, Y.; Wang, J.; Chen, R.; Zhou, D.; Xiang, L. A Review on the Fabrication of Hierarchical ZnO Nanostructures for Photocatalysis Application. *Crystals* **2016**, *6*, 148. [CrossRef]
56. Kovács, Z.; Márta, V.; Gyulavári, T.; Ágoston, Á.; Baia, L.; Pap, Z.; Hernadi, K. Noble metal modified (002)-oriented ZnO hollow spheres for the degradation of a broad range of pollutants. *J. Environ. Chem. Eng.* **2022**, *10*, 107655. [CrossRef]
57. Song, K.-W.; Park, M.-H.; Kim, T.-H.; Lim, S.-H.; Yang, C.-W. UV enhanced synthesis of high density Au coated ZnO nanocomposite. *J. Nanosci. Nanotechnol.* **2014**, *14*, 8766–8770. [CrossRef]
58. Silva, E.; Vilchis-Nestor, A.R.; de La Cruz, W.; Regalado-Contreras, A.; Castro-Beltran, A.; Luque, P.A. Analysis of the structural, morphological, and optical properties of Selaginella lepidophylla-mediated ZnO semiconductor nanoparticles and their influence on the photocatalytic removal of Rhodamine B dye. *J. Mater. Sci. Mater. Electron.* **2023**, *34*, 1330. [CrossRef]
59. Ng, H.T.; Chen, B.; Li, J.; Han, J.; Meyyappan, M.; Wu, J.; Li, S.X.; Haller, E.E. Optical properties of single-crystalline ZnO nanowires on m -sapphire. *Appl. Phys. Lett.* **2003**, *82*, 2023–2025. [CrossRef]
60. Kim, Y.; Kang, S. Calculation of formation energy of oxygen vacancy in ZnO based on photoluminescence measurements. *J. Phys. Chem. B* **2010**, *114*, 7874–7878. [CrossRef] [PubMed]

61. Lipovsky, A.; Tzitrinovich, Z.; Friedmann, H.; Applerot, G.; Gedanken, A.; Lubart, R. EPR Study of Visible Light-Induced ROS Generation by Nanoparticles of ZnO. *J. Phys. Chem. C* **2009**, *113*, 15997–16001. [CrossRef]
62. He, W.; Jia, H.; Cai, J.; Han, X.; Zheng, Z.; Wamer, W.G.; Yin, J.-J. Production of Reactive Oxygen Species and Electrons from Photoexcited ZnO and ZnS Nanoparticles: A Comparative Study for Unraveling their Distinct Photocatalytic Activities. *J. Phys. Chem. C* **2016**, *120*, 3187–3195. [CrossRef]
63. Dumontel, B.; Canta, M.; Engelke, H.; Chiodoni, A.; Racca, L.; Ancona, A.; Limongi, T.; Canavese, G.; Cauda, V. Enhanced biostability and cellular uptake of zinc oxide nanocrystals shielded with a phospholipid bilayer. *J. Mater. Chem. B* **2017**, *5*, 8799–8813. [CrossRef]
64. Srivastav, A.K.; Dhiman, N.; Khan, H.; Srivastav, A.K.; Yadav, S.K.; Prakash, J.; Arjaria, N.; Singh, D.; Yadav, S.; Patnaik, S.; et al. Impact of Surface-Engineered ZnO Nanoparticles on Protein Corona Configuration and Their Interactions With Biological System. *J. Pharm. Sci.* **2019**, *108*, 1872–1889. [CrossRef]
65. Mudunkotuwa, I.A.; Rupasinghe, T.; Wu, C.-M.; Grassian, V.H. Dissolution of ZnO nanoparticles at circumneutral pH: A study of size effects in the presence and absence of citric acid. *Langmuir* **2012**, *28*, 396–403. [CrossRef] [PubMed]
66. James, S.A.; Feltis, B.N.; de Jonge, M.D.; Sridhar, M.; Kimpton, J.A.; Altissimo, M.; Mayo, S.; Zheng, C.; Hastings, A.; Howard, D.L.; et al. Quantification of ZnO nanoparticle uptake, distribution, and dissolution within individual human macrophages. *ACS Nano* **2013**, *7*, 10621–10635. [CrossRef] [PubMed]
67. Li, Y.; Li, J.; Lu, Y.; Ma, Y. ZnO nanomaterials target mitochondrial apoptosis and mitochondrial autophagy pathways in cancer cells. *Cell Biochem. Funct.* **2024**, *42*, e3909. [CrossRef] [PubMed]
68. Giau, V.-V.; Park, Y.-H.; Shim, K.-H.; Son, S.-W.; An, S.-S.A. Dynamic changes of protein corona compositions on the surface of zinc oxide nanoparticle in cell culture media. *Front. Chem. Sci. Eng.* **2019**, *13*, 90–97. [CrossRef]
69. Ickrath, P.; Wagner, M.; Scherzad, A.; Gehrke, T.; Burghartz, M.; Hagen, R.; Radeloff, K.; Kleinsasser, N.; Hackenberg, S. Time-Dependent Toxic and Genotoxic Effects of Zinc Oxide Nanoparticles after Long-Term and Repetitive Exposure to Human Mesenchymal Stem Cells. *Int. J. Environ. Res. Public Health* **2017**, *14*, 1590. [CrossRef]
70. Pandurangan, M.; Veerappan, M.; Kim, D.H. Cytotoxicity of zinc oxide nanoparticles on antioxidant enzyme activities and mRNA expression in the cocultured C2C12 and 3T3-L1 cells. *Appl. Biochem. Biotechnol.* **2015**, *175*, 1270–1280. [CrossRef]
71. Sharma, V.; Anderson, D.; Dhawan, A. Zinc oxide nanoparticles induce oxidative DNA damage and ROS-triggered mitochondria mediated apoptosis in human liver cells (HepG2). *Apoptosis* **2012**, *17*, 852–870. [CrossRef]
72. Jovito, B.L.; Paterno, L.G.; Sales, M.J.A.; Gross, M.A.; Silva, L.P.; de Souza, P.; Báo, S.N. Graphene Oxide/Zinc Oxide Nanocomposite Displaying Selective Toxicity to Glioblastoma Cell Lines. *ACS Appl. Bio Mater.* **2021**, *4*, 829–843. [CrossRef]
73. Heng, B.C.; Zhao, X.; Tan, E.C.; Khamis, N.; Assodani, A.; Xiong, S.; Ruedl, C.; Ng, K.W.; Loo, J.S.-C. Evaluation of the cytotoxic and inflammatory potential of differentially shaped zinc oxide nanoparticles. *Arch. Toxicol.* **2011**, *85*, 1517–1528. [CrossRef]
74. Sahu, D.; Kannan, G.M.; Tailang, M.; Vijayaraghavan, R. In Vitro Cytotoxicity of Nanoparticles: A Comparison between Particle Size and Cell Type. *J. Nanosci.* **2016**, *2016*, 4023852. [CrossRef]
75. Chung, I.-M.; Rahuman, A.A.; Marimuthu, S.; Kirthi, A.V.; Anbarasan, K.; Rajakumar, G. An Investigation of the Cytotoxicity and Caspase-Mediated Apoptotic Effect of Green Synthesized Zinc Oxide Nanoparticles Using Eclipta prostrata on Human Liver Carcinoma Cells. *Nanomaterials* **2015**, *5*, 1317–1330. [CrossRef]
76. Meyer, K.; Rajanahalli, P.; Ahamed, M.; Rowe, J.J.; Hong, Y. ZnO nanoparticles induce apoptosis in human dermal fibroblasts via p53 and p38 pathways. *Toxicol. In Vitro* **2011**, *25*, 1721–1726. [CrossRef]
77. Wahab, R.; Siddiqui, M.A.; Saquib, Q.; Dwivedi, S.; Ahmad, J.; Musarrat, J.; Al-Khedhairy, A.A.; Shin, H.-S. ZnO nanoparticles induced oxidative stress and apoptosis in HepG2 and MCF-7 cancer cells and their antibacterial activity. *Colloids Surf. B Biointerfaces* **2014**, *117*, 267–276. [CrossRef]
78. Wang, J.; Deng, X.; Zhang, F.; Chen, D.; Ding, W. ZnO nanoparticle-induced oxidative stress triggers apoptosis by activating JNK signaling pathway in cultured primary astrocytes. *Nanoscale Res. Lett.* **2014**, *9*, 117. [CrossRef]
79. Fuster, E.; Candela, H.; Estévez, J.; Vilanova, E.; Sogorb, M.A. Titanium Dioxide, but Not Zinc Oxide, Nanoparticles Cause Severe Transcriptomic Alterations in T98G Human Glioblastoma Cells. *Int. J. Mol. Sci.* **2021**, *22*, 2084. [CrossRef]
80. Wiesmann, N.; Kluenker, M.; Demuth, P.; Brenner, W.; Tremel, W.; Brieger, J. Zinc overload mediated by zinc oxide nanoparticles as innovative anti-tumor agent. *J. Trace Elem. Med. Biol.* **2019**, *51*, 226–234. [CrossRef] [PubMed]
81. Yan, L.; Gu, Z.; Zhao, Y. Chemical mechanisms of the toxicological properties of nanomaterials: Generation of intracellular reactive oxygen species. *Chem. Asian J.* **2013**, *8*, 2342–2353. [CrossRef] [PubMed]
82. Kou, L.; Sun, R.; Xiao, S.; Zheng, Y.; Chen, Z.; Cai, A.; Zheng, H.; Yao, Q.; Ganapathy, V.; Chen, R. Ambidextrous Approach To Disrupt Redox Balance in Tumor Cells with Increased ROS Production and Decreased GSH Synthesis for Cancer Therapy. *ACS Appl. Mater. Interfaces* **2019**, *11*, 26722–26730. [CrossRef]

83. Marfavi, Z.H.; Farhadi, M.; Jameie, S.B.; Zahmatkeshan, M.; Pirhajati, V.; Jameie, M. Glioblastoma U-87MG tumour cells suppressed by ZnO folic acid-conjugated nanoparticles: An in vitro study. *Artif. Cells Nanomed. Biotechnol.* **2019**, *47*, 2783–2790. [CrossRef]
84. Sadri, A.; Changizi, V.; Eivazadeh, N. Evaluation of glioblastoma (U87) treatment with ZnO nanoparticle and X-ray in spheroid culture model using MTT assay. *Radiat. Phys. Chem.* **2015**, *115*, 17–21. [CrossRef]
85. Gomaa, S.; Nassef, M.; Tabl, G.; Zaki, S.; Abdel-Ghany, A. Doxorubicin and folic acid-loaded zinc oxide nanoparticles-based combined anti-tumor and anti-inflammatory approach for enhanced anti-cancer therapy. *BMC Cancer* **2024**, *24*, 34. [CrossRef] [PubMed]
86. Wingett, D.; Louka, P.; Anders, C.B.; Zhang, J.; Punnoose, A. A role of ZnO nanoparticle electrostatic properties in cancer cell cytotoxicity. *Nanotechnol. Sci. Appl.* **2016**, *9*, 29–45. [CrossRef]
87. Rajan, S.S.; Chandran, R.; Abrahamse, H. Overcoming challenges in cancer treatment: Nano-enabled photodynamic therapy as a viable solution. *Wiley Interdiscip. Rev. Nanomed. Nanobiotechnol.* **2024**, *16*, e1942. [CrossRef] [PubMed]

nanomaterials

MDPI

Article

Reactive Oxygen Species Yield near Gold Nanoparticles Under Ultrahigh-Dose-Rate Electron Beams: A Monte Carlo Study

Chloe Doen Kim [1] and James C. L. Chow [1,2,3,*]

1 Radiation Medicine Program, Princess Margaret Cancer Centre, University Health Network, Toronto, ON M5G 1X6, Canada; chloe.kim@uhn.ca

2 Department of Radiation Oncology, University of Toronto, Toronto, ON M5T 1P5, Canada

3 Department of Materials Science and Engineering, University of Toronto, Toronto, ON M5S 3E4, Canada

* Correspondence: james.chow@uhn.ca; Tel.: +1-416-946-4501

Abstract

Ultrahigh dose rate (UHDR) radiotherapy, also known as FLASH radiotherapy (FLASH-RT), has shown potential for increasing tumor control while sparing normal tissue. In parallel, gold nanoparticles (GNPs) have been extensively explored as radiosensitizers due to their high atomic number and ability to enhance the generation of reactive oxygen species (ROS) through water radiolysis. In this study, we investigate the synergistic effects of UHDR electron beams and GNP-mediated radiosensitization using Monte Carlo (MC) simulations based on the Geant4-DNA code. A spherical water phantom with embedded GNPs of varying sizes (5–100 nm) was irradiated using pulsed electron beams (100 keV and 1 MeV) at dose rates of 60, 100, and 150 Gy/s. The chemical yield of ROS near the GNPs was quantified and compared to an equivalent water nanoparticle model, and the yield enhancement factor (YEF) was used to evaluate radiosensitization. Results demonstrated that YEF increased with smaller GNP sizes and at lower UHDR, particularly for 1 MeV electrons. A maximum YEF of 1.25 was observed at 30 nm from the GNP surface for 5 nm particles at 60 Gy/s. The elevated ROS concentration near GNPs under FLASH conditions is expected to intensify DNA damage, especially double-strand breaks, due to increased hydroxyl radical interactions within nanometric distances of critical biomolecular targets. These findings highlight the significance of nanoparticle size and beam parameters in optimizing ROS production for FLASH-RT. The results provide a computational basis for future experimental investigations into the combined use of GNPs and UHDR beams in nanoparticle-enhanced radiotherapy.

Keywords: gold nanoparticles; FLASH radiotherapy; reactive oxygen species; Monte Carlo simulation; ultrahigh dose rate; electron beam; nanoparticle-enhanced radiotherapy; dose enhancement; yield enhancement factor; DNA damage; DNA dosimetry; nanodosimetry

1. Introduction

Radiotherapy remains a cornerstone in the treatment of various malignancies, and recent advances have aimed to enhance its efficacy while minimizing damage to healthy tissues. One such advancement is FLASH radiotherapy (FLASH-RT), which involves the delivery of radiation at ultrahigh dose rates (UHDRs), commonly defined as dose rates exceeding 40 Gy/s. This emerging modality has demonstrated the potential to achieve comparable or superior tumor control compared to conventional dose rates while significantly reducing normal tissue toxicity [1–3]. The so-called FLASH effect, a phenomenon wherein normal tissues experience enhanced sparing without compromising tumor control, has been supported by multiple preclinical studies and is currently under active investigation for clinical translation [4].

Several hypotheses have been proposed to explain the mechanisms underlying the FLASH effect. One prevailing theory involves transient oxygen depletion in normal tissues during UHDR exposure, which induces a temporary hypoxic state that renders the tissues more resistant to radiation damage [4–6]. Early studies on bacteria [7] and mammalian cells [8] have laid the foundation for the oxygen depletion hypothesis [9]. Another theory posits a differential biochemical response between normal and cancerous tissues. In this framework, cancer cells, which often have impaired antioxidant defenses, are more vulnerable to oxidative damage induced by ionizing radiation. Conversely, normal cells may exhibit higher resistance to reactive oxygen species (ROS) due to more robust antioxidant systems. Moreover, the limited diffusion of ROS produced during water radiolysis at UHDR may further contribute to the selective targeting of cancerous tissues [5,10]. It is also noted that FLASH-RT can reduce inflammation in normal tissues while enhancing anti-tumor immunity, preserving a greater number of circulating immune cells to survive compared to conventional radiotherapy [4,9].

Preclinical studies using various radiation modalities and delivery techniques have demonstrated that FLASH-RT can significantly spare normal tissues while maintaining therapeutic efficacy against tumors [11–14]. These findings underscore the promise of FLASH-RT as a paradigm-shifting approach in cancer therapy and have catalyzed interest in complementary strategies that further enhance its selectivity and potency [15].

In parallel, the application of engineered nanomaterials (ENMs), notably gold nanoparticles (GNPs), has gained substantial attention in the field of radiation oncology. According to the International Organization for Standardization (ISO), ENMs are materials with at least one dimension in the range of 1 to 100 nm [16,17]. Among them, GNPs have emerged as a particularly attractive class of radiosensitizers due to their biocompatibility, ease of surface modification, and high atomic number (Z = 79), which enhances their ability to absorb radiation [18,19]. Importantly, GNPs are generally considered safer for in vivo applications compared to other high-Z materials, making them suitable for clinical development [18].

Mechanistically, GNPs enhance the biological effect of radiation through both physical and chemical pathways. Physically, they increase local dose deposition via secondary electron emission. Chemically, the interaction of radiation with GNPs in an aqueous environment leads to enhanced ROS production, which amplifies oxidative stress and contributes to DNA damage in cancer cells [19–24]. Specifically, the photoelectric effect plays a dominant role in energy deposition when high-Z materials like gold interact with low-energy photons or electrons. In this process, incident radiation transfers energy to inner-shell electrons of gold atoms, ejecting photoelectrons and initiating a cascade of Auger electrons. These low-energy electrons travel short distances in water, leading to dense ionization tracks that cause water radiolysis and the subsequent formation of ROS (e.g., OH•, H_2O_2, H•). The localized production of these species near DNA significantly enhances the probability of inducing double-strand breaks (DSBs) and other lethal lesions [25]. Among the various forms of radiation-induced DNA lesions, DSBs are considered the most lethal and are a critical determinant of radiotherapeutic efficacy [20]. The potential benefit of GNP through enhanced ROS-induced DNA damage brings new prospects for personalized and targeted cancer treatments as nanotechnology in medicine progresses [26].

Despite significant research into the individual effects of UHDR radiation and GNP-mediated radiosensitization, there are limited studies that explore their combined interaction. This gap highlights the need for a deeper understanding of how GNPs influence ROS dynamics and DNA damage under FLASH-RT conditions. Monte Carlo (MC) simulation techniques offer a powerful tool for investigating these interactions at the nanoscale. In particular, Geant4-DNA, an extension of the Geant4 toolkit developed at the European Organization for Nuclear Research (CERN), provides a versatile platform for simulating radiation interactions with matter at both the physical and chemical stages, including the generation and diffusion of ROS in water [27,28].

Previous MC studies have examined dose enhancement ratios and energy deposition around GNPs using various radiation types, including photons [29–32], electrons [33], and protons [34,35]. In addition, the dependence of ROS enhancement on GNP size has been analyzed with different radiation modalities [35,36]. However, these studies primarily focus on conventional dose rates and do not account for the unique characteristics of UHDR delivery. To the best of our knowledge, few simulation studies have addressed the interplay between GNPs and FLASH-RT, particularly with respect to ROS production during water radiolysis at UHDR.

To address this gap, the present study employs Geant4-DNA-based MC simulations to investigate the influence of GNPs on ROS yield under electron-beam irradiation at UHDR. By modeling the spatial distribution of ROS and comparing it to an equivalent water nanoparticle (WNP) control, we aim to quantify the yield enhancement factor (YEF) as a function of GNP size, dose rate, and distance from the nanoparticle. These insights are expected to provide computational support for experimental studies and help optimize nanoparticle-enhanced FLASH-RT strategies. Previous experimental studies have validated the enhanced production of ROS and increased DNA damage in the presence of GNPs under irradiation, supporting the predictions made by MC-based models. For example, Misawa and Takahashi [21] demonstrated increased ROS generation with GNPs under X-ray and UV exposure, while Butterworth et al. [22] and Hainfeld et al. [23] reported enhanced cytotoxicity and tumor control due to nanoparticle radiosensitization. These findings align with simulation outcomes and underscore the relevance of using MC tools like Geant4-DNA to model nanoscale radiation interactions and optimize radiotherapeutic conditions.

2. Materials and Methods

2.1. MC Simulation Using Geant4-DNA

MC simulations were employed in this study to model the interactions between incident electrons and biological media, including the presence of GNPs. The simulations were conducted using Geant4-DNA, an extension of the general-purpose Geant4 toolkit developed by CERN. Geant4-DNA is specifically designed to simulate the physical, physicochemical, and chemical interactions of ionizing radiation with liquid water at nanometric spatial resolutions, making it particularly well-suited for nanoscale radiobiology studies [37,38].

The simulations were performed using Geant4 version 11.2, which includes an updated chemistry module capable of modeling time-resolved radiolysis processes [35]. The Livermore physics model was adopted to describe particle interactions within the GNP due to its accuracy in representing electromagnetic interactions for high-Z materials at low energies. For the surrounding water medium, the Geant4-DNA physics models were applied to simulate the full cascade of interactions, including energy deposition, track structure, and chemical stage evolution of reactive species.

The simulations aimed to model a GNP embedded within a water environment under UHDR electron-beam irradiation. The system was carefully configured to enable the analysis of both physical dose deposition and chemical yield of ROS around the nanoparticle, providing a multiscale understanding of radiation-induced effects. The chemical stage of the simulation, including ROS formation, was implemented using the

Geant4-DNA chemistry module, which models the time evolution of reactive species following water radiolysis. After energy deposition by primary and secondary electrons, the simulation proceeds through physicochemical processes, such as ionization, excitation, and dissociation, leading to the formation of early species (e.g., H_3O^+, e^-_{aq}, OH•, H•, and H_2O_2). These species are then tracked for up to 10 nanoseconds to account for diffusion and chemical reactions in the aqueous medium. ROS yield was quantified at this time point, consistent with literature indicating peak biological relevance of radical interactions within this timescale [3,39].

2.2. Simulation Geometry and Configuration

To approximate a simplified biological environment, the simulation geometry was designed as a spherical water phantom with a radius of 5 μm, large enough to ensure that secondary electrons and ROS products did not escape the simulation volume yet small enough to maintain computational efficiency [3]. At the center of the sphere, a spherical GNP was embedded, with diameters varied among 5, 10, 50, and 100 nm to study the size-dependent effects.

Incident electron beams were simulated to irradiate the GNP at two different initial energies: 100 keV and 1 MeV. These energy values were selected to approximate the local energy of electrons reaching tumor cells, after accounting for the attenuation and energy loss of clinical electron beams (typically 6–20 MeV) as they traverse tissue. At the cellular or subcellular level, particularly in the presence of GNPs, electrons in the range of 100 keV to 1 MeV are representative of the actual energy spectrum responsible for ionization and ROS production. Therefore, using these energies enables simulation of clinically relevant FLASH RT effects at the nanoscale. Each electron was emitted from the inner surface of the water sphere, directed toward the GNP center. The initial source position for each electron was randomly sampled from the water sphere's surface, mimicking an isotropic irradiation field.

The simulation geometry is illustrated in Figure 1, where electrons (solid red lines) with initial energies of 100 keV or 1 MeV are irradiated toward the center of the water sphere from the surface of the GNP, as shown in Figure 1a. The experimental setup for the MC simulation is shown in Figure 1b, with the spherical water phantom (diameter of 10 μm) embedded with the GNP in varying size (a diameter of 5, 10, 50, and 100 nm). Secondary electrons (red dashed lines) generated from interactions within the GNP, along with ROS represented by multicolored dots, are also visualized. As these secondary electrons propagate along their tracks (solid red lines), they deposit energy into the surrounding water medium, initiating radiolysis, as depicted in Figure 1c.

2.3. Simulation of UHDR Electron-Beam Irradiation

The simulation of UHDR conditions was based on the characteristics of the Oriadtron eRT6 linear accelerator (PMB, Rousset, France), which delivers pulsed electron beams at clinically relevant UHDR parameters [40,41]. Each radiation pulse has a duration of 2 μs with a repetition rate of 100 Hz and a maximum dose delivery of up to 10 Gy per pulse.

The MC simulation used a single pulse out of 100 pulses in a second (100 Hz repetition rate). To realistically model these UHDR conditions in the MC framework, a methodology adapted from Chappuis et al. [3] was employed. During the simulation, electrons were propagated through the water sphere, and the total number of incident electrons was dynamically adjusted such that the deposited dose matched one of three target thresholds: 0.6, 1.0, and 1.5 Gy per pulse. These correspond to dose rates of 60, 100, and 150 Gy/s, respectively, which fall within the FLASH regime. The beam parameters used are summarized in Table 1.

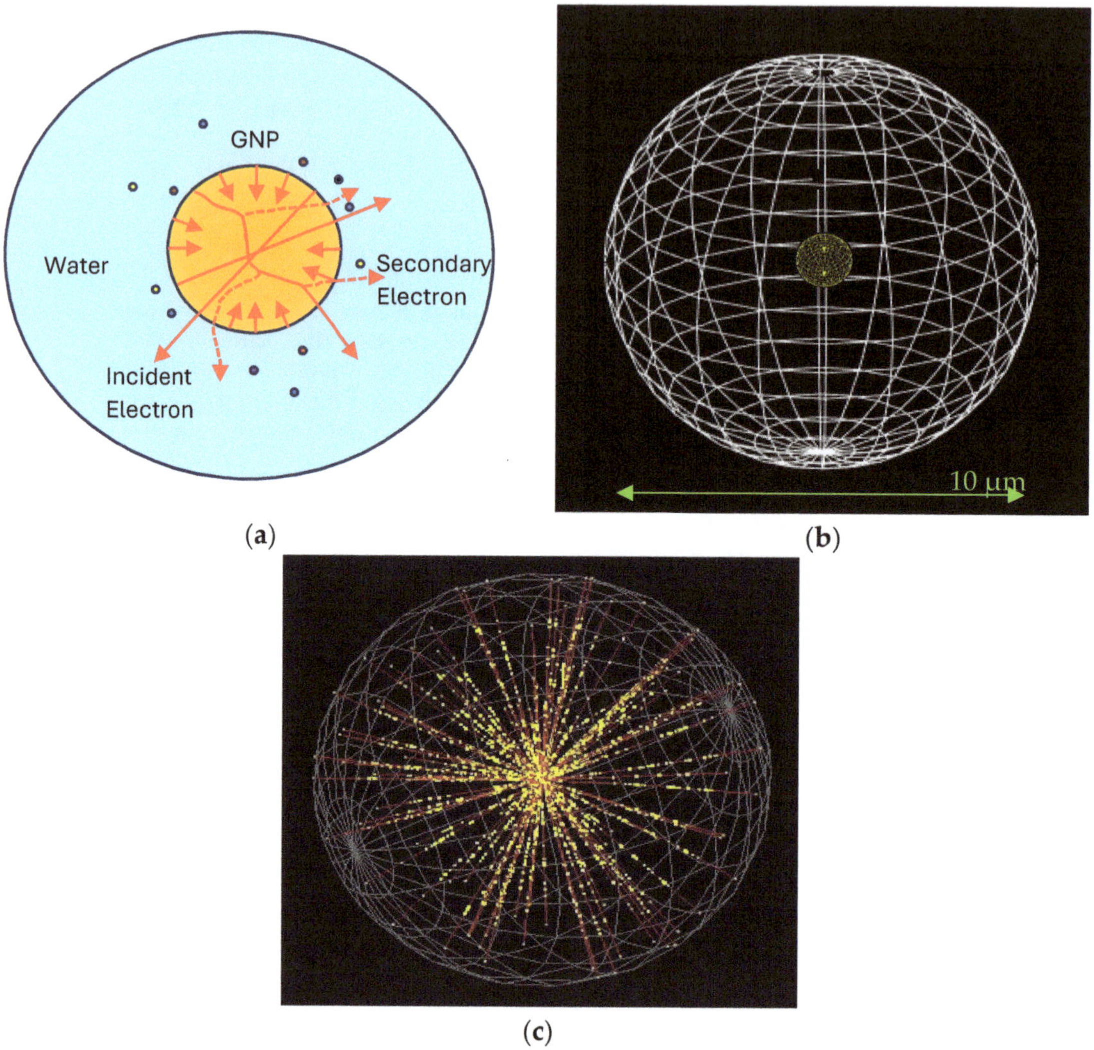

Figure 1. MC geometry modeled in Geant4-DNA. (**a**) A GNP (orange), with diameters of 5, 10, 50, or 100 nm, is positioned at the center of a water sphere (light blue) with a radius of 5 μm. ROS, incident electrons, and secondary electrons are denoted by multicolored dots, solid red lines, and dashed red lines, respectively. (**b**) The experimental setup for the MC simulation consists of the spherical water phantom (diameter of 10 μm) embedded with the GNP in varying size. (**c**) Electrons with an energy of 1 MeV are directed toward the center of the sphere from the surface of the GNP. Secondary electrons (yellow dots), generated along the electron tracks (solid red lines), deposit energy within the water sphere. The primary electron beams used in this simulation model the clinical delivery of electron-beam therapy under UHDR conditions. When these incident electrons interact with the GNPs and surrounding water medium, they undergo inelastic scattering, leading to the emission of low-energy secondary electrons. These secondary electrons are primarily responsible for inducing localized ionization and excitation in water molecules, resulting in the production of ROS through radiolysis. This cascade of electron-induced processes forms the mechanistic basis for radiosensitization in the presence of GNPs.

Table 1. UHDR electron-beam parameters for Geant4-DNA Monte Carlo simulation.

Mode	Dose Per Pulse (Gy)	Dose Rate (Gy/s)
UHDR	0.6	60
UHDR	1	100
UHDR	1.5	150

This configuration allowed the investigation of dose rate-dependent effects on energy deposition and ROS production around the GNP under conditions that replicate preclinical FLASH-RT delivery.

2.4. ROS Quantification and YEF

To quantify the radiosensitizing impact of GNPs under UHDR irradiation, the formation of ROS was analyzed during the chemical stage of water radiolysis. In this phase, short-lived radicals, such as hydroxyl radicals (OH•), play a critical role in biological damage mechanisms, particularly in the formation of DNA DSBs [42].

According to prior studies, the diffusion and interaction of OH• radicals with biomolecules occur predominantly within the first 10 nanoseconds following radiation exposure. Beyond this window, ROS concentrations tend to saturate, and further reactions become secondary. Therefore, ROS concentrations were evaluated at 10 ns to capture the peak biological relevance of ROS-induced damage.

To isolate the effects attributable to GNP radiosensitization, a control simulation was also conducted using a WNP of identical size and position, replacing the GNP within the same simulation geometry. The YEF was then calculated as the ratio of total ROS yield, comprising H_3O^+, OH•, H_2O_2, OH^-, and H• radicals, produced with GNPs to that with WNPs under identical conditions as shown in Equation (1).

$$\mathrm{YEF} = \frac{\mathrm{ROS_{GNP}}}{\mathrm{ROS_{WNP}}} \tag{1}$$

This metric provides a direct measure of the chemical amplification attributable to GNP radiosensitization under FLASH conditions and enables comparative analysis across different nanoparticle sizes, electron energies, and dose rates.

3. Results

3.1. Influence of UHDR on YEF

The effects of different UHDR, namely, 60, 100, and 150 Gy/s, on the YEF were evaluated at varying distances (10–30 nm) from the surface of GNPs of different diameters (5, 10, 50, and 100 nm). Simulations were conducted using two electron energies of 1 MeV and 100 keV. For 1 MeV electrons, the YEF was observed to be highest for small GNP sizes (5 nm and 10 nm) and low UHDR conditions (60 Gy/s). As shown in Figure 2, the YEF increased with increasing distance from the nanoparticle surface, reaching a maximum value of 1.25 at 30 nm from a 5 nm GNP at 60 Gy/s. While there was no change in YEF with an increase in UHDR from 100 Gy/s to 150 Gy/s in a 5 nm GNP, a decrease in YEF was found for the same UHDR change in a 10 nm GNP. This suggests enhanced ROS production farther from the GNP under specific conditions of lower dose rates and smaller particle sizes.

In contrast, for larger GNPs (50 nm and 100 nm), the YEF remained relatively constant across all dose rates, indicating a reduced sensitivity to UHDR in terms of ROS amplification. When examining the results for 100 keV electrons (Figure 3), the dose rate had minimal influence on YEF across most GNP sizes. A slight increase in YEF was noted only in the 10 nm GNP case, where the highest YEF was recorded at 150 Gy/s. However, the overall enhancement was less pronounced compared to the 1 MeV case, and maximum YEF values remained below those obtained with high-energy electrons. Nevertheless, there was also a trend similar to 1 MeV where the YEF increased with increasing distance from the nanoparticle surface.

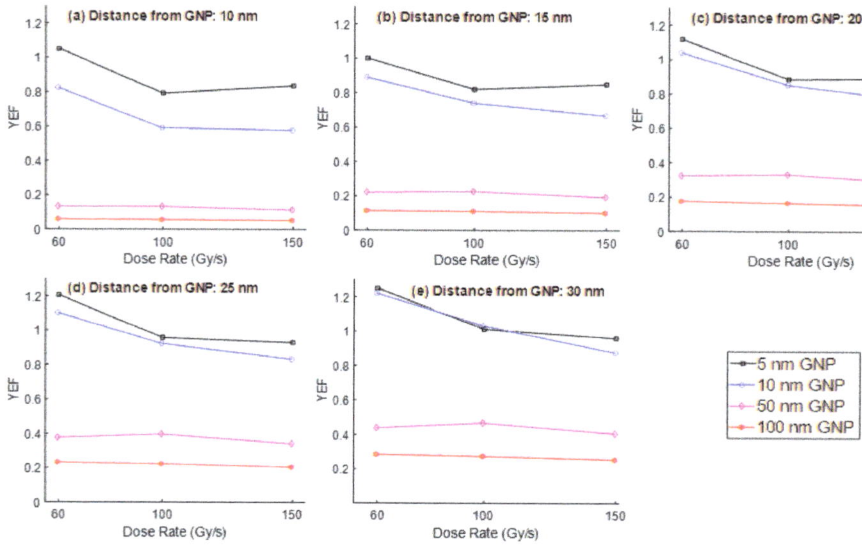

Figure 2. YEF vs. UHDR (60, 100, and 150 Gy/s) at distances of (**a**) 10 nm, (**b**) 15 nm, (**c**) 20 nm, (**d**) 25 nm, and (**e**) 30 nm from the surface of GNPs of different diameters (5, 10, 50, and 100 nm) for the electron energy of 1 MeV.

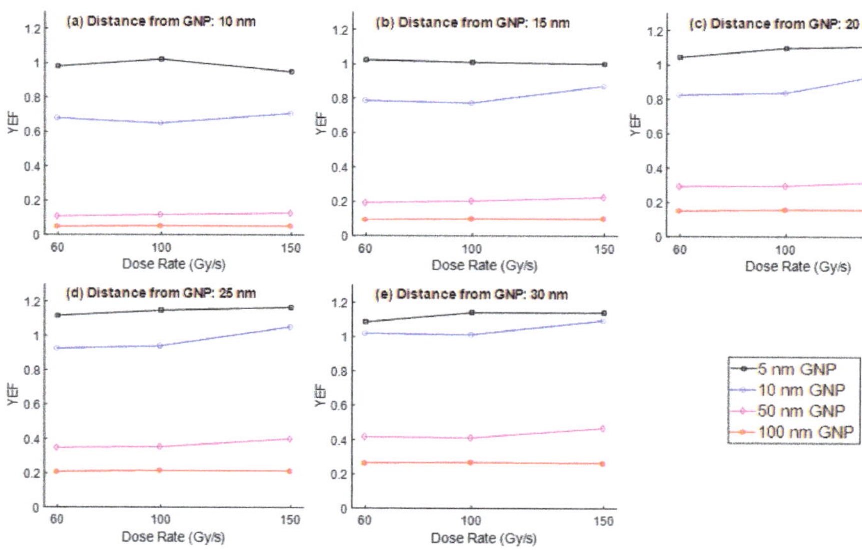

Figure 3. YEF vs. UHDR (60, 100, and 150 Gy/s) at distances of (**a**) 10 nm, (**b**) 15 nm, (**c**) 20 nm, (**d**) 25 nm, and (**e**) 30 nm from the surface of GNPs of different diameters (5, 10, 50, and 100 nm) for the electron energy of 100 keV.

3.2. Effect of GNP Size on YEF

The impact of GNP size on ROS generation was further analyzed across the four selected diameters (5, 10, 50, and 100 nm) at different radial distances (10–30 nm) from the nanoparticle surface. The YEF values for both 1 MeV and 100 keV electrons were plotted in Figures 4 and 5, respectively.

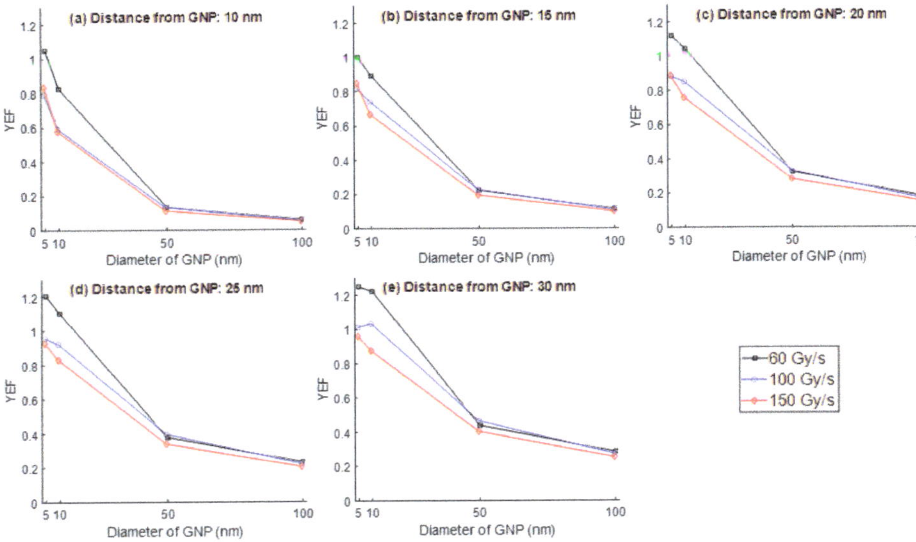

Figure 4. YEF vs. GNP size (5, 10, 50, and 100 nm) at distances of (**a**) 10 nm, (**b**) 15 nm, (**c**) 20 nm, (**d**) 25 nm, and (**e**) 30 nm from the surface of GNP at different UHDR (60, 100, and 150 Gy/s) for the electron energy of 1 MeV.

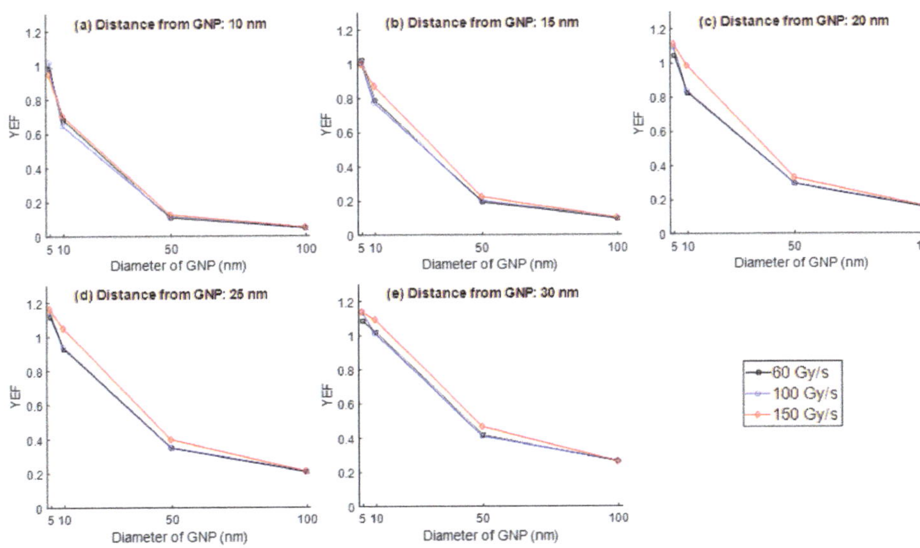

Figure 5. YEF vs. GNP size (5, 10, 50, and 100 nm) at distances of (**a**) 10 nm, (**b**) 15 nm, (**c**) 20 nm, (**d**) 25 nm, and (**e**) 30 nm from the surface of GNP at different UHDR (60, 100, and 150 Gy/s) for the electron energy of 100 keV.

For both electron energies, a clear inverse relationship was observed between GNP size and YEF. Smaller nanoparticles consistently yielded higher ROS enhancement, particularly for 5 nm GNPs, which exhibited peak YEF values under the 1 MeV, 60 Gy/s condition. At all distances, the YEF declined significantly with increasing GNP diameter, especially between 10 nm and 50 nm. For example, at 30 nm from the GNP surface under 1 MeV irradiation with a dose rate equal to 60 Gy/s (Figure 4e), YEF dropped from approximately 1.25 for 5 nm GNPs to less than 0.4 for 100 nm GNPs.

This trend was also present in the 100 keV electron case (Figure 5), although YEF values were generally lower compared to the high-energy scenario. Nevertheless, the results confirm that GNP size is a critical determinant of ROS amplification during FLASH irradiation.

4. Discussion

4.1. Dose Rate Effects on ROS Amplification

The results reveal that the FLASH dose rate affects ROS generation, but only under specific conditions. For 1 MeV electrons, lower dose rates (60 Gy/s) appeared to promote greater ROS yields in proximity to small GNPs, aligning with previous experimental findings that suggest FLASH irradiation can modulate radiolytic chemistry [43]. It was previously reported that as the mean dose rate of UHDR increased, ROS (H_2O_2) production decreased, warranting further optimization of dose rate parameters for UHDR radiotherapy [44].

These observations are also consistent with Chappuis et al. [3], who described dose rate–dependent differences in water radiolysis using Geant4-DNA. The reduced ROS yield at higher dose rates may be attributed to the saturation of radiolytic pathways or suppression of radical diffusion under rapid energy deposition conditions.

In contrast, low-energy electrons (100 keV) exhibited minimal sensitivity to UHDR conditions. In consideration of the fact that ROS can be induced through secondary electron interactions, this could be due to the shorter track lengths and reduced penetration depth of low-energy electrons, which limit their capacity to generate sufficient secondary electrons or promote extended ROS diffusion.

4.2. Role of GNP Size and Self-Absorption

The observed inverse relationship between GNP size and YEF can be attributed to several physical and chemical mechanisms. First, smaller GNPs allow a greater fraction of secondary electrons to escape their surfaces and interact with the surrounding medium. Conversely, larger GNPs exhibit self-absorption, whereby low-energy secondary electrons are reabsorbed or scattered internally, reducing the effective dose enhancement near the GNP surface [45–48].

Moreover, electron shadowing effects in larger GNPs have been reported to contribute to reduced radiolytic yields [49]. While the publications above recorded a reduction in ROS with increasing GNP size, another study found that ROS production increased with GNP size [50]. This discrepancy and the findings in our study highlight the importance of nanoparticle size in optimizing radiosensitization, especially under UHDR delivery, by balancing dose enhancement with energy escape efficiency.

In addition to size-dependent effects, it is important to note that aggregation of GNPs is frequently observed in biological environments due to factors such as protein corona formation, ionic interactions, or endocytosis. Aggregation can significantly alter the effective surface area exposed to radiation and modify the local electron transport conditions. Previous studies have shown that aggregated or clustered nanoparticles may reduce ROS yield compared to isolated particles, as closely packed GNPs can enhance self-absorption of low-energy electrons and limit the escape of secondary electrons into the surrounding medium [45–48]. Although our current simulation models a single, isolated GNP to provide a clean and quantifiable baseline for radiosensitization, future studies could incorporate multi-particle or clustered configurations [46] to better mimic physiological conditions and assess their impact on FLASH-RT outcomes.

4.3. Implications for FLASH Nanoparticle-Enhanced Radiotherapy

This study offers valuable computational insights into the combination of GNP-enhanced radiosensitization with UHDR electron therapy, a relatively unexplored domain in radiation oncology. The enhanced ROS generation observed under low UHDR and small GNP sizes suggests that careful selection of nanoparticle parameters and beam settings may maximize therapeutic benefits while maintaining the normal tissue-sparing effects of FLASH-RT.

However, it is important to note that this work focuses on single-pulse simulations and evaluates ROS yields near GNP (10–30 nm distance from the surface of GNP) at a fixed chemical time point (10 ns). Future studies should consider multiple pulse effects, pulse duration, biological endpoints (e.g., DNA damage), and experimental validation to build a more comprehensive model for clinical translation.

5. Conclusions

This study employed Geant4-DNA MC simulations to evaluate the impact of GNPs on ROS generation under UHDR electron-beam irradiation, mimicking FLASH-RT conditions. By comparing GNPs of various sizes (5–100 nm) with equivalent WNPs, the YEF was quantified as a measure of radiosensitization.

Results showed that small GNPs (5–10 nm) significantly enhanced ROS production under 1 MeV irradiation, particularly at lower dose rates (60 Gy/s), with YEF values reaching up to 1.25. In contrast, larger GNPs demonstrated limited enhancement regardless of dose rates, likely due to self-absorption effects. For 100 keV electrons, the dose rate had minimal influence on YEF.

These findings suggest that GNP size and beam energy play crucial roles in optimizing chemical enhancement during FLASH-RT. Although the MC simulation approximates complicated biological systems, limited to single-pulse irradiation and idealized conditions, this study offers computational insight into nanoparticle-mediated radiosensitization at UHDR.

This study has some limitations. The simulations assumed isolated GNPs, while aggregation is common in biological environments and may affect ROS yield. Only single-pulse irradiation was modeled, and biological endpoints such as DNA damage were not directly evaluated. These aspects will be addressed in future studies to enhance clinical relevance and experimental validation.

Author Contributions: Conceptualization, J.C.L.C.; methodology, C.D.K. and J.C.L.C.; software, C.D.K. and J.C.L.C.; validation, C.D.K. and J.C.L.C.; formal analysis, C.D.K.; investigation, C.D.K. and J.C.L.C.; resources, C.D.K. and J.C.L.C.; data curation, C.D.K.; writing—original draft preparation, C.D.K.; writing—review and editing, C.D.K. and J.C.L.C.; visualization, C.D.K.; supervision, J.C.L.C.; project administration, J.C.L.C. All authors have read and agreed to the published version of the manuscript.

Funding: This research was supported by the Government of Canada's New Frontiers in Research. Fund—Exploration (Grant number: NFRFE-2022-00707), through the three federal research funding agencies (CIHR, NSERC and SSHRC).

Institutional Review Board Statement: Not applicable.

Informed Consent Statement: Not applicable.

Data Availability Statement: No new data were created.

Conflicts of Interest: The authors declare no conflicts of interest.

Abbreviations

The following abbreviations are used in this manuscript:

Abbreviation	Full Term
MC	Monte Carlo
Geant4-DNA	GEometry ANd Tracking version 4—DNA extension
UHDR	Ultrahigh Dose Rate
FLASH-RT	FLASH Radiotherapy
GNP	Gold Nanoparticle
WNP	Water Nanoparticle
ROS	Reactive Oxygen Species
YEF	Yield Enhancement Factor
DNA	Deoxyribonucleic Acid
DSB	Double-Strand Break
ISO	International Organization for Standardization
OH•	Hydroxyl Radical
H_2O_2	Hydrogen Peroxide
H•	Hydrogen Radical
H_3O^+	Hydronium Ion
OH^-	Hydroxide Ion
MeV	Mega Electron Volt
keV	Kilo Electron Volt
CERN	European Organization for Nuclear Research

References

1. Vozenin, M.C.; Bourhis, J.; Durante, M. Towards clinical translation of FLASH radiotherapy. *Nat. Rev. Clin. Oncol.* **2022**, *19*, 791–803. [CrossRef]
2. Favaudon, V.; Labarbe, R.; Limoli, C.L. Model studies of the role of oxygen in the FLASH effect. *Med. Phys.* **2022**, *49*, 2068–2081. [CrossRef]
3. Chappuis, F.; Tran, H.N.; Jorge, P.G.; Zein, S.A.; Kyriakou, I.; Emfietzoglou, D.; Bailat, C.; Bochud, F.; Incerti, S.; Desorgher, L. Investigating ultra-high dose rate water radiolysis using the Geant4-DNA toolkit and a Geant4 model of the Oriatron eRT6 electron linac. *Sci. Rep.* **2024**, *14*, 26707. [CrossRef]
4. Chow, C.L.; Ruda, H.E. Flash Radiotherapy: Innovative Cancer Treatment. *Encyclopedia* **2023**, *3*, 808–823. [CrossRef]
5. Abolfath, R.; Grosshans, D.; Mohan, R. Oxygen depletion in FLASH ultra-high-dose-rate radiotherapy: A molecular dynamics simulation. *Med. Phys.* **2020**, *47*, 6551–6561. [CrossRef]
6. Cao, X.; Zhang, R.; Esipova, T.V.; Allu, S.R.; Ashraf, R.; Rahman, M.; Gunn, J.R.; Bruza, P.; Gladstone, D.J.; Williams, B.B.; et al. Quantification of oxygen depletion during FLASH irradiation in vitro and in vivo. *Int. J. Radiat. Oncol. Biol. Phys.* **2021**, *111*, 240–248. [CrossRef]
7. Dewey, D.L.; Boag, J.W. Modification of the oxygen Effect When bacteria are Given Large pulses of radiation. *Nature* **1959**, *183*, 1450. [CrossRef]
8. Epp, E.R.; Weiss, H.; Djordjevic, B.; Santomasso, A. The radiosensitivity of cultured mammalian cells exposed to single high intensity pulses of electrons in various concentrations of oxygen. *Radiat. Res.* **1972**, *52*, 324–332. [CrossRef]
9. Rosini, G.; Ciarrocchi, E.; D'Orsi, B. Mechanisms of the FLASH effect: Current insights and advances. *Front. Cell Dev. Biol.* **2025**, *13*, 1575678. [CrossRef]
10. Zhou, G. Mechanisms underlying FLASH radiotherapy, a novel way to enlarge the differential responses to ionizing radiation between normal and tumour tissues. *Radiat. Med. Prot.* **2020**, *1*, 35–40. [CrossRef]

11. Favaudon, V.; Caplier, L.; Monceau, V.; Pouzoulet, F.; Sayarath, M.; Fouillade, C.; Poupon, M.-F.; Brito, I.; Hupé, P.; Bourhis, J.; et al. Ultrahigh dose-rate FLASH irradiation increases the differential response between normal and tumor tissue in mice. *Sci. Transl. Med.* **2014**, *6*, 245ra93. [CrossRef]
12. Harrington, K.J. Ultrahigh dose-rate RT: Next steps for FLASH-RT. *Clin. Cancer Res.* **2019**, *25*, 3–5. [CrossRef]
13. Loo, B.W.; Schuler, E.; Lartey, F.M.; Rafat, M.; King, G.J.; Trovati, S.; Koong, A.C.; Maxim, P.G. (P003) Delivery of ultra-rapid flash radiation therapy and demonstration of normal tissue sparing after abdominal irradiation of mice. *Int. J. Radiat. Oncol.* **2017**, *98*, E16. [CrossRef]
14. Vozenin, M.-C.; De Fornel, P.; Petersson, K.; Favaudon, V.; Jaccard, M.; Germond, J.-F.; Petit, B.; Burki, M.; Ferrand, G.; Patin, D.; et al. The advantage of Flash RT confirmed in mini-pig and cat-cancer patients. *Clin. Cancer Res.* **2019**, *25*, 35–42. [CrossRef]
15. Mascia, A.E.; Daugherty, E.C.; Zhang, Y.; Lee, E.; Xiao, Z.; Sertorio, M.; Woo, J.; Backus, L.R.; McDonald, J.M.; McCann, C.; et al. Proton FLASH Radiotherapy for the Treatment of Symptomatic Bone Metastases The FAST-01 Nonrandomized Trial. *JAMA Oncol.* **2023**, *9*, 62–69. [CrossRef]
16. *ISO/TS 80004-1*; Nanotechnologies—Vocabulary—Part 1: Core Terms. International Organization for Standardization: Geneva, Switzerland, 2015; pp. 1–4.
17. Nguyen, N.H.A.; Falagan-Lotsch, P. Mechanistic Insights into the Biological Effects of Engineered Nanomaterials: A Focus on Gold Nanoparticles. *Int. J. Mol. Sci.* **2023**, *24*, 4109. [CrossRef]
18. Chen, Y.; Yang, J.; Fu, S.; Wu, J. Gold nanoparticles as radiosensitizers in cancer radiotherapy. *Int. J. Nanomed.* **2020**, *15*, 9407–9430. [CrossRef]
19. Bardane, A.; Maalej, N.; Chakir, E.M.; Ibrahmi, E.M.A. Gold nanoparticle effect on dose and DNA damage enhancement in the vicinity of gold nanoparticles. *Nucl. Anal.* **2024**, *3*, 100126. [CrossRef]
20. Leenhouts, H.P.; Chadwick, K.H. The crucial role of DNA double-strand breaks in cellular radiobiological effects. In *Advances in Radiation Biology*; Elsevier: Amsterdam, The Netherlands, 1978; Volume 7, pp. 55–101.
21. Misawa, M.; Takahashi, J. Generation of reactive oxygen species induced by gold nanoparticles under X-ray and UV irradiations. *Nanomedicine* **2011**, *7*, 604–614. [CrossRef]
22. Butterworth, K.T.; Coulter, J.A.; Jain, S.; McMahon, S.J.; Schettino, G.; Prise, K.M.; Currell, F.J.; Hirst, D.G. Evaluation of cytotoxicity and radiation enhancement using 1.9 nm gold particles: Potential application for cancer therapy. *Nanotechnology* **2010**, *21*, 295101. [CrossRef]
23. Hainfeld, J.F.; Smilowitz, H.M.; O'Connor, M.J.; Dilmanian, F.A.; Slatkin, D.N. Gold nanoparticle imaging and radiotherapy of brain tumors in mice. *Nanomedicine* **2013**, *8*, 1601–1609. [CrossRef]
24. Hullo, M.; Grall, R.; Perrot, Y.; Mathé, C.; Ménard, V.; Yang, X.; Lacombe, S.; Porcel, E.; Villagrasa, C.; Chevillard, S.; et al. Radiation enhancer effect of platinum nanoparticles in breast cancer cell lines: In vitro and in silico analyses. *Int. J. Mol. Sci.* **2021**, *22*, 4436. [CrossRef]
25. Chow, J.C. Biophysical insights into nanomaterial-induced DNA damage: Mechanisms, challenges, and future directions. *AIMS Biophysics.* **2024**, *11*, 340–369. [CrossRef]
26. Lo, C.-Y.; Tsai, S.-W.; Niu, H.; Chen, F.-H.; Hwang, H.-C.; Chao, T.-C.; Hsiao, I.-T.; Liaw, J.-W. Gold-Nanoparticles-Enhanced Production of Reactive Oxygen Species in Cells at Spread-Out Bragg Peak under Proton Beam Radiation. *ACS Omega* **2023**, *8*, 17922–17931. [CrossRef]
27. Chappuis, F.; Tran, H.N.; Zein, S.A.; Bailat, C.; Incerti, S.; Bochud, F.; Desorgher, L. The general-purpose Geant4 Monte Carlo toolkit and its Geant4-DNA extension to investigate mechanisms underlying the FLASH effect in radiotherapy: Current status and challenges. *Phys. Med.* **2023**, *110*, 102601. [CrossRef]
28. Sakata, D.; Kyriakou, I.; Okada, S.; Tran, H.N.; Lampe, N.; Guatelli, S.; Bordage, M.-C.; Ivanchenko, V.; Murakami, K.; Sasaki, T.; et al. Geant4-DNA track-structure simulations for gold nanoparticles: The importance of electron discrete models in nanometer volumes. *Med. Phys.* **2018**, *45*, 2230–2242. [CrossRef]
29. Engels, E.; Bakr, S.; Bolst, D.; Sakata, D.; Li, N.; Lazarakis, P.; McMahon, S.J.; Ivanchenko, V.; Rosenfeld, A.B.; Incerti, S.; et al. Advances in modelling gold nanoparticle radiosensitization using new Geant4-DNA physics models. *Phys. Med. Biol.* **2020**, *65*, 225017. [CrossRef]
30. Lechtman, E.; Mashouf, S.; Chattopadhyay, N.; Keller, B.M.; Lai, P.; Cai, Z.; Reilly, R.M.; Pignol, J.-P. A Monte Carlo-based model of gold nanoparticle radiosensitization accounting for increased radiobiological effectiveness. *Phys. Med. Biol.* **2013**, *58*, 3075–3087. [CrossRef]
31. Lechtman, E.; Chattopadhyay, N.; Cai, Z.; Mashouf, S.; Reilly, R.; Pignol, J.P. Implications on clinical scenario of gold nanoparticle radiosensitization in regards to photon energy, nanoparticle size, concentration and location. *Phys. Med. Biol.* **2011**, *56*, 4631–4647. [CrossRef]

32. Tsai, M.-Y.; Tian, Z.; Qin, N.; Yan, C.; Lai, Y.; Hung, S.-H.; Chi, Y.; Jia, X. A new open-source GPU-based microscopic Monte Carlo simulation tool for the calculations of DNA damages caused by ionizing radiation-Part I: Core algorithm and validation. *Med. Phys.* **2020**, *47*, 1958–1970. [CrossRef]
33. Chow, J.C.L.; Leung, M.K.K.; Fahey, S.; Chithrani, D.B.; Jaffray, D.A. Monte Carlo simulation on low-energy electrons from gold nanoparticle in radiotherapy. *J. Phys. Conf. Ser.* **2012**, *341*, 012012. [CrossRef]
34. Lin, Y.; McMahon, S.J.; Scarpelli, M.; Paganetti, H.; Schuemann, J. Comparing gold nano-particle enhanced radiotherapy with protons, megavoltage photons and kilovoltage photons: A Monte Carlo simulation. *Phys. Med. Biol.* **2014**, *59*, 7675–7689. [CrossRef]
35. Tran, H.N.; Karamitros, M.; Ivanchenko, V.N.; Guatelli, S.; McKinnon, S.; Murakami, K.; Sasaki, T.; Okada, S.; Bordage, M.C.; Francis, Z.; et al. Geant4 Monte Carlo simulation of absorbed dose and radiolysis yields enhancement from a gold nanoparticle under MeV proton irradiation. *Nucl. Inst. Meth. B.* **2016**, *373*, 126–139. [CrossRef]
36. Peukert, D.; Kempson, I.; Douglass, M.; Bezak, E. Gold nanoparticle enhanced proton therapy: A Monte Carlo simulation of the effects of proton energy, nanoparticle size, coating material, and coating thickness on dose and radiolysis yield. *Med. Phys.* **2020**, *47*, 651–661. [CrossRef]
37. Villagrasa, C.; Francis, Z.; Incerti, S. Physical models implemented in the GEANT4-DNA extension of the GEANT-4 toolkit for calculating initial radiation damage at the molecular level. *Radiat. Prot. Dosim.* **2011**, *143*, 214–218. [CrossRef]
38. Incerti, S.; Suerfu, B.; Xu, J.; Ivantchenko, V.; Mantero, A.; Brown, J.M.C.; Bernal, M.A.; Francis, Z.; Karamitros, M.; Tran, H.N. Simulation of Auger electron emission from nanometer-size gold targets using the Geant4 Monte Carlo simulation toolkit. *Nucl. Instrum. Methods Phys. Res. Sect. B* **2016**, *372*, 91–101. [CrossRef]
39. Tran, H.N.; Archer, J.; Baldacchino, G.; Brown, J.M.C.; Chappuis, F.; Cirrone, G.A.P.; Desorgher, L.; Dominguez, N.; Fattori, S.; Guatelli, S.; et al. Review of chemical models and applications inGeant4-DNA: Report from the ESA BioRad III Project. *Med. Phys.* **2024**, *51*, 5873–5889. [CrossRef]
40. Jorge, P.G.; Grilj, V.; Bourhis, J.; Vozenin, M.-C.; Germond, J.-F.; Bochud, F.; Bailat, C.; Moeckli, R. Technical note: Validation of an ultrahigh dose rate pulsed electron beam monitoring system using a current transformer for FLASH preclinical studies. *Med. Phys.* **2022**, *49*, 1831–1838. [CrossRef]
41. Jaccard, M.; Durán, M.T.; Petersson, K.; Germond, J.-F.; Liger, P.; Vozenin, M.-C.; Bourhis, J.; Bochud, F.; Bailat, C. High dose-perpulse electron beam dosimetry: Commissioning of the Oriatron eRT6 prototype linear accelerator for preclinical use. *Med. Phys.* **2018**, *45*, 863–874. [CrossRef]
42. Zhao, X.; Liu, R.; Zhao, T.; Reynoso, F.J. Quantification of gold nanoparticle photon radiosensitization from direct and indirect effects using a complete human genome single cell model based on Geant4. *Med. Phys.* **2021**, *48*, 8127–8139. [CrossRef] [PubMed]
43. Thomas, W.; Sunnerberg, J.; Reed, M.; Gladstone, D.J.; Zhang, R.; Harms, J.; Swartz, H.M.; Pogue, B.W. Proton and electron UHDR isodose irradiations produce differences in reactive oxygen species yields. *Int. J. Radiat. Ocol. Biol. Phys.* **2023**, *118*, 262–267. [CrossRef] [PubMed]
44. Sunnerberg, J.P.; Zhang, R.; Gladstone, D.J.; Swartz, H.M.; Gui, J.; Pogue, B.W. Mean dose rate in ultra-high dose rate electron irradiation is a significant predictor for O_2 consumption and H_2O_2 yield. *Phys. Med. Biol.* **2023**, *68*, 165014. [CrossRef]
45. Poignant, F.; Charfi, H.; Chan, C.-H.; Dumont, E.; Loffreda, D.; Testa, É.; Gervais, B.; Beuve, M. Monte Carlo simulation of free radical production under keV photon irradiation of gold nanoparticle aqueous solution. Part. I: Global primary chemical boost. *Radiat. Phys. Chem.* **2020**, *172*, 108790. [CrossRef]
46. Santiago, C.A.; Chow, J.C.L. Variations in Gold Nanoparticle Size on DNA Damage: A Monte Carlo Study Based on a Multiple-Particle Model Using Electron Beams. *Appl. Sci.* **2023**, *13*, 4916. [CrossRef]
47. Chow, J.C.L.; Santiago, C.A. DNA Damage of Iron-Gold Nanoparticle Heterojunction Irradiated by kV Photon Beams: A Monte Carlo Study. *Appl. Sci.* **2023**, *13*, 8942. [CrossRef]
48. Jabeen, M.; Chow, J.C.L. Gold Nanoparticle DNA Damage by Photon Beam in a Magnetic Field: A Monte Carlo Study. *Nanomaterials* **2021**, *11*, 1751. [CrossRef]
49. Peukert, D.; Kempson, I.; Douglass, M.; Bezak, E. Gold Nanoparticle Enhanced Proton Therapy: Monte Carlo Modeling of Reactive Species' Distributions Around a Gold Nanoparticle and the Effects of Nanoparticle Proximity and Clustering. *Int. J. Mol. Sci.* **2019**, *20*, 4280. [CrossRef]
50. Antunes, J.; Rabus, H.; Mendes, F.; Paulo, A.; Sampaio, J.M. Chemical mechanism in gold nanoparticles radiosensitization: A Monte Carlo simulation study. *Radiat. Phys. Chem.* **2025**, *232*, 112637. [CrossRef]

nanomaterials

MDPI

Article

Dynamic Susceptibility Contrast Magnetic Resonance Imaging with Carbon-Encapsulated Iron Nanoparticles Navigated to Integrin Alfa V Beta 3 Receptors in Rat Glioma

Agnieszka Stawarska [1,*], Magdalena Bamburowicz-Klimkowska [1], Wojciech Szeszkowski [2] and Ireneusz Piotr Grudzinski [1,*]

1 Department of Toxicology and Food Science, Faculty of Pharmacy, Medical University of Warsaw, Banacha 1, 02-097 Warsaw, Poland; mjbamburowicz@wum.edu.pl

2 Department of Clinical Radiology, Faculty of Medicine, Medical University of Warsaw, Banacha 1A, 02-097 Warsaw, Poland; wojciech.szeszkowski@wum.edu.pl

* Correspondence: agnieszka.stawarska@wum.edu.pl (A.S.); ireneusz.grudzinski@wum.edu.pl (I.P.G.)

Abstract

Overexpression of $\alpha v\beta 3$ integrin is found in a diverse group of tumors originating from glial cells in the brain, making this transmembrane receptor a promising biomarker for molecular MRI diagnosis. In the study, we conjugated a monoclonal antibody against the $\beta 3$ subunit (CD61) of the $\alpha v\beta 3$ integrin receptor with carbon-encapsulated iron nanoparticles to yield Fe@C-$(CH_2)_2$-CONH-anti-CD61 bioconjugates that were used in dynamic susceptibility contrast magnetic resonance imaging (DSC-MRI). Wistar rats bearing C6 gliomas were injected as a single bolus (0.5 mL) through the tail vain with a suspension of Fe@C-$(CH_2)_2$-CONH-anti-CD61 nanoparticles (200 $\mu g\ mL^{-1}$) and the animals were imaged using the T2*-weighted echo planar imaging (T2* EPI) technique. Results showed that intravenously infused nanoparticles targeting $\alpha v\beta 3$ integrin receptors provide strong contrast in rat glioma tissues. No such effects were observed in other rat organs, although some post-contrast effects were also noted in the liver and kidney. The study shows that the as-developed nanoparticles decorated with anti-CD61 monoclonal antibodies might be considered as a novel contrast candidate for noninvasive DSC-MRI diagnosis in CD61-positive gliomas.

Keywords: carbon-encapsulated iron nanoparticles; alfa V beta 3 integrin; glioma; magnetic resonance imaging

1. Introduction

Integrins have critical functions in malignancies, angiogenesis, cell invasion and migration and play vital roles in other cancer-relevant processes, including white blood cell trafficking and activation, chronic inflammation and immune mimicry, which ultimately determine disease state [1]. Since integrins are integral to the process of cell adhesion and migration, these transmembrane receptors have been assessed as potential contributors to glioma invasion, as have the co-operating extracellular matrix (ECM) components [1]. The ECM in the perivascular region of the brain comprises such components as collagen, fibronectin and laminin [2]. The perivascular region is a site of frequent invasion, and it has been shown to be a site where tumor cells are capable of adhering and spreading particularly well [3]. Expression of integrin β1 is enhanced in both high-grade glioma samples and in glioma-derived cell lines, relative to normal brain tissue samples [4,5]. The β1 subunit is also a key integrin component as it is able to partner many subunits capable of interaction with the ECM components present in the perivascular region [4]. These integrins typically mediate epithelial cell adhesion to the basement membrane, but might contribute to migration, proliferation and survival in tumor cells [6]. Both αvβ3 and αvβ5 integrins are expressed in astrocytes and epithelial cells at the tumor–normal tissue margin and together with corresponding ECM components have a possible role in invasion, as they can be highly upregulated in cancer tissue [5,7–9]. Both the αv and to a lesser extent β3 integrin subunits are expressed in numerous glioma-derived cell lines and high-grade glioma biopsies [6,8,10]. Tumor cell expression of αvβ3, and αvβ5 integrins, is correlated with disease progression in glioblastoma [6,9]. Note that αvβ3 and αvβ5 integrins are associated with the tumor vasculature of anaplastic astrocytomas and glioblastoma multiforme specimens but are not expressed in the vasculature of low-grade astrocytomas or in normal tissue [8]. Furthermore, expression of αvβ3 is observed at the periphery of high-grade gliomas, while, in contrast, expression of αvβ5 was greater towards the center of the tumor [4]. The integrin α6β1 plays an important role in the regulation of glioma-initiating cells [11]. This integrin mediates the interaction of glioma-initiating cells with laminin, an extracellular matrix protein expressed in basement membranes, including those supporting endothelial cells [12]. Such interaction provides an anchorage for glioma-initiating cells within the perivascular niche and supports their tumorigenic potential [5]. As integrins are overexpressed in different glioma cells [13], these molecules make them attractive targets for developing novel nanotechnology-based targeted contrast agents used in magnetic resonance imaging (MRI) [14].

Nanoparticle-targeted imaging has emerged as a promising strategy in diagnostic medicine, particularly in the context of cancer. Functionalized nanomaterials play a pivotal role in advancing biomedical imaging and diagnostics by offering enhanced sensitivity, specificity and multiplexing capabilities. These materials, ranging from noble metal-based nanoparticles (e.g., gold, silver), semiconductor quantum dots and iron oxide nanoparticles, to carbon nanomaterials, are engineered at the nanoscale and functionalized with specific ligands to interact with molecular targets relevant to disease processes [15,16]. The surface modification, or functionalization, of these nanomaterials involves the attachment of biological ligands such as antibodies, aptamers, peptides or small molecules such as folates, and even vitamins or carbohydrates that confer target specificity and biocompatibility [17]. In diagnostic applications, the choice of ligand determines the targeting capacity and functional performance of the nanomaterial. Antibodies, with high affinity and specificity, are frequently used for the detection of protein biomarkers such as HER2 in breast cancer or PSA in prostate cancer [18]. Aptamers—short, single-stranded oligonucleotides—are synthetically selected to bind various targets including small molecules, proteins and cells, and offer advantages such as low immunogenicity and ease of chemical modification [19]. Peptides, due to their small size and stability, are employed for targeting receptors like integrins (e.g., αvβ3) overexpressed in tumor vasculature [20]. Small molecules such as folic acid are also used as targeting ligands, exploiting their affinity for receptors (e.g., folate receptor) that are upregulated in many cancers [17]. All these ligands enable nanomaterials to recognize and bind to a wide range of molecular targets, including cell surface receptors, enzymes, nucleic acids and exosomal markers. Such targeting facilitates the use of nanomaterials in various imaging modalities. For instance, gold nanoparticles conjugated with antibodies or peptides enhance contrast in computed tomography (CT) and photoacoustic imaging due to their strong X-ray attenuation and plasmonic properties [21]. Quantum dots functionalized with aptamers or antibodies serve as bright, photostable probes in fluorescence imaging for the detection of cancer cells or circulating tumor markers [22]. Magnetic nanoparticles such as superparamagnetic iron oxide particles, when functionalized with tumor-targeting ligands, improve the resolution and specificity of magnetic resonance imaging, aiding in early cancer detection and localization [20]. Beyond imaging, functionalized nanomaterials are increasingly integrated into biosensors and point-of-care (POC) diagnostics. Their high surface area and customizable surfaces enable the capture and detection of trace amounts of disease biomarkers in blood, saliva or urine. Nanoparticle-based biosensors employing ligands like aptamers or antibodies can detect molecular targets such as microRNAs, cytokines or viral antigens with high sensitivity and rapid turnaround, making them promising tools for early diagnosis and monitoring [23]. However, despite significant preclinical success, major scientific and translational gaps of using targeted nanoparticles remain that limit their routine clinical use. One of the central challenges is the low delivery efficiency of nanoparticles to target tissues. Quantitative studies have demonstrated that less than 1% of the injected dose of nanoparticles typically accumulates at the target site, with the rest being rapidly cleared by the mononuclear phagocyte system, particularly the liver and spleen [24]. Another limitation lies in the biological behavior of nanoparticles, which is influenced by their interaction with blood components. Upon systemic administration, nanoparticles rapidly adsorb plasma proteins, forming a "protein corona" that can mask targeting ligands and alter biodistribution [25]. This phenomenon often leads to reduced targeting specificity and enhanced clearance, diminishing the effectiveness of active targeting strategies. While many nanoparticle formulations show minimal or moderate toxicity in animal models, the long-term biocompatibility and immunogenic potential in humans are less well understood. Inorganic nanoparticles, such as those based on iron oxide or gold, raise concerns about tissue accumulation and degradation, especially in the absence of reliable clearance mechanisms [26]. From a technical standpoint, imaging sensitivity and quantification present further limitations. Many nanoparticle platforms used in magnetic resonance imaging, for example, require relatively high local concentrations to produce a measurable signal, which increases the required dose and potential toxicity [27]. Finally, the complexity of nanoparticle synthesis, surface functionalization and reproducibility creates substantial barriers to clinical translation. Variability in size, surface charge and ligand density not only impacts biological performance but also complicates regulatory approval. Current good manufacturing practices (GMPs) are not yet fully adapted to the nuanced production requirements of multifunctional nanoparticle systems [28]. While nanoparticle-targeted imaging is a powerful tool with significant theoretical and experimental support, its clinical application is hindered by biological, technical and regulatory limitations. Addressing these challenges will require interdisciplinary approaches that combine advances in nanotechnology, imaging science, pharmacology and regulatory policy to enable the safe and effective use of targeted nanoparticles in medical diagnostics.

Recent advances in nanotechnology have opened new avenues for early diagnosis and treatments of cancers due to functionalized next-generation nanoparticles navigating to different molecular targets [29–31]. One of these promising targets are integrins already tested in both preclinical and clinical studies [32–35]. Types of integrin-targeted nanosized contrasts include superparamagnetic iron oxide nanoparticles [14,36,37], gadolinium-based nanoparticles [38–40] and porphyrin-based nanoparticles [37]. Each type has its own advantages and disadvantages. Magnetic nanoparticles (MNPs) have emerged as promising candidates for cancer diagnosis and therapy due to their unique physical properties, such as their small size, magnetic properties and biocompatibility [41–43]. Moreover, functionalization of MNPs with ligands, such as monoclonal antibodies, tumor-penetrating peptides and aptamers can enhance their specificity and selectivity towards cancer cells that could be used as targeting contrast agents in MRI [43–45].

Magnetic resonance imaging is the most commonly used technique for gliomas due to its superior soft tissue contrast and ability to differentiate between normal and tumor tissues [46,47]. Key features of MRI for glioma diagnosis include T1- and T2-weighted imaging, fluid-attenuated inversion recovery (FLAIR) imaging, diffusion-weighted imaging (DWI), perfusion-weighted imaging (PWI) and magnetic resonance (MRS) spectroscopy [37,39,48,49]. Here, we developed a novel MRI contrast agent composed of carbon-encapsulated iron nanoparticles functionalized with monoclonal antibodies against CD61 (Fe@C-$(CH_2)_2$-CONH-anti-CD61) to target the beta 3 subunit (CD61) of the integrin $\alpha v\beta 3$ transmembrane receptor, which is highly overexpressed on glioma cells. Dynamic susceptibility-weighted contrast-enhanced (DSC-MRI) imaging was performed on Wistar rats bearing C6 glioma tumors using monoclonal antibody-navigated nanoparticles as T2*-weighted contrast agents.

2. Materials and Methods

Carbon-encapsulated iron nanoparticles were synthesized by a carbon arc discharge route [50,51] and fully characterized for the size, shape, composition and physicochemical properties as described elsewhere in detail [52]. More bioconjugation studies on carbon-encapsulated magnetic nanoparticles using polyclonal and monoclonal antibodies to yield Fe@C-$(CH_2)_2$-CONH-anti-CD61 bioconstructs have been completed and characterized in our previous studies [20,53]. The magnetic and relaxometric properties of carbon-encapsulated iron nanoparticles as a negative contrast agent for MRI have been recently described [54].

In animal studies, anti-CD61 monoclonal antibody-functionalized carbon-encapsulated iron nanoparticles (Fe@C-$(CH_2)_2$-CONH-anti-CD61) were tested as contrast agents using mouse IgG1 anti-rat CD61 monoclonal antibodies (BD Biosciences, San Jose, CA, USA) as navigating ligands. Male Wistar rats (6–7 weeks old, n = 5) purchased from the Mossakowski Medical Research Institute Polish Academy of Science were used in the experiments following the protocol approved by the local ethical committee for experimental animals (No 3/2009). Rat glioma C6 cells (ATCC-CCL-107) suspended in phosphate buffered saline (Sigma-Aldrich, St. Louis, MO, USA) were administered (10^5 cells per 100 μL^{-1}) subcutaneously into the right flank of the animals and the rats bearing C6 tumors were examined three weeks after post-implantation. A 1.5T scanner (Magnetom Avanto, Siemens Medical Solutions, Erlangen, Germany) and a 4-channel surface receiver coil "flex" (Siemens Medical Solutions, Erlangen, Germany) were used in the imaging studies of rats. The imaging protocols were conducted using the T2*-weighted echo planar imaging (T2* EPI) with spoiler in the coronal plane (TR/TE 1250/58 ms, flip angle 90°, FoV 200 mm, slices 8, slice thickness 3 mm, base resolution 128, phase resolution 100%, bandwidth 1502 Hz/Px, EPI factor 128, gradient mode Fast*). Regardless of the adopted EPI regime and acquisition, the T2-weighted (anatomical) turbo spin echo (TSE) sequence in the coronal orientation (TR/TE 3000/80 ms, flip angle 150°, FoV 200 mm, slices 23, slice thickness 2 mm, base resolution 250, phase resolution 100%, bandwidth 191 Hz/Px) was also applied in the studies. All imaging data was analysed at a commercially available workstation (Leonardo workstation for MRI, Siemens, Erlangen, Germany) and BrainMagix version 2.0.1 (Imagilys SPRL, Belgium). The animals bearing C6 glioma under ketamine/xylosin (Sigma-Aldrich, St. Louis, MO, USA) anesthesia (45/3 mg kg^{-1}) were placed in the MR scanner (1.5T). The MRI examination began with setting the so-called localizer sequence and precise positioning of the animal in the 3D planes. In order to precisely dose the nanoparticle suspension to the tail vein of rats, automatic administration was applied from a dispenser programmed in a defined time cycle. In the studies, the nanoparticle suspension was prepared in a physiological sterile phosphate buffered saline (PBS) solution, to which carboxymethylcellulose (CMC, Sigma-Aldrich, St. Louis, MO,

USA), as a surfactant, was added at a concentration of 0.1 mg mL^{-1}. The as-prepared suspension of Fe@C-$(CH_2)_2$CONH-anti-CD61 nanoparticles (200 μg mL^{-1}) was administered in a single rapid bolus (0.5 mL) through a cannulated catheter into the tail vein of rats bearing C6 glioma with acquisitions of T2*-weighted EPI images performed in the first 12.5 min (750 s) post-dosing. The anatomical T2-weighted imaging was also performed using TSE sequences. The superimposition of T2*- and T2-weighted images was performed by reconstructing perpendicular sections from the volume in a common coordinate system using image data interpolation. The mutual position and geometry of the volumes as well as the coefficients correcting the shifts and rotation between the collected image series were taken into account. Regions of interest (ROI) were marked on the obtained images, from which an averaged curve of relative arbitrary units of nanoparticle concentration over time was calculated, specifying the time to reach the maximum amplitude on the pharmacokinetic curves. The MR signal intensity was converted to a relative concentration of nanoparticles expressed as arbitrary units according to the formula [55]:

$$C_t(t) = k_t \Delta R_2^*(t) = -\frac{k_t}{TE} ln\left(\frac{S(t)}{S_0}\right) \quad (1)$$

where:

$C_t(t)$—the relative concentration of the contrast agent (nanoparticles) in the tissue at time t;

K_t—a proportionality constant that depends on the tissue, the contrast agent (nanoparticles), the field strength and the pulse sequence parameters;

R_2*—the relaxation rate (R_2* = 1/T_2*);

TE—the echo time of the MRI sequence;

$S(t)$—the signal intensity in the tissue at time t;

S_0—the signal intensity during the baseline period before the arrival of the contrast agent (nanoparticles).

3. Results and Discussion

Dynamic susceptibility contrast magnetic resonance imaging (DSC-MRI), commonly referred to as the "bolus tracking" method, monitors the first pass of an intravascular, non-diffusible contrast agent through tissue. It relies on susceptibility-induced signal loss observed in T2*-weighted sequences, which results from a bolus of a contrast passing through the capillary bed [56]. This method involves intravenous injection of a paramagnetic or superparamagnetic contrast agent in a bolus, i.e., a single rapid administration, and measurement of the signal during its direct flow [57]. Since the flow time of the contrast agent through the given tissue usually lasts several seconds, the use of fast MR imaging techniques is required for correct characterization of the signal intensity. Echo planar imaging (EPI) and fast low tilt angle sequences are commonly used in such studies [58]. The T2*-weighted sequences based on perfusion measurement are used in the case of contrast agent movement only in the lumen of the blood vessel [59]. For this reason, they commonly use blood pool contrast agents to modulate the signals [60]. Although the vascular space is a small part of the total tissue volume (e.g., for the human brain it is ca. 3–5%), the effect of contrast susceptibility of the flowing agent in the bolus causes a strong momentary decrease in the signal in this space [37]. This type of phenomenon was considered appropriate to analyze the possibility of using innovative core-shell type carbon-encapsulated iron nanoparticles, which in light of the conducted previous studies was classified as a typical negative contrast agent, i.e., reducing the MRI signal intensity on T2- and T2*-weighted magnetic resonance images [52].

The DSC-MRI study was performed on Wistar rats bearing C6 tumors growing on a flank that were intravenously administered into the tail vain with a single bolus of Fe@C-$(CH_2)_2$-CONH-anti-CD61 nanoparticles conjugated with monoclonal antibodies against the beta 3 subunit (CD61) of the integrin $\alpha v\beta 3$ receptors. Note that the sedimentation conditions of the as-used nanoparticle suspension were discussed in the publication by Grudzinski et al. [61]. Based on the performed studies, the targeted deposition of nanoparticles with a functionalized antibody against the CD61 integrin in tumor tissues was demonstrated. This was noted as a strong post-contrast signal change in magnetic susceptibility visible as increased red color on T2*-weighted images (Figure 1). Results shown in Figure 1 provide T2* EPI-weighted images expressed as blood volume parametric maps superimposed with T2-weighted images to better localize the tumor tissues. Creating a blood volume parametric map (Figure 1) from DSC-MRI involved a structured series of image-processing and signal-modeling steps. The process begins with the acquisition of DSC-MRI data during the injection of a nanoparticle-based contrast agent. Once the data were acquired, preprocessing was performed to ensure the accuracy of the signal analysis. This included motion correction to align all volumes over time, and registration to anatomical images. Following preprocessing, the MRI signal was converted into a concentration-time curve by calculating the change in transverse relaxation rate (please see the formula). The relative blood volume was then calculated by integrating the $\Delta R_2{}^*(t)$ curve over time. This step was performed using numerical methods. Finally, the calculated blood volume values were used to generate a parametric map that matches the spatial dimensions of the original MRI data. The result is a voxel-wise representation of blood volume, useful for assessing C6 glioma perfusion in Wistar rats (Figure 1).

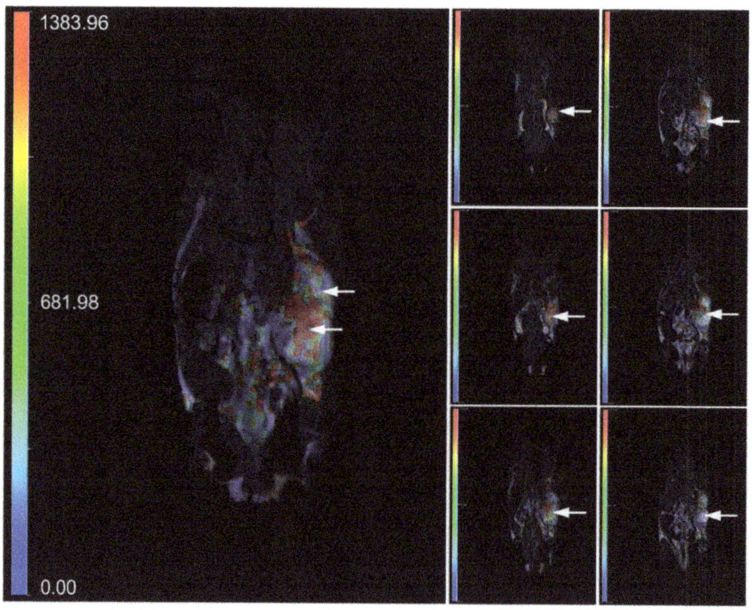

Figure 1. Representative parametric maps of blood volume, overlaid on T2-weighted images, are shown across different slices of a Wistar rat bearing a C6 glioma. The animal underwent dynamic susceptibility contrast MR imaging (DSC-MRI) following intravenous injection of Fe@C-$(CH_2)_2$CONH-anti-CD61 nanoparticles (200 µg mL^{-1}) administered as a single bolus (0.5 mL) into the tail vein. DSC-MRI was conducted over 12.5 min using T2*-weighted EPI sequences (TR/TE = 1250/58 ms).

Anatomical T2 images were performed using TSE sequences (TR/TE = 3000/80 ms). The left large image displays a parametric map from a single slice, while the right small images combine maps from multiple slices of the same rat. The accumulation of nanoparticles targeting the beta 3 subunit (CD61) on the integrin αvβ3 receptors in glioma tissue leads to localized magnetic susceptibility changes, visualized as areas of increased red intensity on the arbitrary unit (a.u.) color scale (refer to the scale for interpretation). Both the left and right images use the same color scale ranged from 0 to 1383.96 a.u. The C6 glioma is indicated by arrows.

In the studies, we observed a pronounced increase in the concentration of Fe@C-$(CH_2)_2$CONH-anti-CD61 nanoparticles in glioma tissues and this could be associated with enhanced permeability and retention (EPR) effects (Figure 2). Such a significant increase (arbitrary units) was not found in the case of the liver or kidney (Figure 2). Please note that vascular extravasation in tumors refers to the leakage of blood components, including nanoparticles, from the tumor vasculature into the surrounding tumor tissue due to the abnormal structure and permeability of tumor blood vessels [62]. This phenomenon is often enhanced in tumors, allowing for greater penetration of nanoparticles, which can accumulate at higher concentrations in the tumor microenvironment, making them potential vehicles for targeted drug delivery [63]. The EPR effect, a hallmark of tumors, further facilitates the preferential accumulation of nanoparticles in tumor tissues, improving diagnostic and therapeutic outcomes [64,65]. This process is influenced by factors like nanoparticle size, tumor characteristics and the presence of leaky vasculature. The presence of beta 3 integrin receptors in glioma tissues and glioma vasculature may significantly enhance the deposition of these nanoparticles in glioma as compared to other analyzed organs (Figure 2). Our previous relaxometric measurements indicate the temporary deposition of Fe@C-$(CH_2)_2$CONH-anti-CD61 in glioma and melanoma tumors examined in murine syngeneic mice models [20,54]. The current results therefore confirm the observed effects, indicating the deposition of nanoparticles functionalized with antibodies recognizing CD61 integrin receptors. Of course, in the case of nanoparticles, we cannot rule out the nanoparticle extravasation effects.

Quantification of bolus-tracking data using an intravascular agent is based on the principles of tracer kinetics for non-diffusible tracers [55]. These are used to model the time-dependent concentration of the nanoparticle-based contrast agent in the tissue as a function of the injected bolus, the blood flow and the fraction of the contrast agent remaining in the tissue. One of the differences of MRI compared with other imaging modalities is that MRI does not measure the concentration of the contrast agent directly. In order to apply pharmacokinetic models of contrast distribution to imaging-based data, the first essential step is to use the signal changes observed in the dynamic acquisition to calculate quantitative parametric images of contrast concentration at each time point. The concentration is related to its effect on the relaxation time and the change (compared with the pre-injection baseline value) in the relaxation rate ($R_2^* = 1/T_2^*$) is linearly proportional to the concentration of the contrast agent (please see the formula) [55]. Pharmacokinetic studies evidenced a rapid increase in the concentration of Fe@C-$(CH_2)_2$CONH-anti-CD61 nanoparticles in the tumor environment at the first minutes following saturation of this process in the next minutes (Figure 2). Minimal post-contrast effects were also observed in the liver and kidney, but the contrast effects were significantly weaker compared to glioma tissues (Figure 2). Basically, no such changes in the resonance signal were observed in the gastrocnemius muscle of the rat, which served as a control for the pharmacokinetic study. When analyzing the concentration profile of antibody-conjugated carbon-encapsulated iron nanoparticles over time in rat C6 glioma tissues (Figure 2), it should be emphasized that no typical curve pattern was observed, such as the one characteristic of gadolinium-based chelate agents routinely used in clinical DSC-MRI studies with glioma patients [66,67]. When gadolinium passes through the blood vessels, it causes changes in the magnetic properties between the vessels and the nearby tissue. Even though blood vessels make up only a small percentage of the total glioma tissue volume, the presence of gadolinium inside them disrupts the uniformity of the local magnetic field. This disturbance spreads beyond just the blood vessels, affecting nearby areas of tissue. As a result, both the spins inside the blood vessels and those in the surrounding tissue experience a drop in T2*, leading to a noticeable temporary decrease in signal. Following a standard intravenous bolus injection, gadolinium typically exhibits a concentration-time curve marked by an initial rise in concentration, followed by a rapid decline as the agent exits the tumor vasculature [68]. In contrast, the nanoparticle concentration profile in our study showed a markedly different pattern (Figure 2). In the glioma tissue, we observed a sharp initial increase in concentration, which then stabilized, forming a sustained plateau lasting up to 750 s. This result clearly shows that nanoparticles are capable of causing blood vessels to leak, a process linked to the phenomenon of extravasation (Figure 3). Additionally, their ability to bind to integrin $\alpha v\beta 3$ receptors due to navigating antibodies (anti-CD61) on both the endothelial cells of glioma blood vessels and glioma cells enhances their accumulation within the tumor site. This accumulation may lead to localized decreases in signal intensity on T2*-weighted MRI scans (Figure 1). Notably, this behavior was not observed in the liver or kidneys. Please note that the liver acts as a major site for the capture, processing and elimination of particles, including nanomaterials, which can be taken up by hepatocytes, Kupffer cells and endothelial cells in the liver sinusoids. The fenestrated endothelial architecture of hepatic sinusoids further facilitates nanoparticle entry into the liver parenchyma. This unique permeability allows nanoparticles—particularly those under 100 nm—to diffuse into the space of Disse and be taken up by hepatocytes and stellate cells. Over time, this can lead to nanoparticle accumulation and potential hepatic toxicity, especially for inorganic nanomaterials that are not easily degraded or excreted [69]. Kupffer cells comprise the largest population of tissue macrophages in the organism and play a crucial role in the immune response. Iron nanoparticles are taken up by Kupffer cells and this is used in imaging the primary liver tumor [70]. Interestingly, uptake of iron oxide nanoparticles by Kupffer cells is used as a biomarker in MRI distinguishing benign from malignant liver lesions [71,72]. Renal clearance is another significant pathway for nanoparticle elimination, especially for particles with hydrodynamic diameters below ~5–8 nm. These ultrasmall nanoparticles can pass through the glomerular filtration barrier and be excreted in the urine. However, this is not always beneficial. In many cases, renal clearance competes with the accumulation of nanoparticles at disease sites, shortening circulation time and reducing imaging efficacy. Moreover, repeated exposure or improper surface functionalization can result in tubular reabsorption or nephrotoxicity, particularly if the particles contain heavy metals or reactive surface chemistries [73]. Importantly, the balance between hepatic and renal clearance is strongly influenced by the physicochemical properties of nanoparticles, including size, shape, surface charge and coating composition. For example, larger particles (>100 nm) tend to be retained in the liver and spleen, while smaller, neutrally charged and hydrophilic particles are more likely to be filtered by the kidneys. These competing clearance mechanisms create a narrow design window for optimizing targeted imaging agents that can circulate long enough to reach target tissues, yet avoid premature clearance and off-target accumulation [74]. The challenge, therefore, is not simply reducing liver and kidney uptake, but engineering nanoparticles that can evade these organs long enough to engage disease targets while maintaining biocompatibility and eventual excretion.

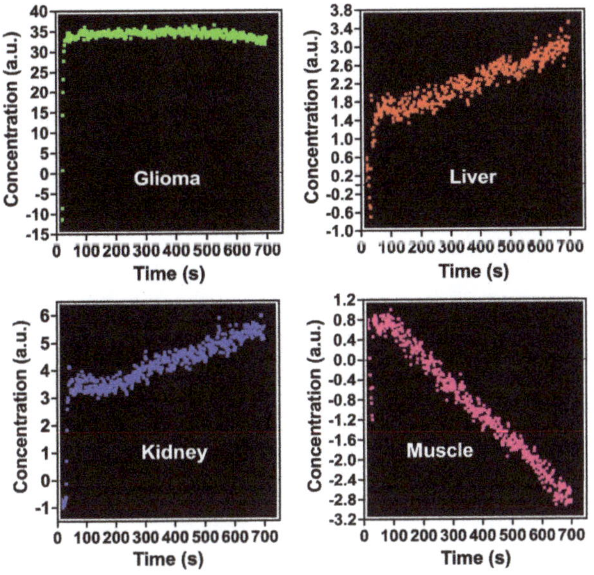

Figure 2. Pharmacokinetic analysis of antibody-conjugated carbon-encapsulated iron nanoparticles (Fe@C-$(CH_2)_2$CONH-anti-CD61) in rat C6 glioma and other organs. The concentration of nanoparticles is reported in arbitrary units (a.u.). A single bolus dose (0.5 mL) containing 200 μg mL^{-1} of the nanoparticles was administered via a cannulated tail vein catheter to rats implanted with C6 glioma. MRI scans were conducted for up to 12.5 min (750 s) post-injection. Dynamic susceptibility contrast MRI (DSC-MRI) was performed using a T2*-weighted EPI sequence with parameters TR/TE = 1250/58 ms.

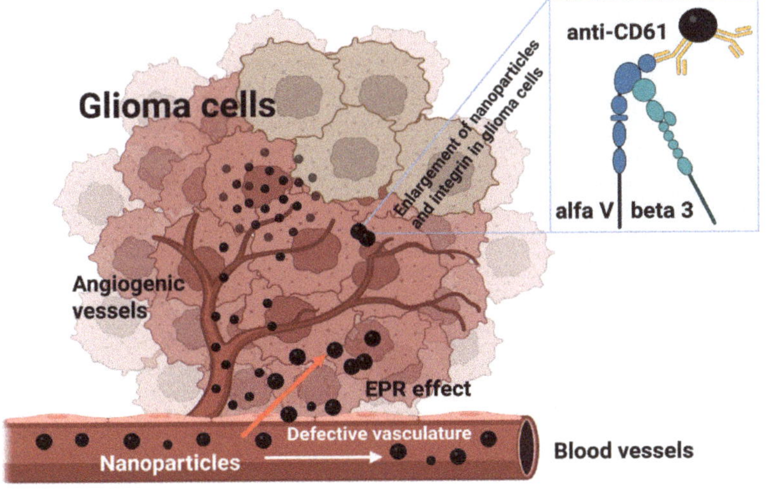

Figure 3. Schematic diagram of nanoparticle extravasation in rat C6 glioma tissues. A single bolus of carbon-encapsulated iron nanoparticles conjugated with a monoclonal antibody (anti-CD61) was injected into the tail vein of Wistar rats bearing C6 glioma. These nanoparticles exited the glioma vasculature through an EPR (enhanced permeability and retention)-mediated effect. Upon reaching the glioma cells, they bound to the αvβ3 integrin, resulting in a detectable change in the magnetic resonance signal (refer to Figure 1). As a result, the concentration of nanoparticles within the glioma tissue increased (refer to Figure 2).

Monoclonal antibody-conjugated carbon-encapsulated iron nanoparticles targeting the beta 3 subunit (CD61) of the integrin αvβ3 receptors represent a promising class of nanomaterials for targeted MRI diagnostics. In the context of MRI, these nanoparticles act as effective T2 contrast agents due to their high magnetic susceptibility, significantly improving image contrast and sensitivity [52]. Their superparamagnetic behavior at room temperature minimizes remanence, reducing the risk of particle aggregation—a key factor in safe in vivo applications [52]. Additionally, the carbon shell offers a versatile platform for functionalization with targeting ligands (e.g., CD61), enabling specific binding to biomarkers or tumor cells, thereby enhancing diagnostic precision [20,54]. In bolus-tracking methods such as DSC-MRI, the T2*-based techniques for measuring perfusion using contrast agents are primarily used in cases where there is a significant compartmentalization of the contrast agent (to observe a significant decrease in T2* relaxation). In the present studies, we evidenced nanoparticle extravasation effects, which refers to the movement of nanoparticles from blood vessels into surrounding glioma tissues (Figure 3). This effect is primarily influenced by the EPR effect, where leaky tumor vasculature and poor lymphatic drainage allow nanoparticles to accumulate in tumor tissues more than in normal tissues. While this mechanism enhances nanoparticle accumulation, it may also cause off-target effects and plausibly toxicity in non-tumor tissues. Therefore, understanding and optimizing extravasation is crucial for improving the efficacy and safety of nanoparticle-based diagnostics, and ongoing research will seek nanoparticles for controlled and targeted extravasation.

4. Conclusions

The performed preclinical DSC-MRI studies indicate local T2*-weighted contrast changes in magnetic susceptibility in rat C6 glioma tissues after administration of carbon-encapsulated iron nanoparticles functionalized with a monoclonal antibody (anti-CD61) against the beta 3 subunit of the rat integrin αvβ3 receptors. The pharmacokinetic studies of nanoparticles in rats bearing C6 tumors indicate the possibility of nanoparticle accumulation in the glioma tissue. Considering the intravenous administration of nanoparticles in rats, it cannot be ruled out that hepatic deposition also occurred after administration. The studies indicate the importance of surface functionalization of carbon-encapsulated iron nanomaterials with monoclonal antibodies selectively recognizing the beta 3 subunit of the αvβ3 integrin receptor in targeted dynamic susceptibility contrast magnetic resonance imaging in CD61-positive glioma cells. Therefore, the obtained results open new avenues in the field of the use of targeted core-shell type magnetic nanoparticles in both glioma and plausibly liver tumors, because the uptake of these nanoparticles in the hepatic macrophage system may probably be a beneficial element in the case of hepatotropic imaging, including primary liver tumors with overexpression of the integrin receptor αvβ3 [70,75]. Overall, monoclonal antibody (anti-CD61)-conjugated carbon-encapsulated iron nanoparticles hold significant promise in advancing MRI-based diagnostics, offering a combination of high contrast capability and targetability. Further research will determine their preclinical safety and viability in clinical settings. Challenges will also remain in large-scale synthesis with controlled size and surface properties, as well as in thorough long-term stability, extravasation and clearance studies.

Author Contributions: Conceptualization, I.P.G.; methodology, A.S., M.B.-K., W.S. and I.P.G.; formal analysis, A.S., M.B.-K., W.S. and I.P.G.; investigation, A.S., M.B.-K., W.S. and I.P.G.; data curation, A.S., W.S. and I.P.G.; writing—original draft preparation, A.S., M.B.-K. and I.P.G.; writing—review and editing, A.S., M.B.-K., W.S. and I.P.G.; visualization, W.S. and I.P.G.; project administration,

I.P.G.; funding acquisition, I.P.G. All authors have read and agreed to the published version of the manuscript.

Funding: This study was supported by the National Science Center Grant N N518 381737.

Institutional Review Board Statement: The animal study protocol was approved by the Ethics Committee of Local ethical committee for experimental animals (protocol code No 3/2009 and date of approval 27 January 2009).

Data Availability Statement: The original contributions presented in this study are included in the article. Further inquiries can be directed to the corresponding author(s).

Acknowledgments: Graphical abstract and schematic diagram (Figure 3) was created with BioRender (academic license agreement, Open Access Agreement number: YR28NAGA84 (graphical abstract) and CU28NADYLQ (Figure 3)).

Conflicts of Interest: The authors declare no conflicts of interest.

Abbreviations

The following abbreviations are used in this manuscript:

MRI	magnetic resonance imaging
DSC-MRI	dynamic susceptibility contrast magnetic resonance imaging
T2* EPI	T2*-weighted echo planar imaging
DWI	diffusion-weighted imaging
PWI	perfusion-weighted imaging
TSE	turbo spin echo
ROI	regions of interest
CD61	beta 3 subunit of the $\alpha v\beta 3$ integrin receptor
ECM	extracellular matrix
MNPs	magnetic nanoparticles
TPP	tumor penetrating peptides
EPR	enhanced permeability and retention effect
CT	computed tomography
POC	point-of-care
GMP	good manufacturing practice

References

1. Liu, F.; Wu, Q.; Dong, Z.; Liu, K. Integrins in cancer: Emerging mechanisms and therapeutic opportunities. *Pharmacol. Ther.* **2023**, *247*, 108458. [CrossRef] [PubMed]
2. Soles, A.; Selimovic, A.; Sbrocco, K.; Ghannoum, F.; Hamel, K.; Moncada, E.L.; Gilliat, S.; Cvetanovic, M. Extracellular Matrix Regulation in Physiology and in Brain Disease. *Int. J. Mol. Sci.* **2023**, *24*, 7049. [CrossRef]
3. Biermann, J.; Melms, J.C.; Amin, A.D.; Wang, Y.; Caprio, L.A.; Karz, A.; Tagore, S.; Barrera, I.; Ibarra-Arellano, M.A.; Andreatta, M.; et al. Dissecting the treatment-naive ecosystem of human melanoma brain metastasis. *Cell* **2022**, *185*, 2591–2608.e30. [CrossRef] [PubMed]
4. D'abaco, G.M.; Kaye, A.H. Integrins: Molecular determinants of glioma invasion. *J. Clin. Neurosci.* **2007**, *14*, 1041–1048. [CrossRef]
5. Tabatabai, G.; Weller, M.; Nabors, B.; Picard, M.; Reardon, D.; Mikkelsen, T.; Ruegg, C.; Stupp, R. Targeting integrins in malignant glioma. *Target. Oncol.* **2010**, *5*, 175–181. [CrossRef]
6. Ellert-Miklaszewska, A.; Poleszak, K.; Pasierbinska, M.; Kaminska, B. Integrin Signaling in Glioma Pathogenesis: From Biology to Therapy. *Int. J. Mol. Sci.* **2020**, *21*, 888. [CrossRef]
7. Bello, L.; Francolini, M.; Marthyn, P.; Zhang, J.; Carroll, R.S.; Nikas, D.C.; Strasser, J.F.; Villani, R.; Cheresh, D.A.; Black, P.M. $\alpha v\beta 3$ and $\alpha v\beta 5$ Integrin Expression in Glioma Periphery. *Neurosurgery* **2001**, *49*, 380–390. [CrossRef]
8. Paolillo, M.; Serra, M.; Schinelli, S. Integrins in glioblastoma: Still an attractive target? *Pharmacol. Res.* **2016**, *113*, 55–61. [CrossRef]
9. Hamidi, H.; Ivaska, J. Every step of the way: Integrins in cancer progression and metastasis. *Nat. Rev. Cancer* **2018**, *18*, 533–548. [CrossRef]

10. Tabatabai, G.; Tonn, J.-C.; Stupp, R.; Weller, M. The Role of Integrins in Glioma Biology and Anti-Glioma Therapies. *Curr. Pharm. Des.* **2011**, *17*, 2402–2410. [CrossRef] [PubMed]
11. Delamarre, E.; Taboubi, S.; Mathieu, S.; Bérenguer, C.; Rigot, V.; Lissitzky, J.-C.; Figarella-Branger, D.; Ouafik, L.; Luis, J. Expression of Integrin $\alpha 6\beta 1$ Enhances Tumorigenesis in Glioma Cells. *Am. J. Pathol.* **2009**, *175*, 844–855. [CrossRef] [PubMed]
12. Ishikawa, T.; Wondimu, Z.; Oikawa, Y.; Gentilcore, G.; Kiessling, R.; Brage, S.E.; Hansson, J.; Patarroyo, M. Laminins 411 and 421 differentially promote tumor cell migration via $\alpha 6\beta 1$ integrin and MCAM (CD146). *Matrix Biol.* **2014**, *38*, 69–83. [CrossRef]
13. Wagle, N.; Nguyen, M.; Carrillo, J.; Truong, J.; Dobrawa, L.; Kesari, S. Characterization of molecular pathways for targeting therapy in glioblastoma. *Chin. Clin. Oncol.* **2020**, *9*, 77. [CrossRef]
14. Lin, X.; Beringhs, A.O.; Lu, X. Applications of Nanoparticle-Antibody Conjugates in Immunoassays and Tumor Imaging. *AAPS J.* **2021**, *23*, 43. [CrossRef]
15. Gupta, A.K.; Gupta, M. Synthesis and surface engineering of iron oxide nanoparticles for biomedical applications. *Biomaterials* **2005**, *26*, 3995–4021. [CrossRef]
16. Hou, Z.; Liu, Y.; Xu, J.; Zhu, J. Surface engineering of magnetic iron oxide nanoparticles by polymer grafting: Synthesis progress and biomedical applications. *Nanoscale* **2020**, *12*, 14957–14975. [CrossRef]
17. Bamburowicz-Klimkowska, M.; Poplawska, M.; Grudzinski, I.P. Nanocomposites as biomolecules delivery agents in nanomedicine. *J. Nanobiotechnol.* **2019**, *17*, 48. [CrossRef] [PubMed]
18. Medintz, I.L.; Uyeda, H.T.; Goldman, E.R.; Mattoussi, H. Quantum dot bioconjugates for imaging, labelling and sensing. *Nat. Mater.* **2005**, *4*, 435–446. [CrossRef]
19. Lv, M.; Zhou, W.; Tavakoli, H.; Bautista, C.; Xia, J.; Wang, Z.; Li, X. Aptamer-functionalized metal-organic frameworks (MOFs) for biosensing. *Biosens. Bioelectron.* **2021**, *176*, 112947. [CrossRef] [PubMed]
20. Stawarska, A.; Bamburowicz-Klimkowska, M.; Bystrzejewski, M.; Kasprzak, A.; Grudzinski, I.P. Carbon-Encapsulated Iron Nanoparticles Seeking Integrins in Murine Glioma. *Int. J. Nanomed.* **2025**, *20*, 5475–5488. [CrossRef]
21. Jain, P.K.; Huang, X.; El-Sayed, I.H.; El-Sayed, M.A. Noble Metals on the Nanoscale: Optical and Photothermal Properties and Some Applications in Imaging, Sensing, Biology, and Medicine. *Accounts Chem. Res.* **2008**, *41*, 1578–1586. [CrossRef]
22. Michalet, X.; Pinaud, F.F.; Bentolila, L.A.; Tsay, J.M.; Doose, S.; Li, J.J.; Sundaresan, G.; Wu, A.M.; Gambhir, S.S.; Weiss, S. Quantum Dots for Live Cells, in Vivo Imaging, and Diagnostics. *Science* **2005**, *307*, 538–544. [CrossRef]
23. Maduraiveeran, G.; Sasidharan, M.; Ganesan, V. Electrochemical sensor and biosensor platforms based on advanced nanomaterials for biological and biomedical applications. *Biosens. Bioelectron.* **2018**, *103*, 113–129. [CrossRef]
24. Wilhelm, S.; Tavares, A.J.; Dai, Q.; Ohta, S.; Audet, J.; Dvorak, H.F.; Chan, W.C.W. Analysis of nanoparticle delivery to tumours. *Nat. Rev. Mater.* **2016**, *1*, 16014. [CrossRef]
25. Blanco, E.; Shen, H.; Ferrari, M. Principles of nanoparticle design for overcoming biological barriers to drug delivery. *Nat. Biotechnol.* **2015**, *33*, 941–951. [CrossRef] [PubMed]
26. Danhier, F. To exploit the tumor microenvironment: Since the EPR effect fails in the clinic, what is the future of nanomedicine? *J. Control. Release* **2016**, *244*, 108–121. [CrossRef] [PubMed]
27. Dutta, T.; Burgess, M.; McMillan, N.A.; Parekh, H.S. Dendrosome-based delivery of siRNA against E6 and E7 oncogenes in cervical cancer. *Nanomed. Nanotechnol. Biol. Med.* **2010**, *6*, 463–470. [CrossRef] [PubMed]
28. Anselmo, A.C.; Mitragotri, S. Nanoparticles in the clinic: An update. *Bioeng. Transl. Med.* **2019**, *4*, e10143. [CrossRef]
29. Wang, B.; Hu, S.; Teng, Y.; Chen, J.; Wang, H.; Xu, Y.; Wang, K.; Xu, J.; Cheng, Y.; Gao, X. Current advance of nanotechnology in diagnosis and treatment for malignant tumors. *Signal Transduct. Target. Ther.* **2024**, *9*, 200. [CrossRef]
30. Mosleh-Shirazi, S.; Abbasi, M.; Moaddeli, M.R.; Vaez, A.; Shafiee, M.; Kasaee, S.R.; Amani, A.M.; Hatam, S. Nanotechnology Advances in the Detection and Treatment of Cancer: An Overview. *Nanotheranostics* **2022**, *6*, 400–423. [CrossRef] [PubMed]
31. Nasir, A.; Khan, A.; Li, J.; Naeem, M.; Khalil, A.A.K.; Khan, K.; Qasim, M. Nanotechnology, A Tool for Diagnostics and Treatment of Cancer. *Curr. Top. Med. Chem.* **2021**, *21*, 1360–1376. [CrossRef] [PubMed]
32. Xie, H.; Diagaradjane, P.; Deorukhkar; Goins, B.; Bao, A.; Phillips, W.T.; Wang, Z.; Schwartz, J.; Krishnan, S. Integrin $\alpha v\beta 3$-targeted gold nanoshells augment tumor vasculature-specific imaging and therapy. *Int. J. Nanomed.* **2011**, *6*, 259–269. [CrossRef]
33. Sugahara, K.N.; Teesalu, T.; Karmali, P.P.; Kotamraju, V.R.; Agemy, L.; Girard, O.M.; Hanahan, D.; Mattrey, R.F.; Ruoslahti, E. Tissue-Penetrating Delivery of Compounds and Nanoparticles into Tumors. *Cancer Cell* **2009**, *16*, 510–520. [CrossRef] [PubMed]
34. Ahmad, K.; Lee, E.J.; Shaikh, S.; Kumar, A.; Rao, K.M.; Park, S.-Y.; O Jin, J.; Han, S.S.; Choi, I. Targeting integrins for cancer management using nanotherapeutic approaches: Recent advances and challenges. *Semin. Cancer Biol.* **2021**, *69*, 325–336. [CrossRef]
35. Pang, X.; He, X.; Qiu, Z.; Zhang, H.; Xie, R.; Liu, Z.; Gu, Y.; Zhao, N.; Xiang, Q.; Cui, Y. Targeting integrin pathways: Mechanisms and advances in therapy. *Signal Transduct. Target. Ther.* **2023**, *8*, 1. [CrossRef]

36. Dammes, N.; Peer, D. Monoclonal antibody-based molecular imaging strategies and theranostic opportunities. *Theranostics* **2020**, *10*, 938–955. [CrossRef] [PubMed]
37. Zhao, W.; Yu, X.; Peng, S.; Luo, Y.; Li, J.; Lu, L. Construction of nanomaterials as contrast agents or probes for glioma imaging. *J. Nanobiotechnol.* **2021**, *19*, 125. [CrossRef]
38. Byrne, H.L.; Le Duc, G.; Lux, F.; Tillement, O.; Holmes, N.M.; James, A.; Jelen, U.; Dong, B.; Liney, G.; Roberts, T.L.; et al. Enhanced MRI-guided radiotherapy with gadolinium-based nanoparticles: Preclinical evaluation with an MRI-linac. *Cancer Nanotechnol.* **2020**, *11*, 9. [CrossRef]
39. Li, A.Y.; Iv, M. Conventional and Advanced Imaging Techniques in Post-treatment Glioma Imaging. *Front. Radiol.* **2022**, *2*, 883293. [CrossRef]
40. Hu, L.S.; Hawkins-Daarud, A.; Wang, L.; Li, J.; Swanson, K.R. Imaging of intratumoral heterogeneity in high-grade glioma. *Cancer Lett.* **2020**, *477*, 97–106. [CrossRef]
41. Farzin, A.; Etesami, S.A.; Quint, J.; Memic, A.; Tamayol, A. Magnetic Nanoparticles in Cancer Therapy and Diagnosis. *Adv. Healthc. Mater.* **2020**, *9*, e1901058. [CrossRef]
42. Kolosnjaj-Tabi, J.; Wilhelm, C. Magnetic Nanoparticles in Cancer Therapy: How Can Thermal Approaches help? *Nanomedicine* **2017**, *12*, 573–575. [CrossRef] [PubMed]
43. He, M.; Chen, S.; Yu, H.; Fan, X.; Wu, H.; Wang, Y.; Wang, H.; Yin, X. Advances in nanoparticle-based radiotherapy for cancer treatment. *iScience* **2024**, *28*, 111602. [CrossRef]
44. Rezaei, B.; Yari, P.; Sanders, S.M.; Wang, H.; Chugh, V.K.; Liang, S.; Mostufa, S.; Xu, K.; Wang, J.; Gómez-Pastora, J.; et al. Magnetic Nanoparticles: A Review on Synthesis, Characterization, Functionalization, and Biomedical Applications. *Small* **2023**, *20*, e2304848. [CrossRef]
45. Hola, K.; Markova, Z.; Zoppellaro, G.; Tucek, J.; Zboril, R. Tailored functionalization of iron oxide nanoparticles for MRI, drug delivery, magnetic separation and immobilization of biosubstances. *Biotechnol. Adv.* **2015**, *33*, 1162–1176. [CrossRef]
46. Fatania, K.; Mohamud, F.; Clark, A.; Nix, M.; Short, S.C.; O'cOnnor, J.; Scarsbrook, A.F.; Currie, S. Intensity standardization of MRI prior to radiomic feature extraction for artificial intelligence research in glioma—A systematic review. *Eur. Radiol.* **2022**, *32*, 7014–7025. [CrossRef] [PubMed]
47. Li, G.; Li, L.; Li, Y.; Qian, Z.; Wu, F.; He, Y.; Jiang, H.; Li, R.; Wang, D.; Zhai, Y.; et al. An MRI radiomics approach to predict survival and tumour-infiltrating macrophages in gliomas. *Brain* **2022**, *145*, 1151–1161. [CrossRef] [PubMed]
48. Pope, W.B.; Brandal, G. Conventional and advanced magnetic resonance imaging in patients with high-grade glioma. *Q. J. Nucl. Med. Mol. Imaging* **2018**, *62*, 239–253. [CrossRef]
49. Verburg, N.; Hamer, P.C.d.W. State-of-the-art imaging for glioma surgery. *Neurosurg. Rev.* **2020**, *44*, 1331–1343. [CrossRef]
50. Borysiuk, J.; Grabias, A.; Szczytko, J.; Bystrzejewski, M.; Twardowski, A.; Lange, H. Structure and magnetic properties of carbon encapsulated Fe nanoparticles obtained by arc plasma and combustion synthesis. *Carbon* **2008**, *46*, 1693–1701. [CrossRef]
51. Bystrzejewski, M.; Huczko, A.; Lange, H. Arc plasma route to carbon-encapsulated magnetic nanoparticles for biomedical applications. *Sens. Actuators B Chem.* **2005**, *109*, 81–85. [CrossRef]
52. Grudzinski, I.P.; Bystrzejewski, M.; Bogorodzki, P.; Cieszanowski, A.; Szeszkowski, W.; Poplawska, M.; Bamburowicz-Klimkowska, M. Comprehensive magnetic resonance characteristics of carbon-encapsulated iron nanoparticles: A new frontier for the core-shell–type contrast agents. *J. Nanoparticle Res.* **2020**, *22*, 82. [CrossRef]
53. Poplawska, M.; Bystrzejewski, M.; Grudziński, I.P.; Cywińska, M.A.; Ostapko, J.; Cieszanowski, A. Immobilization of gamma globulins and polyclonal antibodies of class IgG onto carbon-encapsulated iron nanoparticles functionalized with various surface linkers. *Carbon* **2014**, *74*, 180–194. [CrossRef]
54. Bamburowicz-Klimkowska, M.; Bystrzejewski, M.; Kasprzak, A.; Cieszanowski, A.; Grudzinski, I.P. Monoclonal antibody-navigated carbon-encapsulated iron nanoparticles used for MRI-based tracking integrin receptors in murine melanoma. *Nanomed. Nanotechnol. Biol. Med.* **2023**, *55*, 102721. [CrossRef]
55. Calamante, F. Quantification of Dynamic Susceptibility Contrast T2* MRI in Oncology. In *Dynamic Contrast-Enhanced Magnetic Resonance Imaging in Oncology*; Jackson, A., Buckley, D.L., Parker, G.J.M., Eds.; Springer: Berlin/Heidelberg, Germany, 2005; pp. 53–67.
56. Arvidsson, J.; Starck, G.; Lagerstrand, K.; Ziegelitz, D.; Jalnefjord, O. Effects of bolus injection duration on perfusion estimates in dynamic CT and dynamic susceptibility contrast MRI. *Magn. Reson. Mater. Phys. Biol. Med.* **2022**, *36*, 95–106. [CrossRef]
57. Günther, M. Perfusion Imaging. *J. Magn. Reson. Imaging* **2013**, *40*, 269–279. [CrossRef]
58. Halder, A.; Harris, C.T.; Wiens, C.N.; Soddu, A.; Chronik, B.A. Optimization of Gradient-Echo Echo-Planar Imaging for T_2* Contrast in the Brain at 0.5 T. *Sensors* **2023**, *23*, 8428. [CrossRef]
59. Sachdev, P.; Wen, W.; Shnier, R.; Brodaty, H. Cerebral Blood Volume in T2-Weighted White Matter Hyperintensities Using Exogenous Contrast Based Perfusion MRI. *J. Neuropsychiatry* **2004**, *16*, 83–92. [CrossRef]

60. Zhuang, D.; Zhang, H.; Hu, G.; Guo, B. Recent development of contrast agents for magnetic resonance and multimodal imaging of glioblastoma. *J. Nanobiotechnol.* **2022**, *20*, 284. [CrossRef] [PubMed]
61. Grudzinski, I.P.; Bystrzejewski, M.; Cywinska, M.A.; Kosmider, A.; Poplawska, M.; Cieszanowski, A.; Ostrowska, A. Cytotoxicity evaluation of carbon-encapsulated iron nanoparticles in melanoma cells and dermal fibroblasts. *J. Nanoparticle Res.* **2013**, *15*, 1835. [CrossRef] [PubMed]
62. Fukumura, D.; Duda, D.G.; Munn, L.L.; Jain, R.K. Tumor Microvasculature and Microenvironment: Novel Insights Through Intravital Imaging in Pre-Clinical Models. *Microcirculation* **2010**, *17*, 206–225. [CrossRef]
63. Torchilin, V. Tumor delivery of macromolecular drugs based on the EPR effect. *Adv. Drug Deliv. Rev.* **2011**, *63*, 131–135. [CrossRef]
64. Fang, J.; Islam, W.; Maeda, H. Exploiting the dynamics of the EPR effect and strategies to improve the therapeutic effects of nanomedicines by using EPR effect enhancers. *Adv. Drug Deliv. Rev.* **2020**, *157*, 142–160. [CrossRef]
65. Maeda, H. Toward a full understanding of the EPR effect in primary and metastatic tumors as well as issues related to its heterogeneity. *Adv. Drug Deliv. Rev.* **2015**, *91*, 3–6. [CrossRef]
66. Quan, G.; Zhang, K.; Liu, Y.; Ren, J.-L.; Huang, D.; Wang, W.; Yuan, T.; Chauncey, T.R. Role of Dynamic Susceptibility Contrast Perfusion MRI in Glioma Progression Evaluation. *J. Oncol.* **2021**, *2021*, 1696387. [CrossRef] [PubMed]
67. Wongsawaeng, D.; Schwartz, D.; Li, X.; Muldoon, L.L.; Stoller, J.; Stateler, C.; Holland, S.; Szidonya, L.; Rooney, W.D.; Wyatt, C.; et al. Comparison of dynamic susceptibility contrast (DSC) using gadolinium and iron-based contrast agents in high-grade glioma at high-field MRI. *Neuroradiol. J.* **2024**, *37*, 473–482. [CrossRef] [PubMed]
68. Lipiński, S. Creation of a Simulated Sequence of Dynamic Susceptibility Contrast—Magnetic Resonance Imaging Brain Scans as a Tool to Verify the Quality of Methods for Diagnosing Diseases Affecting Brain Tissue Perfusion. *Computation* **2024**, *12*, 54. [CrossRef]
69. Howorka, S. Building membrane nanopores. *Nat. Nanotechnol.* **2017**, *12*, 619–630. [CrossRef]
70. Nowak-Jary, J.; Machnicka, B. In vivo Biodistribution and Clearance of Magnetic Iron Oxide Nanoparticles for Medical Applications. *Int. J. Nanomed.* **2023**, *18*, 4067–4100. [CrossRef]
71. Lee, D.; Sohn, J.; Kirichenko, A. Quantifying Liver Heterogeneity via R2*-MRI with Super-Paramagnetic Iron Oxide Nanoparticles (SPION) to Characterize Liver Function and Tumor. *Cancers* **2022**, *14*, 5269. [CrossRef]
72. Li, W.; Jia, X.; Yin, L.; Yang, Z.; Hui, H.; Li, J.; Huang, W.; Tian, J.; Zhang, S. Advances in magnetic particle imaging and perspectives on liver imaging. *iLIVER* **2022**, *1*, 237–244. [CrossRef] [PubMed]
73. Longmire, M.; Choyke, P.L.; Kobayashi, H. Clearance Properties of Nano-Sized Particles and Molecules as Imaging Agents: Considerations and Caveats. *Nanomedicine* **2008**, *3*, 703–717. [CrossRef]
74. Ding, T.; Guan, J.; Wang, M.; Long, Q.; Liu, X.; Qian, J.; Wei, X.; Lu, W.; Zhan, C. Natural IgM dominates in vivo performance of liposomes. *J. Control. Release* **2020**, *319*, 371–381. [CrossRef] [PubMed]
75. Li, Y.; Hu, P.; Wang, X.; Hou, X.; Liu, F.; Jiang, X. Integrin αvβ3-targeted polydopamine-coated gold nanostars for photothermal ablation therapy of hepatocellular carcinoma. *Regen. Biomater.* **2021**, *8*, rbab046. [CrossRef] [PubMed]

nanomaterials

MDPI

Article

Hydrophobic Silicon Quantum Dots for Potential Imaging of Tear Film Lipid Layer

Sidra Sarwat [1], Fiona Stapleton [1], Mark D. P. Willcox [1], Peter B. O'Mara [2] and Maitreyee Roy [1,*]

[1] School of Optometry and Vision Science, University of New South Wales (UNSW), Sydney 2052, Australia
[2] Electron Microscope Unit, School of Chemistry, Mark Wainwright Analytical Centre, University of New South Wales (UNSW), Sydney 2052, Australia
* Correspondence: maitreyee.roy@unsw.edu.au

Abstract: The tear film, consisting of the aqueous and lipid layers, maintains the homeostasis of the ocular surface; therefore, when disturbed, it can cause dry eye, which affects millions of people worldwide. Understanding the dynamics of the tear film layers is essential for developing efficient drug delivery systems for dry eye disease. Quantum dots (QDs) offer the potential for real-time monitoring of tear film and evaluating its dynamics. Hydrophilic silicon QDs (Si-QDs) have already been optimised to image the aqueous layer of the tear film. This study was conducted to optimise hydrophobic Si-QDs to image the lipid layer of the tear film. Si-QDs were synthesised in solution and characterised by transmission electron microscope and spectrofluorophotometry. The fluorescence emission of Si-QDs was monitored in vitro when mixed with artificial tears. The cytotoxicity was assessed in cultured human corneal epithelial cells using an MTT assay following 24 h of exposure. Si-QDs were 2.65 ± 0.35 nm in size and were non-toxic at <16 µg/mL. Si-QDs emitted stable green fluorescence for 20 min but demonstrated aggregation at higher concentrations. These findings highlight the potential of hydrophobic Si-QDs as a biomarker for the real-time imaging of the tear film lipid layer. However, further research on surface functionalisation and preclinical evaluations are recommended for enhanced solubility and biocompatibility in the ocular surface.

Keywords: tear film lipid layer; dry eye disease; hydrophobic quantum dots; fluorescence imaging

1. Introduction

Dry eye disease (DED) is a multifactorial disease of the ocular surface characterised by a loss of homeostasis of the tear film [1]. DED is accompanied by ocular symptoms in which tear film instability and hyperosmolarity, ocular surface inflammation and damage, and neurosensory abnormalities play etiological roles [1,2]. It is among the most prevalent ocular conditions for people seeking eye care [3–5]. The prevalence of dry eyes ranges from 5 to 50% [6], affecting the quality of vision and life [7]. An altered/deficient tear film is important in the pathophysiology of the DED [8].

The tear film is a complex dynamic fluid that covers the anterior ocular surface [1]. It consists of an outer lipid layer and an underlying muco-aqueous layer, which play an important role in protecting and lubricating the ocular surface [5]. Any change/lack in the aqueous or lipid layer of the tear film can lead to aqueous-deficient or evaporative DED, respectively. The prevalence of evaporative DED is significantly higher than that of aqueous-deficient DED [9]. Any change in the composition or production of tear film lipids may lead to the development of evaporative DED [2].

A range of diagnostic tests and assessments are utilised to evaluate the presence and severity of evaporative DED. These tests focus on tear film stability, lipid layer integrity, and ocular surface health [10]. One of the primary diagnostic tools for evaporative DED is measuring tear film break-up time (TBUT), which assesses the stability of the tear film [11]. However, this test does not accurately assess tear dynamics over time. In addition, ocular surface staining with fluorescein or lissamine green is used to examine the cornea, conjunctiva, and eyelid margins. These dyes disrupt the tear film during the examination [12] and are susceptible to autofluorescence and photobleaching [13]. The Schirmer test assesses tear production by inserting a strip of paper into the lower conjunctival sac [14]. Interferometry [15] and optical coherence tomography [16] have been commonly used in clinical practice to assess lipid layer thickness; however, the impact of dynamic changes and tear film stability is still not well understood [5].

The tear film is subject to dynamic interfacial interactions between its layers [17] and continuous blinking [12]. The interfacial properties impede the investigation of the phase dynamics of the tear film [12]. The information gathered from diagnostics tests does not provide insight into the tear film's layers [12]. Therefore, the knowledge of the interfacial dynamics of individual tear film layers remains inadequate [5,18]. The simultaneous imaging of tear film layers with high-contrast biomarker may address this issue and provide more insight into tear film dynamics [12].

Quantum dots (QDs) offer a promising alternative to traditional imaging techniques and organic dyes and provide strong fluorescence for tear film bioimaging due to their excitation range of around 500 nm [13]. These nano-sized particles (2–20 nm) emit stable, bright fluorescence with reduced photobleaching, making them ideal for bioimaging [19,20]. While cadmium-based QDs are widely used, their toxicity raises concerns, especially for ocular applications [21]. However, silicon QDs (Si-QDs) present a biocompatible option with flexible surface modification, producing either a hydrophilic or hydrophobic charge, making them suitable for imaging tear film layers [22]. Silicon nanomaterials in the 50–150 nm range are compatible with human corneal cells, highlighting their potential as safer alternatives for ocular surface bioimaging [22–27].

High-contrast biological biomarkers for the precise labelling of tear film layers may enhance the knowledge of tear film dynamics [12]. In our previous study, amine-terminated hydrophilic Si-QDs have been optimised for bioimaging tear film aqueous layer [22]. However, the transition to hydrophobic systems in this study was required to image the lipid layer of the tear film, which plays a crucial role in maintaining tear film stability and preventing evaporation. Hydrophobic QDs offer several advantages, including improved compatibility with lipid-rich environments, enhanced stability at the air–liquid interface, and stronger interactions with the tear film's outermost layer. This shift addresses the limitations observed in aqueous systems, where hydrophilic QDs primarily interact with the aqueous layer, making it challenging to study lipid-layer dynamics. Using hydrophobic QDs, we aim to expand the potential of QD-based imaging for a more comprehensive analysis of tear film structure and function.

This study presents the optimisation of hydrophobic scandium-doped Si-QDs (Sc-Si-QDs), which may eventually serve as a biomarker for imaging the tear film lipid layer.

2. Results

2.1. *Characteristics of Si-QDs*

Si-QDs were 2.65 nm (±0.35) in size and had spherical morphology (Figure 1).

Figure 2 shows the photoluminescence emission spectra of Si-QDs at various excitation wavelengths (360–480 nm). The *y*-axis represents the arbitrary units of photoluminescence intensity of emission wavelengths. Si-QDs were excited remarkedly by 360, 380 and 400 nm.

The emission peak (130 RFU) transitioned from 400 nm with a narrow emission bandwidth within the visible light spectrum, to 570 nm. The intensity of emitted light diminished at excitation wavelengths of 500 nm and onwards. Si-QDs were prepared using the same synthesis method; their photoluminescence quantum yield was 14.5%, and full width at half maximum was 90 nm [28,29].

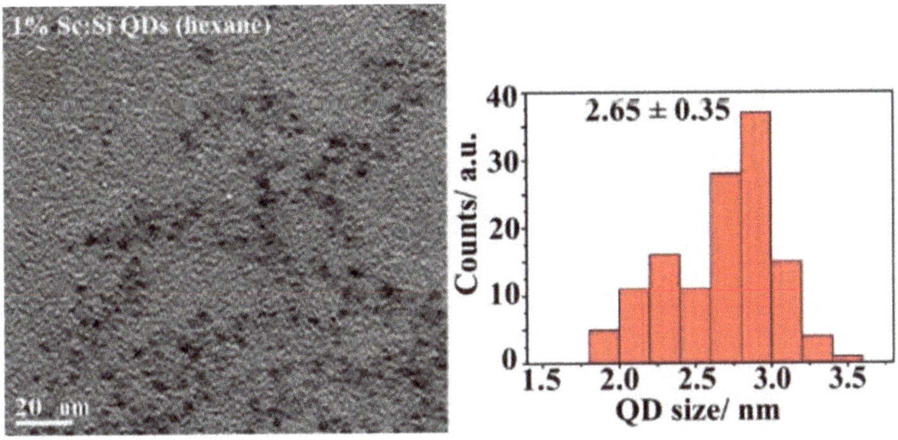

(**a**) (**b**)

Figure 1. (**a**) TEM (Transmission electron microscopy) image of Si-QDs; (**b**) size distribution of Si-QDs.

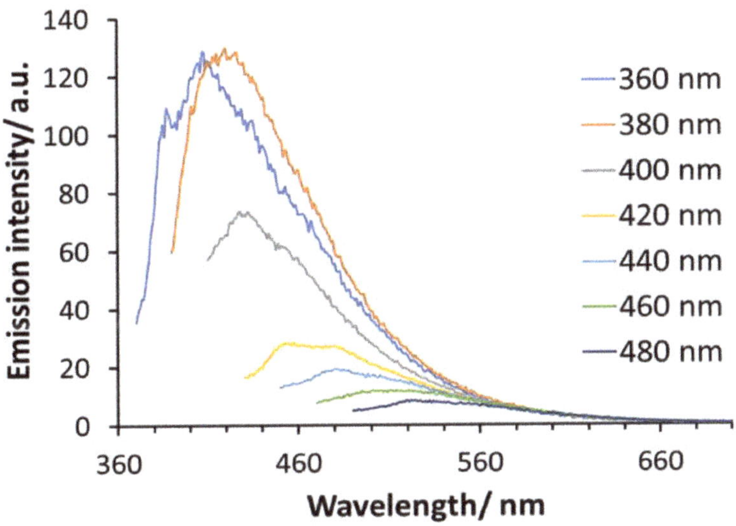

Figure 2. The photoluminescence emission spectra of Si-QDs at different excitation wavelengths (400–520 nm).

2.2. Cytotoxicity of Si-QDs

Figure 3 shows the percentage viability of HCECs following exposure to different concentrations of Si-QDs. Si-QDs showed an average cell viability of >95% at 1 µg/mL and reduced to 5% at 250 µg/mL, compared to the positive control. Si-QDs exhibited no significant reduction in cell viability at 16 µg/mL and below, relative to the negative control ($p = 0.15$).

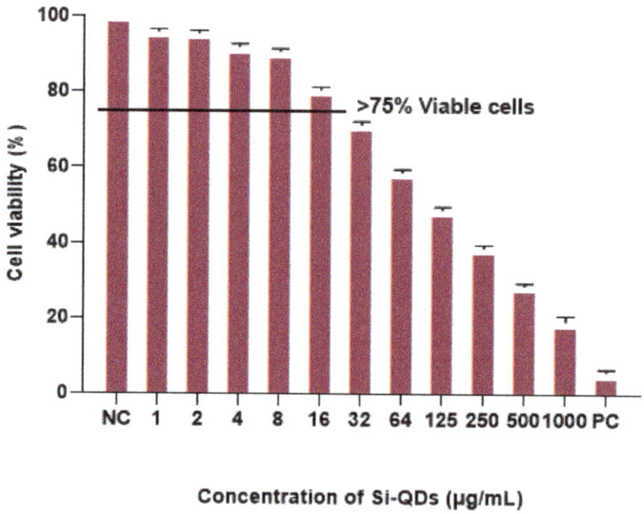

Figure 3. MTT assay: Cell viability (%) of human corneal epithelial cells (HCECs) after exposure to different concentrations of Si-QDs. The horizontal line indicates a cut-off cell viability of more than 75%. Negative control, NC positive control (PC).

2.3. *In Vitro Fluorescence Imaging of TheraTears® with Si-QDs*

The control solution (TheraTears®), devoid of Si-QDs, exhibited no detectable fluorescence signal (Figure 4a). Si-QDs emitted green fluorescence at all given concentrations with scattered fluorescence emission at 16 µg/mL and above. However, Si-QDs appeared to be aggregated at higher concentrations resulting in dispersed fluorescence (Figure 4d).

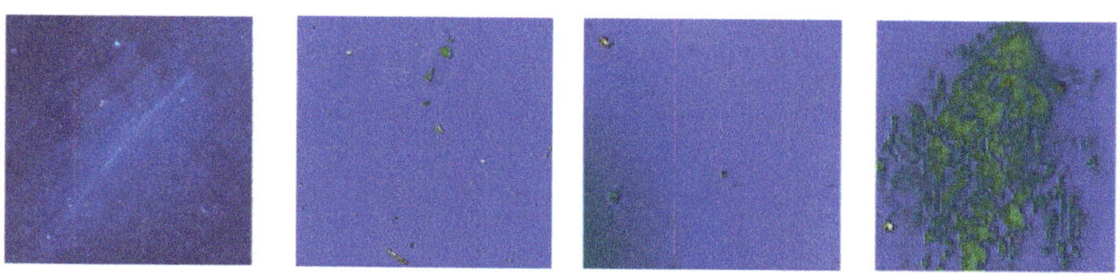

(**a**) Control (**b**) 1 µg/mL (**c**) 16 µg/mL (**d**) 2 mg/mL

Figure 4. Fluorescence emissions of Sc-Si-QDs in TheraTears® at different concentrations. NC = negative control (TheraTears® alone).

Figure 5 illustrates the emission of green fluorescence from Si-QDs at 1 min (Figure 5a) and 20 min (Figure 5b) following their addition to TheraTears®. The average fluorescence intensity was 183 RFU at 1 min and 176 RFU at 20 min. The fluorescence intensity of Si-QDs was stable over the 20 min observation time ($p < 0.05$).

Figure 6 illustrates the detection of fluorescence emission wavelength from artificial tears and Si-QDs using SOLIS software. Si-QDs-free TheraTears showed a broader visible light emission (300–700 nm), while Si-QDs emitted at a discrete peak of 530 nm.

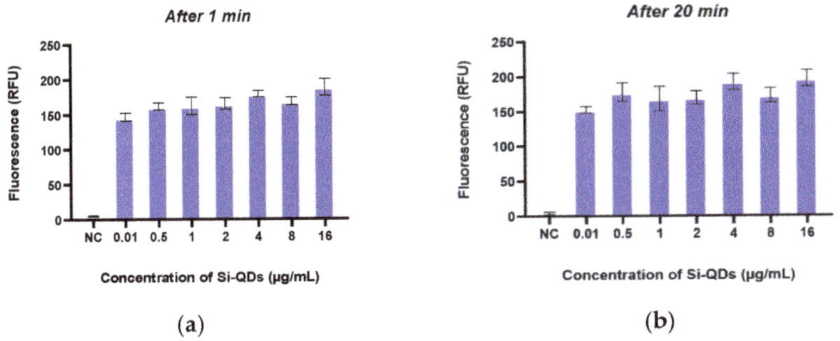

Figure 5. The fluorescence emission of the Sc-Si-QDs at two different time points: after (**a**) 1 min and (**b**) 20 min. RFU: Relative Fluorescence Unit.

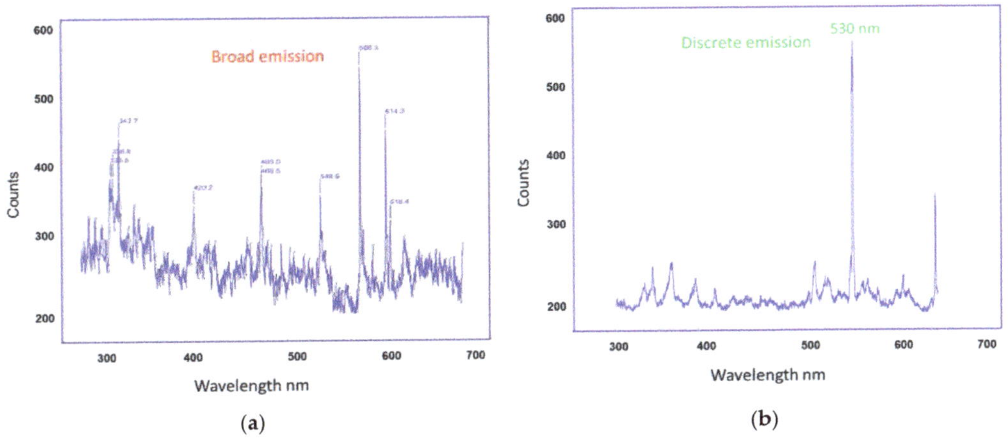

Figure 6. Signal detection from (**a**) Control TheraTears and (**b**) TheraTears + Si-QDs.

2.4. Solubility of Si-QDs

Figure 7 illustrates the solubility of Si-QDs in MilliQ water and Chloroform. In MilliQ water, Si-QDs exhibited minimum solubility as particles aggregated and adhered to the glass walls, and remained at the top surface of the solution. In contrast, Si-QDs showed higher solubility in Chloroform, forming a homogeneous solution.

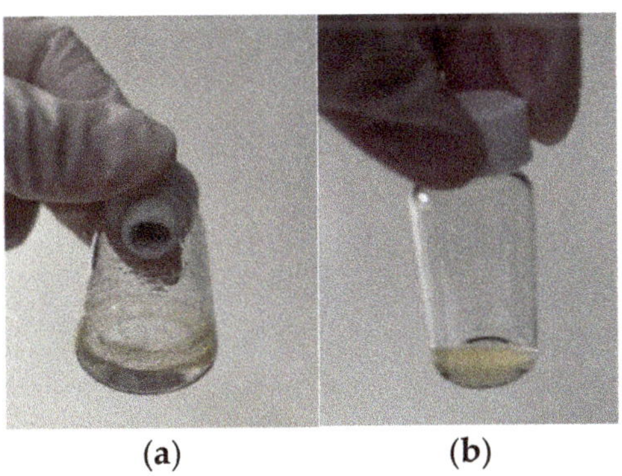

Figure 7. The solubility of Si-QDs in (**a**) MilliQ water and (**b**) Chloroform.

3. Discussion

DED is among the most widespread ocular diseases globally, affecting the quality of vision and daily life [6]. The tear film is essential for developing treatment modalities for DED; however, fundamental knowledge about dynamics remains incomplete [3,5]. QDs provide a novel approach to studying the tear film dynamics [12,22] and an alternative to traditional diagnostic dyes and imaging techniques due to their fluorescence properties and surface functionalisation [13]. The current study demonstrated the optimal size and investigated the fluorescence emission and cytotoxicity of hydrophobic Si-QDs, which could potentially be used as a bioimaging agent to study tear film lipid layer dynamics.

Hydrophobic Si-QDs were synthesised in an aqueous phase [30]. To achieve the quantum yield effect ideal for bright fluorescence emission, a strong reducing agent such as LiAlH4 was used to generate small size particles [31]. This size allows particles to absorb light within a broad spectrum while emitting a narrow wavelength of light [28]. In addition, the surface of Si-QDs was modified with the hexane group to be hydrophobic, making Si-QDs suitable for specific labelling of the tear film's lipid layer [12]. As visible light is commonly employed in ophthalmic imaging devices for safety purposes, it is important to use a light source to excite QDs while delivering a discrete emission [12].

Si-QDs have strong resistance to photobleaching, a major drawback of the currently used organic fluorescent dyes [30]. This characteristic is specifically important in the tear film imaging, which requires prolonged imaging [12]. Si-QDs exhibited stable fluorescence emission at all given concentrations for 20 min, consistent with prior studies [22,28,30,32]. In addition, Si-QDs emit a suitable wavelength of light, which is desirable for specific bioimaging. Therefore, it may be feasible to dispense hydrophobic Si-QDs into the lipid component of the tear film and capture their dynamics using a fluorescence emission signal. In contrast, it is difficult to monitor individual layers of tear film using fluorescein for a more extended period as it destabilises the tear film during examination [33]. Fluorescein cannot differentiate between aqueous-deficient and evaporative dry eyes [34]. QDs could label specifically the tear film lipid layer and help develop a targeted drug delivery system. Nonetheless, the chance of reduced fluorescence in biological environments still remains; hence, studies on in vivo imaging models are recommended for further optimisation of Si-QDs.

In contrast to hydrophilic Si-QDs, hydrophobic Si-QDs aggregated at certain concentrations, as previously reported [35]. This aggregation likelyresulted from the hydrophobic nature of Si-QDs, which was not sufficiently accommodated by the surfactants present in artificial tear formulations [36]. Under specific conditions, hydrophobic QDs with minimal cosolvents, such as chloroform and hexane, can penetrate membranes non-invasively, suggesting the potential for novel nanoprobes [37]. Therefore, hydrophobic QDs could be solubilised with cosolvents to image tear film lipid layers without aggregation. Polymer encapsulation of hydrophobic particles facilitates solubility in aqueous environments, making QDs suitable for bioimaging [38]. This encapsulation enhances the stability of quantum dots in physiological conditions without compromising their optical properties [38]. Another option is surface functionalisation with polyethylene glycol to increase the solubility of Si-QDs and their biocompatibility in an aqueous environment [39].

One prominent advantage of Si-QDs is their biocompatibility compared to traditional QDs, which often constitute toxic elements such as cadmium and lead [25,27,40,41]. Therefore, this biocompatibility enhances the potential of Si-QDs for use in tear film bioimaging, where direct contact with biological tissue is required [23]. In this study, Si-QDs did not significantly reduce the cell viability of HCECs at 16 μg/mL and below. A concentration-dependent reduction in cell viability was seen at concentrations beyond 16 μg/mL. Molybdenum sulfide QDs exhibited a cell viability of >77% at 250 μg/mL and a dose-dependent reduction in cell viability of Hela cells at higher concentrations [42].

Similarly, copper and carbon hydrophobic QDs also did not exhibit cytotoxic effects on mouse fibroblast cells [43,44]. Cadmium-based QDs have been shown to induce significant cytotoxic effects due to the release of reactive oxygen species [45]. Limited studies indicate no toxicity associated with Si-QDs [23,25], while others showed higher toxicity even with small-sized particles (<6 nm) [46]. No cytotoxic effect was observed with large 50 nm QDs [47]. Hydrophobic QDs show significant potential for ocular applications, but their cytotoxicity varies depending on composition, surface modification, and concentration. Gaining a deeper understanding of these factors is essential for developing safer and more efficient quantum dot-based technologies.

4. Materials and Methods

4.1. Synthesis and Characterisation of Hydrophobic Sc-Si-QDs

Si-QDs doped with scandium were synthesised in a solution phase [29]. Solution phase synthesis uses surfactants to interact with nanoparticles and limit their growth to reach the optimal size [48]. This synthesis process was conducted under argon atmosphere in a glove box with oxygen levels below 10 ppm and surface functionalisation with Zinc Sulphide [49] to prevent the oxidation of silicon in a biological environment [48]. Therefore, tetraoctyl ammonium bromide in toluene is used. QDs were analysed for their size and photoluminescence by TEM (Olympus Life Science, Notting Hill, VIC, Australia) and spectrofluorophotometer (RF-5301PC Shimadzu, Rydalmere, NSW, Australia). TEM images were taken at an acceleration voltage of 200 kV.

Briefly, as shown in Figure 8 Si-QDs were doped with four scandium atoms per QDs molecule by adding 0.5 g of tetraoctylammonium bromide and 0.026 mmol of $ScCl_3$ to a Schlenk tube. The Schenk tube underwent three evaluation cycles along the Schlenk line, followed by 5 min of Nitrogen purging per cycle. A total of 50 mL of anhydrous toluene was added, stirring the solution for 24 h. Silicon tetrachloride was added, and the solution was again stirred for an hour. Lithium aluminium hydride (reducing agent) was added, allowing the solution to react for 3 h. This resulted in hydride Si-QDs doped with Sc. Hydrophobic quantum dots were created by modifying silicon hydrogen bonds by adding 0.1 mol hexachloroplatinic acid in isopropyl alcohol and 1-heptene [48]. The samples for Transmission Electron Microscopy (TEM) were prepared by drop-casting doped Si-QDs in 0.5–1.0 mL of hexane onto carbon-coated copper grids. The emission spectrum is used to obtain the most significant emission wavelengths.

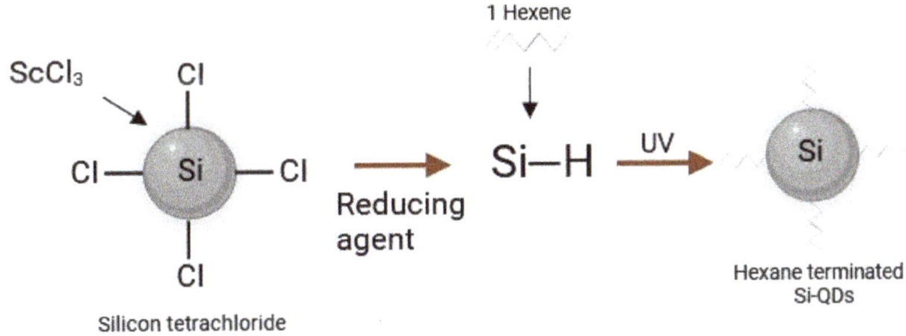

Figure 8. The schematic representation of hydrophobic Si-QDs.

4.2. MTT Assay

Human Corneal Epithelial Cells (HCECs) were cultured in Dulbecco's Modified Eagle Medium (DMEM: Thermo Fisher Scientific, Sydney, Australia) supplemented with 10% Fetal Bovine Medium, 2 ng/mL human recombinant epidermal growth factor, and 1% ITS at 37 °C with 5% CO_2. The MTT assay was used to evaluate the cytotoxicity of hydrophobic

Si-QDs against HCECs after 24 h of exposure [27]. This assay is based on reducing the 3(4,5-dimethylthiazol-2-yl)-2,5-diphenyl tetrazolium bromide (MTT) to dark blue formazan by viable cells in a 96-well plate, incubated at 37 °C in a 5% humidified CO_2 chamber. HCECs were exposed to different concentrations (2000 μg/mL, 1000 μg/mL, 500 μg/mL, 250 μg/mL, 125 μg/mL, 62.6 μg/mL, 32 μg/mL, 16 μg/mL, 8 μg/mL, 4 μg/mL, 2 μg/mL, 1 μg/mL) of Si-QDs in triplicates for 24 h. DMSO and DMEM were used as positive and negative controls, respectively. After incubation, 100 μL of 5 mg MTT was added to each well containing HCECs and three blank wells, and the plate was incubated for 2–4 h until purple precipitates appeared. The supernatants were removed, and 100 μL of DMSO was added to solubilise the MTT. The absorbance was measured at 570 nm using a spectrophotometer (BMG LABTECH, Ortenberg, Germany). The percentage viability was calculated using absorbance values and plotted against concentrations of Si-QDs. The non-toxic concentrations were further analysed for fluorescence emission.

4.3. In Vitro Imaging of Si-QDs

This study used an optimised imaging system based on a previously published procedure to detect fluorescence from Si-QDs [50]. The fluorescence emission of the hydrophobic Si-QDs was detected by mixing a 10 μL aliquot of non-toxic concentrations of Si-QDs (16 μg/mL, 8 μg/mL, 4 μg/mL, 2 μg/mL, 1 μg/mL) with artificial tears (TheraTears®, Akorn, Inc., Ann Arbor, MI, USA). TheraTears®, a balanced electrolyte formula, was used because it closely mimics the physiology of the tear film. The ultimate goal of Si-QDs is to image the tear film; therefore, optimising their fluorescence in an environment closely resembling the tear film is crucial. In addition, in vitro imaging can provide accurate results without the risk of degradation caused by toxic substances in organic dyes [22]. A high-resolution sCMOS camera (Zyla 5.5 megapixel, Andor Oxford Instruments Group, Belfast, Northern Ireland) attached to a slit lamp biomicroscope (Carl Zeiss, Dublin, CA, USA) was used for in vitro imaging of Si-QDs. Images were taken at a frame rate of 25 per second using Slit Lamp (30×) at time intervals of 1 min, 5 min, 10 min, 15 min, and 20 min. The built-in excitation filter of the slit lamp biomicroscope was used while the emission filter (MF530 FITC Emission Filter; Thorlabs Inc., Newton, New Jersey, USA) was placed in front of the slit lamp objective lens to facilitate discrete fluorescence emission from Si-QDs. The background signal was subtracted using SOLIS software. Three repeated single scans were taken using an exposure time of 1.99 s. A clear microscope slide was used as a negative control.

4.4. Assessment of Si-QDs Solubility

The solubility of Si-QDs has been evaluated by their dispersion in aqueous and organic solvents. Si-QDs were added to the MilliQ water and chloroform to achieve the safest concentration of 16 μg/mL. The solution was then vortexed (Vortex Genie 2-Mixer 230 V, 50 Hz) at 1000 RMP/min for 1 min to allow the particles to dissolve. Finally, the resulting solution was observed for its solubility in a solvent.

5. Conclusions

Hydrophobic Si-QDs demonstrated optimal size, biocompatibility, and discrete fluorescence emission, suggesting their potential application in imaging the tear film lipid layer. However, unlike hydrophilic Si-QDs, hydrophobic Si-QDs may aggregate on the ocular surface due to nonpolar functional groups and solvents. Based on the results, surface functionalisation and the use of cosolvent are recommended to enhance their solubility using cosolvent. Future research should also involve preclinical ex vivo and in vivo cytotoxic evaluations for further optimisation.

Author Contributions: Conceptualisation, S.S., F.S., M.D.P.W., P.B.O. and M.R.; data curation, S.S.; formal analysis, S.S.; funding acquisition, M.R.; investigation, S.S., M.R. and P.B.O.; methodology, S.S., F.S., M.D.P.W. and M.R.; project administration, F.S., M.D.P.W. and M.R.; resources, M.R.; software, S.S. and M.R.; supervision, F.S., M.D.P.W. and M.R.; validation, S.S., F.S., M.D.P.W. and M.R.; visualisation, F.S., M.D.P.W. and M.R.; writing—original draft, S.S.; writing—review and editing, S.S., F.S., M.D.P.W. and M.R. All authors have read and agreed to the published version of the manuscript.

Funding: This project was funded by a Faculty Research Grant and Faculty of Science Interdisciplinary Grant (FRG-PS43717), University of New South Wales, Sydney, Australia.

Institutional Review Board Statement: Not applicable.

Informed Consent Statement: Not applicable.

Data Availability Statement: The raw data supporting the conclusions of this article will be made available by the corresponding author upon request.

Conflicts of Interest: The authors report no conflicts of interest and have no proprietary interest in any of the materials mentioned in this article.

References

1. Craig, J.P.; Nichols, K.K.; Akpek, E.K.; Caffery, B.; Dua, H.S.; Joo, C.K.; Liu, Z.; Nelson, J.D.; Nichols, J.J.; Tsubota, K.; et al. TFOS DEWS II definition and classification report. *Ocul. Surf.* **2017**, *15*, 276–283. [CrossRef] [PubMed]
2. Tsubota, K.; Pflugfelder, S.C.; Liu, Z.; Baudouin, C.; Kim, H.M.; Messmer, E.M.; Kruse, F.; Liang, L.; Carreno-Galeano, J.T.; Rolando, M.; et al. Defining dry eye from a clinical perspective. *Int. J. Mol. Sci.* **2020**, *21*, 9271. [CrossRef] [PubMed]
3. Rolando, M.; Zierhut, M. The ocular surface and tear film and their dysfunction in dry eye disease. *Surv. Ophthalmol.* **2001**, *45*, 203–210. [CrossRef]
4. Farrand, K.F.; Fridman, M.; Stillman, I.O.; Schaumberg, D.A. Prevalence of diagnosed dry eye disease in the united states among adults aged 18 years and older. *Am. J. Ophthalmol.* **2017**, *182*, 90–98. [CrossRef]
5. Willcox, M.D.P.; Argüeso, P.; Georgiev, G.A.; Holopainen, J.M.; Laurie, G.W.; Millar, T.J.; Papas, E.B.; Rolland, J.P.; Schmidt, T.A.; Stahl, U.; et al. TFOS DEWS II tear film report. *Ocul. Surf.* **2017**, *15*, 366–403. [CrossRef]
6. Stapleton, F.; Alves, M.; Bunya, V.Y.; Jalbert, I.; Lekhanont, K.; Malet, F.; Na, K.-S.; Schaumberg, D.; Uchino, M.; Vehof, J.; et al. TFOS DEWS II epidemiology report. *Ocul. Surf.* **2017**, *15*, 334–365. [CrossRef]
7. Craig, J.P.; Alves, M.; Wolffsohn, J.S.; Downie, L.E.; Efron, N.; Galor, A.; Gomes, J.A.P.; Jones, L.; Markoulli, M.; Stapleton, F.; et al. TFOS lifestyle report executive summary: A lifestyle epidemic-ocular surface disease. *Ocul. Surf.* **2023**, *30*, 240–253. [CrossRef]
8. Potvin, R.; Makari, S.; Rapuano, C.J. Tear Film osmolarity and dry eye disease: A review of the literature. *Clin. Ophthalmol.* **2015**, *2015*, 2039. [CrossRef]
9. Rouen, P.A.; White, M.L. Dry eye disease: Prevalence, assessment, and management. *Home Healthc. Now* **2018**, *36*, 74–83.
10. Wolffsohn, J.S.; Arita, R.; Chalmers, R.; Djalilian, A.; Dogru, M.; Dumbleton, K.; Gupta, P.K.; Karpecki, P.; Lazreg, S.; Pult, H.; et al. TFOS DEWS II diagnostic methodology report. *Ocul. Surf.* **2017**, *15*, 539–574. [CrossRef]
11. Shapiro, A.; Merin, S. Schirmer test and break-up time of tear film in normal subjects. *Am. J. Ophthalmol.* **1979**, *88*, 752–757. [CrossRef] [PubMed]
12. Khanal, S.; Millar, T.J. Nanoscale phase dynamics of the normal tear film. *Nanomed. Nanotechnol.* **2010**, *6*, 707–713. [CrossRef] [PubMed]
13. Resch-Genger, U.; Grabolle, M.; Cavaliere-Jaricot, S.; Nitschke, R.; Nann, T. Quantum dots versus organic dyes as fluorescent labels. *Nat. Methods* **2008**, *5*, 763. [CrossRef] [PubMed]
14. Serruya, L.G.; Nogueira, D.C.; Hida, R.Y. Schirmer test performed with open and closed eyes: Variations in normal individuals. *Arq. Bras. Oftalmol.* **2009**, *72*, 65–67. [CrossRef]
15. Doane, M.G. An instrument for in vivo tear film interferometry. *Optom. Vis. Sci.* **1989**, *66*, 383–388. [CrossRef]
16. Wang, J.; Aquavella, J.; Palakuru, J.; Chung, S.; Feng, C. Relationships between central tear film thickness and tear menisci of the upper and lower eyelids. *Investig. Ophthalmol. Vis. Sci.* **2006**, *47*, 4349–4355. [CrossRef]
17. Holly, F.J. Physical chemistry of the normal and disordered tear film. *Trans. Ophthalmol. Soc. UK* **1985**, *104*, 374–380.
18. Papas, E.B. Diagnosing dry-eye: Which tests are most accurate? *Contactlens Anterior Eye* **2023**, *46*, 102048. [CrossRef]
19. Walling, M.A.; Novak, J.A.; Shepard, J.R.E. Quantum dots for live cell and in vivo imaging. *Int. J. Mol. Sci.* **2009**, *10*, 441–491. [CrossRef]
20. Arya, H.; Kaul, Z.; Wadhwa, R.; Taira, K.; Hirano, T.; Kaul, S.C. Quantum dots in bio-imaging: Revolution by the small. *Biochem. Biophys. Res. Commun.* **2005**, *329*, 1173–1177. [CrossRef]

21. Dabbousi, B.O.; RodriguezViejo, J.; Mikulec, F.V.; Heine, J.R.; Mattoussi, H.; Ober, R.; Jensen, K.F.; Bawendi, M.G. (CdSe)ZnS core-shell quantum dots: Synthesis and characterization of a size series of highly luminescent nanocrystallites. *J. Phys. Chem. B* **1997**, *101*, 9463–9475. [CrossRef]
22. Sarwat, S.; Stapleton, F.J.; Willcox, M.D.P.; O'Mara, P.B.; Tilley, R.D.; Gooding, J.J.; Roy, M. Feasibility of silicon quantum dots as a biomarker for the bioimaging of tear film. *Nanomaterials* **2022**, *12*, 1965. [CrossRef] [PubMed]
23. Park, J.-H.; Jeong, H.; Hong, J.; Chang, M.; Kim, M.; Chuck, R.S.; Lee, J.K.; Park, C.-Y. The effect of silica nanoparticles on human corneal epithelial cells. *Sci. Rep.* **2016**, *6*, 37762. [CrossRef] [PubMed]
24. Kim, J.Y.; Park, J.H.; Kim, M.; Jeong, H.; Hong, J.; Chuck, R.S.; Park, C.Y. Safety of nonporous silica nanoparticles in human corneal endothelial cells. *Sci. Rep.* **2017**, *7*, 14566. [CrossRef]
25. Yim, B.; Park, J.H.; Jeong, H.; Hong, J.; Shin, Y.J.; Chuck, R.S.; Park, C.Y. The effects of nonporous silica nanoparticles on cultured human keratocytes. *Investig. Ophthalmol. Vis. Sci.* **2017**, *58*, 362–371. [CrossRef]
26. Korhonen, E.; Rönkkö, S.; Hillebrand, S.; Riikonen, J.; Xu, W.; Järvinen, K.; Lehto, V.-P.; Kauppinen, A. Cytotoxicity assessment of porous silicon microparticles for ocular drug delivery. *Eur. J. Pharm. Biopharm.* **2016**, *100*, 1–8. [CrossRef]
27. Fan, J.-W.; Vankayala, R.; Chang, C.-L.; Chang, C.-H.; Chiang, C.-S.; Hwang, K.C. Preparation, cytotoxicity and in vivo bioimaging of highly luminescent water-soluble silicon quantum dots. *Nanotechnology* **2015**, *26*, 215703. [CrossRef]
28. McVey, B.F.P.; König, D.; Cheng, X.; O'Mara, P.B.; Seal, P.; Tan, X.; Tahini, H.A.; Smith, S.C.; Gooding, J.J.; Tilley, R.D. Synthesis, optical properties and theoretical modelling of discrete emitting states in doped silicon nanocrystals for bioimaging. *Nanoscale* **2018**, *10*, 15600–15607. [CrossRef]
29. McVey, B.F.P.; Butkus, J.; Halpert, J.E.; Hodgkiss, J.M.; Tilley, R.D. Solution synthesis and optical properties of transition-metal-doped silicon nanocrystals. *J. Phys. Chem. Lett.* **2015**, *6*, 1573–1576. [CrossRef]
30. McVey, B.F.P.; Tilley, R.D. Solution synthesis, optical properties, and bioimaging applications of silicon nanocrystals. *Acc. Mater. Res.* **2014**, *47*, 3045–3051. [CrossRef]
31. Shiohara, A.; Prabakar, S.; Faramus, A.; Hsu, C.-Y.; Lai, P.-S.; Northcote, P.T.; Tilley, R.D. Sized controlled synthesis, purification, and cell studies with silicon quantum dots. *Nanoscale* **2011**, *3*, 3364–3370. [CrossRef] [PubMed]
32. Warner, J.H.; Hoshino, A.; Yamamoto, K.; Tilley, R.D. Water-soluble photoluminescent silicon quantum dots. *Angew. Chem. Int. Ed.* **2005**, *44*, 4550–4554. [CrossRef] [PubMed]
33. King-Smith, P.E.; Ramamoorthy, P.; Braun, R.; Nichols, J.J. Tear film images and breakup analyzed using fluorescent quenching. *Investig. Opthalmol. Vis. Sci.* **2013**, *54*, 6003. [CrossRef] [PubMed]
34. Sivaraman, G.; Padma, M. Clinical profile of patients with aqueous deficient dry eye. *Indian J. Clin. Exp. Ophthalmol.* **2022**, *8*, 340–344. [CrossRef]
35. Matvienko, O.O.; Savin, Y.N.; Kryzhanovska, A.S.; Vovk, O.M.; Dobrotvorska, M.V.; Pogorelova, N.V.; Vashchenko, V.V. Dispersion and aggregation of quantum dots in polymer–inorganic hybrid films. *Thin Solid Film.* **2013**, *537*, 226–230. [CrossRef]
36. Poderys, V.; Matulionytė, M.; Selskis, A.; Rotomskis, R. Interaction of water-soluble CdTe quantum dots with bovine serum albumin. *Nanoscale Res. Lett.* **2010**, *6*, 9. [CrossRef]
37. Xie, J.; Mei, L.; Sun, Y.; Yong, X.; Han, N.; Dai, J.; Yang, X.; Ruan, G. Direct and noninvasive penetration of bare hydrophobic quantum dots through live cell membranes. *ACS Biomater. Sci. Eng.* **2019**, *5*, 468–477. [CrossRef]
38. Zrazhevskiy, P.; Dave, S.R.; Gao, X. Addressing key technical aspects of quantum dot probe preparation for bioassays. *Part. Part. Syst. Charact.* **2014**, *31*, 1291–1299. [CrossRef]
39. Singh, N.P.; Charan, S.; Sanjiv, K.; Huang, S.-H.; Hsiao, Y.C.; Kuo, C.-N.; Chien, F.C.; Lee, T.C.; Chen, P. Synthesis of tunable and multifunctional ni-doped near-infrared qds for cancer cell targeting and cellular sorting. *Bioconjugate Chem.* **2012**, *23*, 421–430. [CrossRef]
40. Chen, N.; He, Y.; Su, Y.; Li, X.; Huang, Q.; Wang, H.; Zhang, X.; Tai, R.; Fan, C. The cytotoxicity of cadmium-based quantum dots. *Biomaterials* **2012**, *33*, 1238–1244. [CrossRef]
41. Zhao, Y.; Sun, X.; Zhang, G.; Trewyn, B.G.; Slowing, I.I.; Lin, V.S.Y. Interaction of Mesoporous Silica Nanoparticles with Human Red Blood Cell Membranes: Size and Surface Effects. *ACS Nano* **2011**, *5*, 1366–1375. [CrossRef] [PubMed]
42. Xu, S.; Li, D.; Wu, P. One-pot, facile, and versatile synthesis of monolayer MoS quantum dots as bioimaging probes and efficient electrocatalysts for hydrogen evolution reaction. *Adv. Funct. Mater.* **2015**, *25*, 1127–1136. [CrossRef]
43. Stanković, N.K.; Bodík, M.; Šiffalovič, P.; Kotlár, M.; Mičušík, M.; Špitálský, Z.; Danko, M.; Milivojević, D.; Kleinová, A.; Kubát, P.; et al. Antibacterial and antibiofouling properties of light triggered fluorescent hydrophobic carbon quantum dots langmuir–blodgett thin films. *Acs Sustain. Chem. Eng.* **2018**, *6*, 4154–4163. [CrossRef]
44. Amna, T.; Ba, H.V.; Vaseem, M.; Hassan, M.S.; Khil, M.-S.; Hahn, Y.B.; Lee, H.-K.; Hwang, I. Apoptosis induced by copper oxide quantum dots in cultured C2C12 cells via caspase 3 and caspase 7: A study on cytotoxicity assessment. *Appl. Microbiol. Biotechnol.* **2013**, *97*, 5545–5553. [CrossRef]
45. Bai, C.; Wei, T.; Zou, L.; Liu, N.; Huang, X.; Tang, M. The apoptosis induced by CdTe quantum dots through the mitochondrial pathway in dorsal root ganglion cell line ND7/23. *J. Appl. Toxicol.* **2022**, *42*, 1218–1229. [CrossRef]

46. Kuo, T.-R.; Lee, C.-F.; Lin, S.-J.; Dong, C.-Y.; Chen, C.-C.; Tan, H.-Y. Studies of intracorneal distribution and cytotoxicity of quantum dots: Risk assessment of eye exposure. *Chem. Res. Toxicol.* **2011**, *24*, 253–261. [CrossRef]
47. Parak, W.J.; Pellegrino, T.; Plank, C. Labelling of cells with quantum dots. *Nanotechnology* **2005**, *16*, R9–R25. [CrossRef]
48. Tilley, R.D.; Yamamoto, K. The microemulsion synthesis of hydrophobic and hydrophilic silicon nanocrystals. *Adv. Mater.* **2006**, *18*, 2053–2056. [CrossRef]
49. Akerman, M.E.; Chan, W.C.W.; Laakkonen, P.; Bhatia, S.N.; Ruoslahti, E. Nanocrystal targeting in vivo. *Proc. Natl. Acad. Sci. USA* **2002**, *99*, 12617–12621.
50. Sarwat, S.; Stapleton, F.; Willcox, M.; Roy, M. Quantum dots in ophthalmology: A literature review. *Curr. Eye Res.* **2019**, *44*, 1037–1046. [CrossRef]

nanomaterials

Article

Magnetically Controlled Transport of Nanoparticles in Solid Tumor Tissues and Porous Media Using a Tumor-on-a-Chip Format

Tatiana Zimina [1], Nikita Sitkov [2,*], Ksenia Brusina [1], Viacheslav Fedorov [2], Natalia Mikhailova [2], Dmitriy Testov [1], Kamil Gareev [2], Konstantin Samochernykh [2], Stephanie Combs [3] and Maxim Shevtsov [2,3,*]

[1] Department of Micro and Nanoelectronics, St. Petersburg Electrotechnical University "LETI" (ETU "LETI"), Prof. Popova Str., 5, 197022 St. Petersburg, Russia; tmzimina@gmail.com (T.Z.); kebrusina@gmail.com (K.B.); dtestov@bk.ru (D.T.)

[2] Personalized Medicine Centre, Almazov National Medical Research Centre, Akkuratova Str. 2, 197341 St. Petersburg, Russia; fedorovvs.biotech@gmail.com (V.F.); natashashed1@gmail.com (N.M.); kggareev@yandex.ru (K.G.); neurobaby12@gmail.com (K.S.)

[3] Department of Radiation Oncology, Technische Universität München (TUM), Klinikum Rechts der Isar, Ismaninger Str. 22, 81675 Munich, Germany; stephanie.combs@tum.de

* Correspondence: sitkov93@yandex.ru (N.S.); maxim.shevtsov@tum.de (M.S.)

Abstract: This study addresses issues in developing spatially controlled magnetic fields for particle guidance, synthesizing biocompatible and chemically stable MNPs and enhancing their specificity to pathological cells through chemical modifications, developing personalized adjustments, and highlighting the potential of tumor-on-a-chip systems, which can simulate tissue environments and assess drug efficacy and dosage in a controlled setting. The research focused on two MNP types, uncoated magnetite nanoparticles (mMNPs) and carboxymethyl dextran coated superparamagnetic nanoparticles (CD-SPIONs), and evaluated their transport properties in microfluidic systems and porous media. The original uncoated mMNPs of bimodal size distribution and the narrow size distribution of the fractions (23 nm and 106 nm by radii) were demonstrated to agglomerate in magnetically driven microfluidic flow, forming a stable stationary web consisting of magnetic fibers within 30 min. CD-SPIONs were demonstrated to migrate in agar gel with the mean pore size equal to or slightly higher than the particle size. The migration velocity was inversely proportional to the size of particles. No compression of the gel was observed under the magnetic field gradient of 40 T/m. In the brain tissue, particles of sizes 220, 350, 820 nm were not penetrating the tissue, while the compression of tissue was observed. The particles of 95 nm size penetrated the tissue at the edge of the sample, and no compression was observed. For all particles, movement through capillary vessels was observed.

Keywords: tumor; magnetic nanoparticles; magnetically controlled transport; microfluidic systems; organic porous systems; tissue engineering

1. Introduction

The development of methods for the selection of efficient and low-toxicity drugs for target delivery and the suppression of tumors growth locally, thus minimizing damage to healthy organs, is an urgent and acute biomedical problem [1–5]. An intensively developing approach is the utilization of magnetic nanoparticles (MNPs) as vehicles for the targeted delivery of anticancer drugs. The realization of this approach requires solving several problems: the development of external fields providing the spatially determined transport of particles within an organism, the preparation of biocompatible and chemically stable MNPs and their chemical modification with targeting components (e.g., antibodies, peptides, etc.), ensuring their presence at the surface of pathological cells, and releasing antitumor preparations at the target cells [5–9].

References to magnetic nanoparticles (MNPs) for biomedical applications have been around for quite some time. Thus, in [10], the authors discuss applications of magnetic materials in medicine, including applications of the technology of fine particle iron for the treatment of cancer and mental disease, and for diagnostic purposes. Although the real explosion of nanoparticle technologies for biomedicine occurred with the introduction of nanotechnology [11], various applications of MNPs started to intensively develop, particularly in the field of targeted drug delivery [5,12–16]. The most recent efforts are concentrated on the development of various types of coatings for magnetic nanoparticle core [5,17–22]. So, at present, the family of biomedical magnetic nanoparticles includes the following principal types: maghemite (γ-Fe_2O_3) [23] and magnetite (Fe_3O_4) [11], used as a magnetic core, SPIONs [6,7,9,24], magnetic liposomes–MNPs covered with a phospholipid layer [25], and others. Among the biogenic magnetic nanoparticles, magnetosomes should be mentioned, as they are generated by magnetotactic bacteria. These MNPs demonstrate a magnetic core with an outstanding ordered structure and are covered with a layer like a cell wall [5]. It is necessary to note that the isolation of such MNPs is a very difficult and labor-intensive process [26–28].

Contemporary studies in the field of MNPs, particularly in the most current field of development, target-oriented drug delivery particles, are focused on the following directions [5]: the regular structure of MNPs, the biocompatibility and chemical neutrality of the shell, and, at the same time, the possibility of its chemical modification [29], the absence of particle aggregation effects, small size, and the ability to penetrate living tissues.

Solving the above problems at the stage of pharmaceutical preparation development, as well as at the stage of dosage adjustment in personalized medicine and target delivery challenges, is labor-intensive, expensive, and requires the use of laboratory animals, volunteers, etc. A promising solution is the preparation of tumor models on a chip [30–37]. This could be used in extensive experiments on drugs efficiency adjustments, as well as with the samples of patients' tumor cell culture to adjust personal biochemistry and the dosage of the antitumor drug.

The ultimate goals of solving these problems involve creating a topology of microchannels that imitate the vascular system, compartments for the culturing of tumor cells with interfaces with the vascular system and that allow access for tested drugs, special cells for studying the migration of drugs into the culture of tumor cells, and models based on polymer gels similar to living tissues.

The current technological level of hybrid-integrated microfluidic analytical devices fabrication allows the implementation of a wide variety of topologies, thick layer or printing technologies, sheet and nanostructured materials, integrated sensors. and many other possibilities. The precision of realization approaches 1–5 μm [38–41]. However, the integration of live tissues, including 2D, 3D, or spheroid types, represents a specific challenge [38].

A variety of approaches using models of tumors have been described [5] the advances which have been realized within the tumor-on-a-chip (TuOCh) concept design. But despite their advantages, such as simple culture growth, low cost, and the technological availability of formation and assembly methods, these approaches involve subjecting cancer cells to artificially restricted growth conditions and lack key components of the tumor microenvironment that affect cancer biology and drug response, including tumor mechanical properties and intertumoral gradients of compounds. In healthy organs, the vasculature is solid, while in tumors, the vasculature is leaky, with poor lymphatic drainage [36]. Nanoparticles of <200 nm in size are reported to be selectively captured in a tumor by an effect known as enhanced permeability and retention (EPR) [36,37]. The EPR effect in human tumors is of great interest, since it opens a way of drug delivery when the permeability of tissues is normally poor. Thus, the study of tumor models on a chip can help to find ways to increase the permeability of nanoparticles delivering antitumor drugs to targets by using vasculature [42], the EPR effect, or some other mechanisms for the targeted delivery of drugs in nanomedicine. The effect of magnetically guided target delivery was demonstrated in [43] for dextran-coated MNPs of 50 nm size on a murine model. Recently, a dual-mode targeting implementing EPR and ligand–antibody interaction have shown great prospects for the in vitro use of magnetic nanoparticles in therapy. As an example, single core SPIONs carrying siRNA and HER2 antibody fragments have demonstrated high biocompatibility and remained stable and resistant to enzymatic degradation in the models of biological fluids. These particles demonstrated improved intake and retention on HER2-overexpressing breast cancer cells compared to onloaded particles even in the magnetic field. Similarly, <100 nm size core SPIONs functionalized with RGD and glucosamine, aside from biocompatibility, expressed low macrophage activation. Utilizing EPR and ligand specificity, the authors confirmed high SPION retention and imaging effectiveness in tumor tissue.

At the current stage of work, the aim was to study the behavior of a variety of magnetic nanoparticles, including the "bare" magnetite nanoparticles (mMNPs) and carboxymethyl dextran-coated SPIONs (CD-SPIONs) as regards the penetrability into various porous tissues, including synthetic gels and murine brain tissue under the magnetic field drive. The results are important for the development of tumor-on-a-chip models and the target-oriented transport of anticancer medicines in organisms.

2. Materials and Methods

2.1. Magnetic Nanoparticles

Two types of magnetic nanoparticles (MNPs) were synthesized and studied, namely (1) MNPs without a shell (mMNPs) [44–47] and (2) carboxymethyl dextran-coated superparamagnetic iron oxide nanoparticles (CD-SPIONs) [48–51]. The MNPs were synthesized by authors as described below.

2.1.1. Materials for Preparation of Magnetite Nanoparticles

The following reagents were used for the preparation of mMNPs: ferric chloride hexahydrate $FeCl_3 \cdot 6H_2O$, ammonium hydroxide, NH_3OH and sodium tetrahydridoborate, $NaBH_4$—all are of puriss grade, ≥98%, obtained from LLC "Vecton" (St. Petersburg, Russia).

2.1.2. A Procedure of Magnetite Nanoparticles Formation

The mMNPs were synthesized by the proprietary method based on the reduction principle at ETU "LETI" (St. Petersburg, Russia).

The 1.25% *w*/*v* solution of $FeCl_3 \cdot 6H_2O$ was prepared in deionized water under intensive vertical stirring. The 3.5% *w*/*v* solution of NH_3OH was prepared in deionized water. Then, 2 g of $NaBH_4$ was dissolved in 100 mL of 3.5% solution of NH_3OH. The solution of $FeCl_3 \cdot 6H_2O$ heated up to T = 40 °C and the solution of $NaBH_4$+NH_3OH was added to it under intensive stirring. Then, the system was heated to achieve thermodynamic equilibrium at T = 95 °C. Under these conditions, the reduction reaction was completed in 60–120 min. The magnetic nanoparticles were formed during the reduction process, being settled at the bottom of the vessel under the action of a magnetic field. In two hours, the system cooled down to room temperature, and the magnetic phase was isolated using the magnetic separation technique. As a result, a highly stable, black magnetic powder in the form of two fractions of Fe_3O_4 was settled at the bottom of the reaction vessel.

2.1.3. Materials for Preparation of Carboxymethyl Dextran-Coated Superparamagnetic Iron Oxide Particles (CD-SPIONs)

The reagents used—$FeSO_4$ and $FeCl_3$ aqueous solutions, NH_4OH, CsCl, and carboxymethyl dextran sodium salt—all are of puriss grade and were obtained from Sigma-Aldrich (St. Louis, MO, USA).

2.1.4. Procedure of CD-SPIONs Preparation

Carboxymethyl dextran-coated superparamagnetic iron oxide nanoparticles (CD - SPIONs) were synthesized by the microemulsion method at Institute of Cytology RAS as described earlier [52,53] and more specifically by the coprecipitation of Fe^{2+} and Fe^{3+}

salts solutions (at pH = 10, inert atmosphere and T = 80 °C) and followed by electrosteric stabilization by a negatively charged dextran coating. The resulting core of CD-SPIONs are 30 nm size.

2.2. Microfluidic Chips

2.2.1. Materials for Microfluidic Chips Fabrication

The layers of microfluidic chip sandwich structures were prepared by laser ablation from a poly(methylmethacrylate) (PMMA) sheet, 1.5 mm thick, polyethylene terephthalate (PET) sheet 0.5 mm thick, polyethylene (PE) sheet 0.3 mm thick, polypropylene (PP) sheet (Astroplastica, St. Petersburg, Russia). All layers are optically transparent in the visible range of the spectrum and sufficiently chemically and biologically inert (Table 1). In certain cases, 3M™ Scotch® Transparent Film Tape 600 (St. Paul, MN, USA) (polyvinylchloride (PVC) film/acrylic adhesive) was used for hermitization [39,40] (Table 1).

Table 1. Properties of polymer films used in formation of microfluidic chips.

Material Name	Thickness, mm	Density, g/cm³ at 20 °C	Melting Point, °C	MW, kDa	Structural Formula
Poly(methylmeth-acrylate) (PMMA) [39,40]	1.5	1.18	>100	100–150	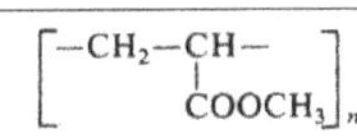
Polyethylene terephthalate (PET) [54–56]	0.5	1.38	>245	150	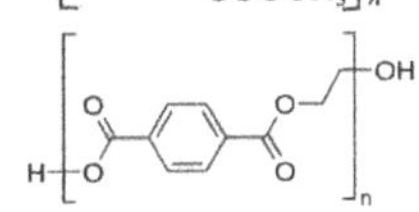
Polyethylene (PE) [57–59]	0.3	0.88–0.96	115–135	<200	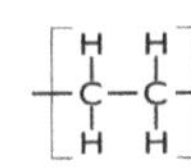
Polypropylene (PP) [60,61]	0.3	-	130–171	-	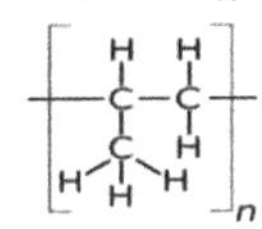
3M™ Scotch® Trans-parent Film Tape 600 (polyvinylchloride (PVC) film/acrylic adhesive) [62]	0.52 0.38 (PVC base)	1.35–1.43	150–220	100–170	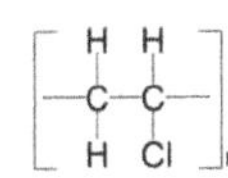

2.2.2. Microfluidic Chips Fabrication Technology

The microfluidic chips were fabricated as a sandwich structure bonded by the thermo-compression technique [39,63–66]. The surfaces were treated by isopropyl alcohol prior to bonding procedure. The temperature of bonding was 120 °C, which was applied under gradual heating stepwise with the intervals of 20 °C and exposition time at each step of 20 min. The pressure was about 200 g/cm^2.

The topologies of planar capillary patterns were formed in polymer sheets of 0.3–0.5 mm thickness using the laser ablation technique with a laser machine Zareff M2, 300 × 200 mm, 40 W CO_2 laser (LLC "Zareff", Moscow, Russia). The sandwich structures were composed by adjusting the base layer and cover layer (with corresponding openings) with the intermediate layer comprising through in-sheet channel structures. The assembly was placed into the steel press (200 g/cm^2) and heated in oven Industrial Lab DZG-6020 25L Electro-Thermostatic Hot Air Circulating Drying Oven (Zhejiang NADE Scientific Instruments Co. Ltd., Hangzhou, China) at 120 °C for 60 min.

2.3. *Porous Media of Differing Mean Pore Diameter*

2.3.1. Materials for Preparation of Porous Media

Porous media as a model of biological tissues in experiments on magnetically mediated transport of nanoparticles were prepared using agarose gel (Agar-Agar 900, LLC MolecularMeal, Moscow, Russia) of differing concentrations of initial suspension.

2.3.2. Procedure of Porous Media Preparation

The mean pore size concentration dependence was selected according to the data presented in earlier publications [67,68]. The mean pore size of the gel varied by the selection of polymer concentration.

To prepare agarose gels, the powders were suspended in water of ambient temperature and left to swell for 20 min. After swelling, the suspended polymer was slowly heated under continuous stirring up to 90 °C for 30 min. Then, the gel was poured into cells of the hybrid microsystem using a dispenser. Agar gels were prepared at various concentrations within the range from 0.25% w/v to 0.5% w/v. The gel models were optically sufficiently transparent.

2.4. *Testing Materials, Buffer Solutions*

Phosphate buffer solution was prepared from PBS tablets (Biotechnology grade, Am-E404-100, Helicon, Moscow, Russia) composed of 0.01 M PBS, 0.137 M NaCl, 0.002 M KCl, and pH 7.4 dissolved in 100 ml of deionized water. Albumin bovine serum (BSA) and heat shock isolation (Biotechnology grade. Amresco Inc., Framingham, MA, USA) were used in preparation of the blood serum model.

2.5. *Methods*

2.5.1. Collection of Laboratory Mouse Brain Tissue Sample

The mouse was humanely sacrificed by severing the brainstem under anesthesia. Anesthesia was induced with propofol; then, the mouse was placed in a chamber where the inhalation of anesthesia with sevoflurane was performed [69]. Under sterile conditions, a linear incision of soft tissues in the projection of the sagittal suture was made. The parietal, frontal and occipital bones were skeletonized, and a craniotomy was made using a diamond bur. Then, the brainstem was transected with a scalpel, a frontal slice of the right hemisphere of the brain was taken in the parietal lobe area, and the brain tissue material was placed in a container filled with saline.

The animals (weighing 250–300 g) were purchased in an animal nursery "Rappolovo" RAMN (St. Petersburg, Russia). All animals were bred and kept under specific pathogen-free conditions in accordance with the guidelines of the Federation of European Laboratory Science Association (FELASA). All animal experiments have been approved by the Ethics Committee of the Almazov National Medical Research Centre of the Ministry of Health of the Russian Federation (Protocol number 05112019 dated 11 November 2019) and were in accordance with institutional guidelines for the welfare of animals.

2.5.2. Application of Magnetic Field in the Microfluidic System

The neodymium magnet (class N35; with a nickel protective coating) with a magnetic field (MF) strength at the surface of 0.27 T (Figure 1a) and magnetic field gradient in the near field of 0.04 T/mm (Figure 1b) was used, which, according to the literature data [70], corresponds to the MF forces of magnets used in medicine. The form applied in the experiments' magnetic field strength distribution, i.e., the relationship of the magnetic field strength versus distance along the chip channel from the magnet, is illustrated in Figure 1b, and the arrangement of the magnet at the top of the microchip compartment with the gel is presented in Figure 1c.

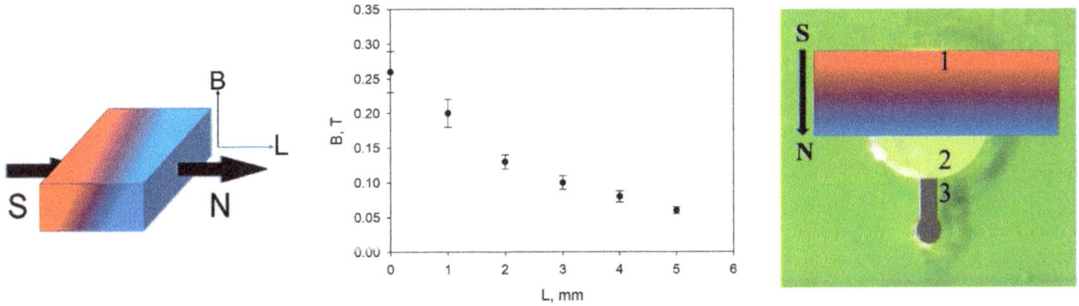

(a) (b) (c)

Figure 1. Geometry of Nd-Fe-B magnet of 30 mm × 10 mm × 4 mm size strength of magnetic field (MF) at the surface value of 0.27 T (**a**); MF gradient versus sample length of ~0.04 T/mm (**b**); positioning of magnet at the chip with gel and suspension of nanoparticles compartment (**c**); 1—neodymium magnet, 2—agar gel, R = 5 mm, h = 0.3 mm, 3—aqueous suspension of MNPs.

3. Results

3.1. Features of Manufacturing Microfluidic Systems

The channels of microfluidic systems (MFS) were prepared by laser ablation. Their quality depends on the velocity of the laser head movement and the laser power. As presented in Figure 2a, the velocity of 10 mm/s gives better results, as regards the quality of the cut edge, than the ablation at 15 mm/s (Figure 2b). In the second case, the wall of the channel is ribbing, and the border has ridges and sagging, which scatter light.

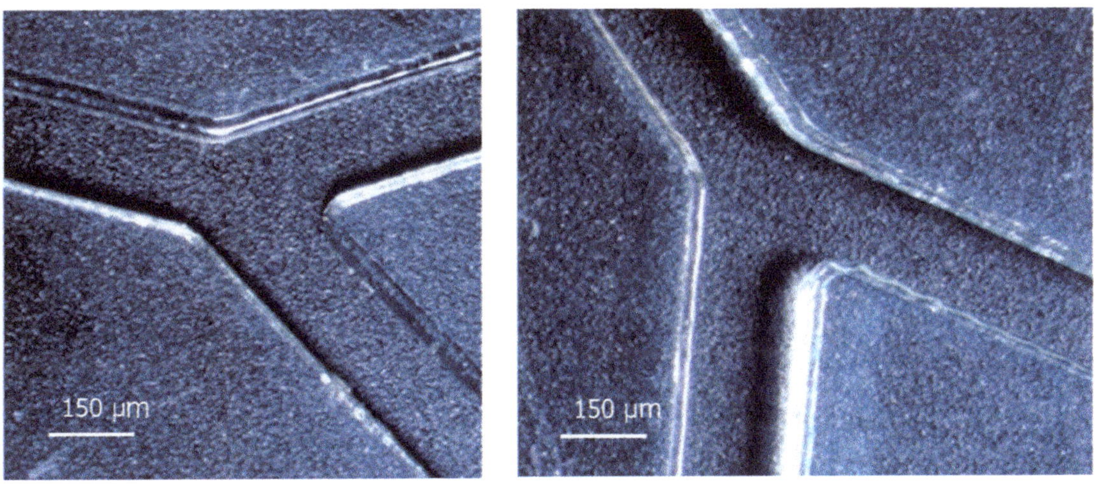

(**a**) (**b**)

Figure 2. Microscopic image of channels 150 μm wide and 300 μm deeply fabricated using the laser ablation technique at different laser head speed values: 10 mm/s (**a**) and 15 mm/s (**b**).

The power of the laser beam in both cases was 8 W. At lower power, it was not possible to achieve a through-cut ablation in the material 0.3 mm thick, while at higher speed, the cracking of PMMA was observed along the edges due to the occurrence of mechanical stresses around high-temperature gradients in polymer material.

The assembly of the multilayer polymer microfluidic systems was performed via thermocompression bonding of the thermoplastic polymers, including PMMA, PET, PE and PP (Table 1). The thermocompression technique makes it possible not only to seal the layers of the sandwich-like system but also to form certain 3D elements, such as inlet basins, lenses etc. within the same technological cycle [71–73]. In this work, two types of three-layer polymer assemblies of microfluidic systems have been developed: (1) PMMA-PET-PMMA with capillaries for analyzing nanoparticles in liquid media made from the 0.3 mm thick PET functional layer (Figure 3a,b), and (2) PMMA-PMMA-PET with basins for gels (Figure 3c,d). Sealing of the second type of microfluidic systems was carried out using a transparent PET film 140 μm thick with a deposited glue layer.

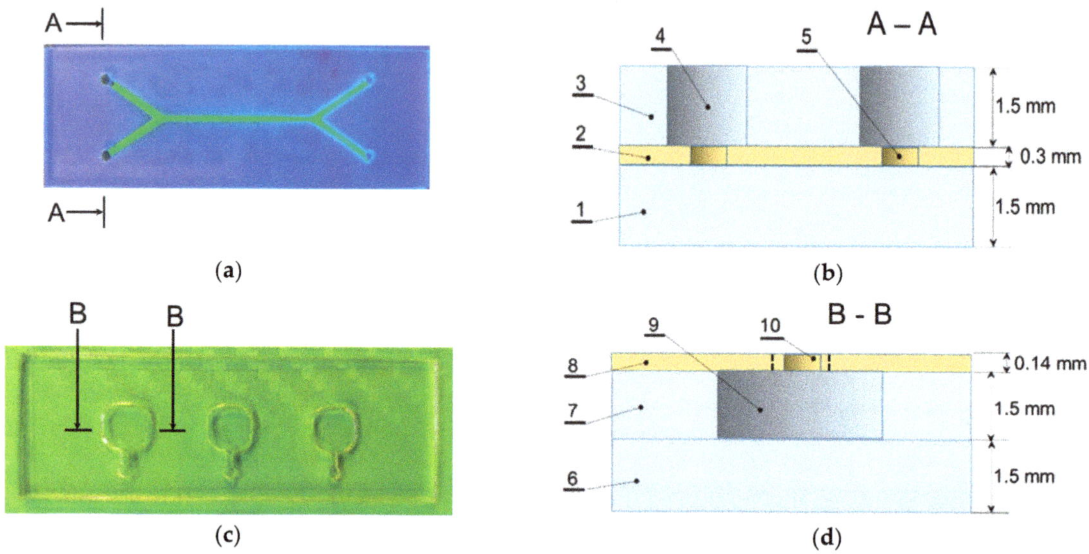

Figure 3. Photographs of cross-sections of microfluidic chips. (**a**)—a schematic of type assembly and (**b**) photographs of MFS-1, where 1—base layer, 2—microfluidic layer, 3—casing layer, 4—inlet, 5—microfluidic channel and MFS-2 (**c**,**d**) for MNP analysis in liquid media; (**c**)—a schematic of type assembly and photograph (**d**) of MFS-2 for tissue-on-a-chip type studies, where 6—base layer, 7—layer for tissue sample, 8—casing layer, 9—tissue sample cell, 10—inlet.

3.2. *Features of Magnetic Nanoparticles Migrating in Aqueous Media Patterns*

Two types of MNPs were investigated: (1) magnetite (Fe_3O_4) nanoparticles without a shell, mMNPs, aqueous suspension of 2 mg/mL concentration and (2) MNPs of iron oxide (Fe_3O_4) in carboxymethyl-dextran shell, CD-SPIONs, aqueous suspension of 2.2–4 mg/mL concentration (Table 2).

Table 2. Samples of magnetic nanoparticles: mMNPs and CD-SPIONs.

	Sample Name	Particle Size, nm	Particle Concentration, mg/mL
1	mMNP	46; 212	2
2	CD-SPION-1	95 ± 3	2.2–4.0
3	CD-SPION-2	220 ± 10	2.2–4.0
4	CD-SPION-3	350 ± 15	2.2–4.0
5	CD-SPION-4	820 ± 200	2.2–4.0

The hydrodynamic radii (R_H) of fractions of mMNPs in the sample were estimated by the laser light scattering using a Photocor Mini (Photocor LLC, Moscow, Russia) as 23 nm and 106 nm, (Figure 4). The concentration of MNPs in suspension is 2 mg/mL.

The experiments on the mobility of the mMNPs in aqueous suspension in capillaries demonstrated intensive agglomeration phenomena, which was also observed earlier elsewhere [40,54,55]. Figure 5a,b show parallel flows of a suspension of mMNPs and aqueous medium in MFS-1 at the initial time (Figure 5a) and 5 min after the start of the experiment (Figure 5b), when the formation of agglomerates of an approximate size range of 0.5–2.0 μm was observed within several minutes. When BSA solution of 7% *w*/*v* was used as the suspension medium, agglomeration was accelerated. Further development of the process led to the formation of agglomerates and appearance of laminar vortices (Figure 5c). After some time of exposition to a static magnetic field (MF) gradient of 40 T/m, the formation of spatial grid-like structures of the type shown in Figure 5d were observed. These intensive agglomeration effects obviously make it impossible to study the diffusion of this type of MNPs and use them as delivery particles in organisms.

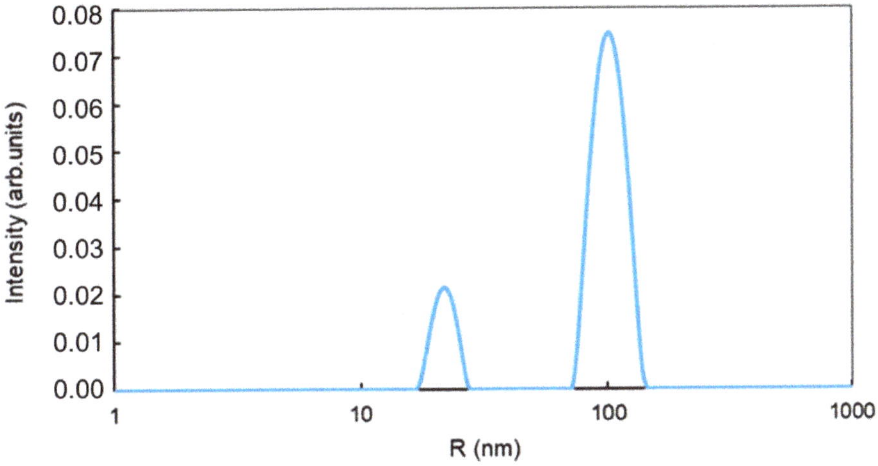

Figure 4. Size distribution of mMNPs obtained by dynamic light scattering technique.

CD-SPIONs were synthesized by the microemulsion method at Institute of Cytology RAS as described earlier [23–26] and more specifically by coprecipitation from Fe^{2+} and Fe^{3+} salt solutions (pH = 10, t = 80 °C) and stabilization by a cross-linked dextran coating. The hydrodynamic size and zeta potential of CD-SPIONs were assessed by dynamic light scattering (DLS) using a Zetasizer Nano (Malvern Instruments, Malvern, UK). The concentration of iron in the nanosuspension was determined by the thiocyanate method as 0.2% *w*/*v* [6].

DLS results for the sizes of CD-SPIONs are shown in Figure 6a. The investigated nanoparticles were sufficiently monodisperse in size with the main distribution peak corresponding to the expected nanocluster diameter. The major peak for the surface charge of all samples was approximately 0 mV, since non-functionalized dextran lack charged sites (Figure 6b). Lower monodispersity, especially for the 820 nm sample, can be explained by the uneven polymer distribution on the surface of the Fe core. TEM images of CD-SPIONs of different sizes are given in the Supplementary Materials.

The resulting suspensions have a high [Fe^{3+}, Fe^{2+}] concentration of ~4 mg/mL and contain superparamagnetic (as was evident from their susceptibility for external magnetic field) particles with dominating fractions of mean size 95 nm, 220 nm, 350 nm and 820 nm (highly heterogeneous), as presented in Figure 6. The concentration of iron in the nanosuspensions was determined by the thiocyanate method as described in [6]. The CD-SPIONs are described in Table 2.

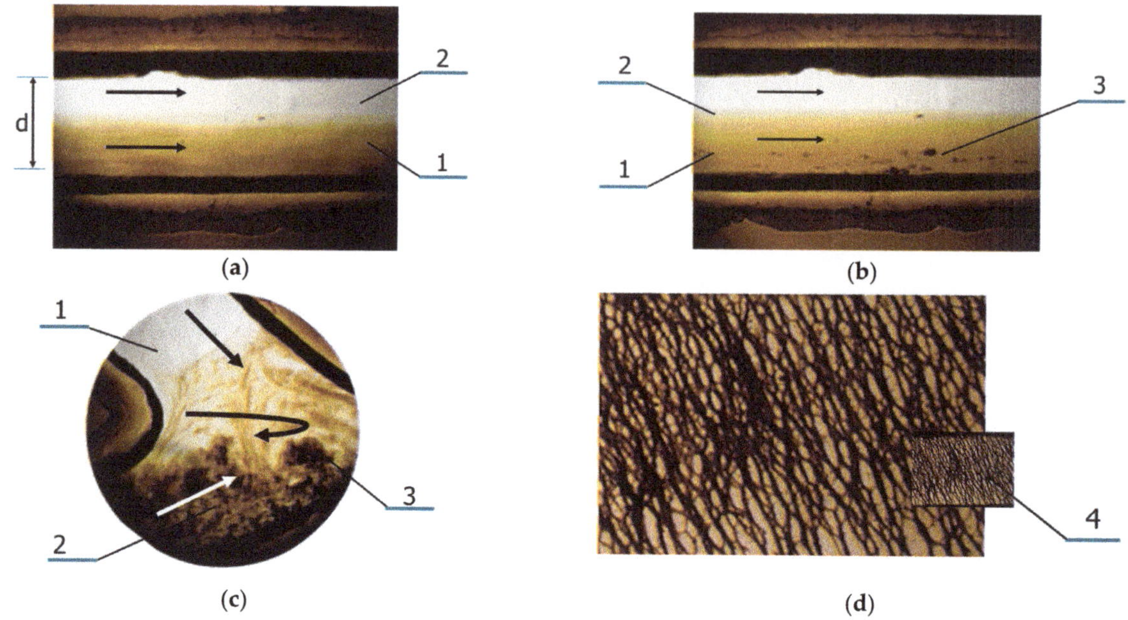

Figure 5. Peculiarities of microfluidic flow for mixture of magnetite nanoparticles (mMNPs) of 23 nm and 106 nm size: (**a**) parallel flow at $V = 1$ mm/s of mMNPs aqueous suspension of 2 mg/mL concentration (1), and water (2) in MFS-1 under MF gradient of 40 T/m in capillary channel 200 µm wide (**d**) and 300 µm deep at an initial time (**a**); in 5 min, when considerable agglomeration (3) is observed (**b**); in 10 min when laminar vortices of mMNP suspension are observed (**c**) and in 20 min when the total grid of magnetic fibers (4) is observed (presented also accelerated image) (**d**).

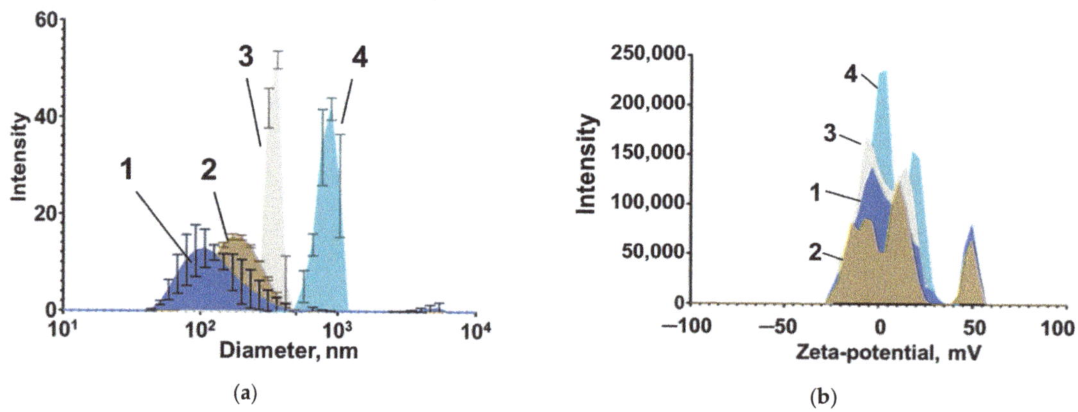

Figure 6. Hydrodynamic diameter (**a**) and zeta-potential (**b**) distribution for CD-SPIONs of various diameters (1—95 nm, 2—220 nm, 3—350 nm, 4—820 nm).

3.3. Study of Magnetically Controlled Migration of SPIONs in Porous Media Modeling Life Tissues

The magnetic field makes it possible to deliver MNPs via blood vessels to the tumor and concentrate MNPs near its surface. The localization of MNPs near the tumor surface greatly simplifies the delivery of drugs into the affected tissues. The migration of MNPs in tissues is a rather slow process, the rate of which depends on the parameters of the tissue, such as density and pore size, on the characteristics of MNPs, their size, shape, shell type, magnetic moment of MNPs and on the strength of the magnetic field (MF) [71]. In this work, the migration of CD-SPIONs of different sizes in the porous media of agar gel samples of various pore sizes was studied.

Three types of agar gel were selected as defined by the concentration of the dry agar: 0.5 *w*/*v*%, 0.4 *w*/*v*% and 0,25 *w*/*v*%, which corresponds to the pore sizes of 100–150 nm (G1), 200–300 nm (G2), and 350–400 nm (G3) [72,73]. In the agar gels G1 and G2 with the mean pore sizes of 100–150 nm and 200–300 nm, correspondingly, no migration of CD-SPIONs of the sizes 95 nm, 220 nm, 350 nm and 820 nm was observed. The cross-section of the experimental chip is presented in Figure 7.

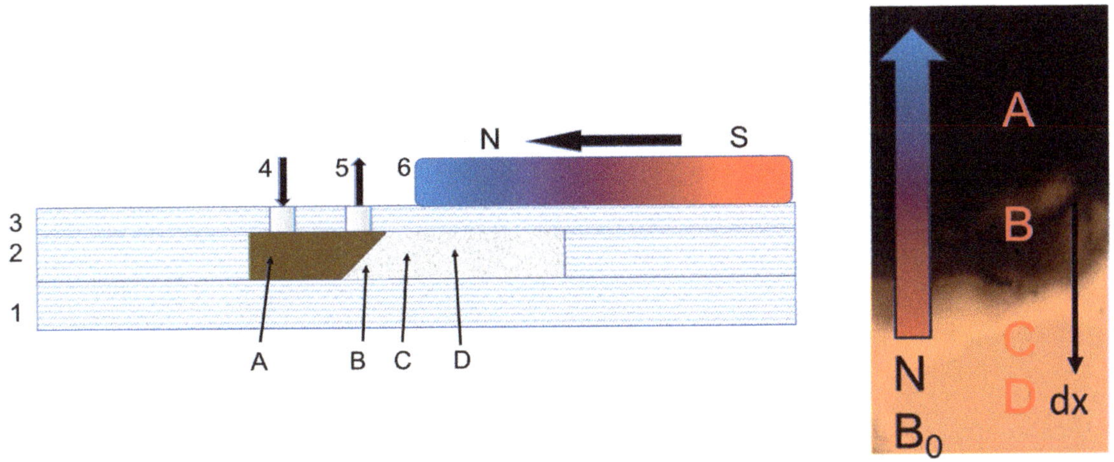

(a) (b)

Figure 7. Experimental setup of microchamber (**a**) for magnetically controlled migration study and the position of the magnet above the sample (**b**): 1—base, prepared of PMMA layer, 1 mm thick; 2—layer with microfluidic system profile, PMMA, 1 mm thick; 3—PP layer, 0.3 mm thick; 4—inlet for CD-SPION suspension; 5—outlet of CD-SPIONs suspension; 6—neodymium magnet. A—MNP suspension zone, B—interface zone, slope of gel massive, C—zone accessible for CD-SPIONs, D—gel.

In the gel G3, the following patterns were detected (Figure 8).

The relationships of magnetically controlled transport of CD-SPIONs in agar gel versus time are presented in Figure 9.

The migration characteristics of CD-SPIONs in brain tissue were studied using particles of 350, 220 and 95 nm hydrodynamic size. The schematic of zones monitoring are presented in Figure 10a–e. The brain tissue was sliced and placed into the large basin of MFS-3 with adjustment of the sample height with the depth of the basin. The CD-SPIONs suspension was delivered into the microfluidic channel and small basin as shown in Figure 10c. The pattern developed under the action of the magnetic field was observed via a microscope. The CD-SPIONs of 350 nm mean diameter on the tissue were displaced under the Lorentz force for 2 min (Figure 10e,f).

The experiments were completed for the particles of the 220 nm and 95 nm hydrodynamic size. The results are presented in Figure 11.

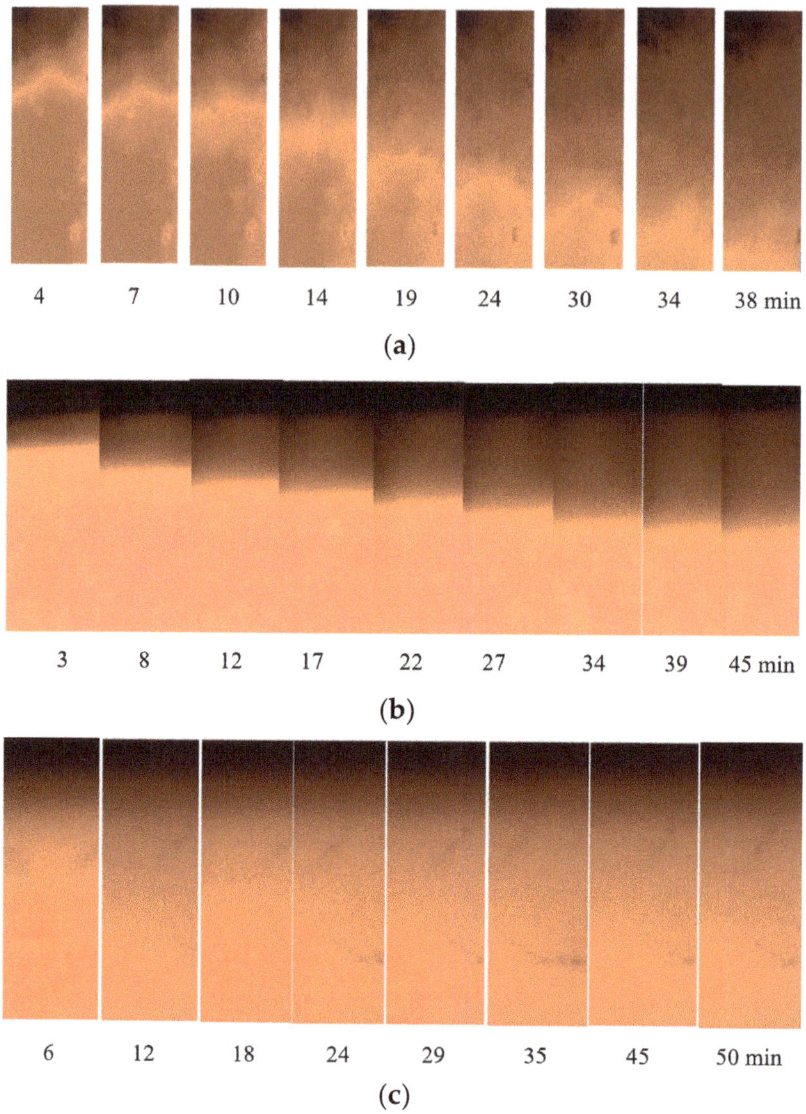

Figure 8. Magnetically controlled fronts of CD-SPIONs (r = 95 nm (**a**), 220 nm (**b**), 350 nm (**c**)) in agar gel of 350–400 nm mean pore size, under 40 T/m gradient of MF in microfluidic chip.

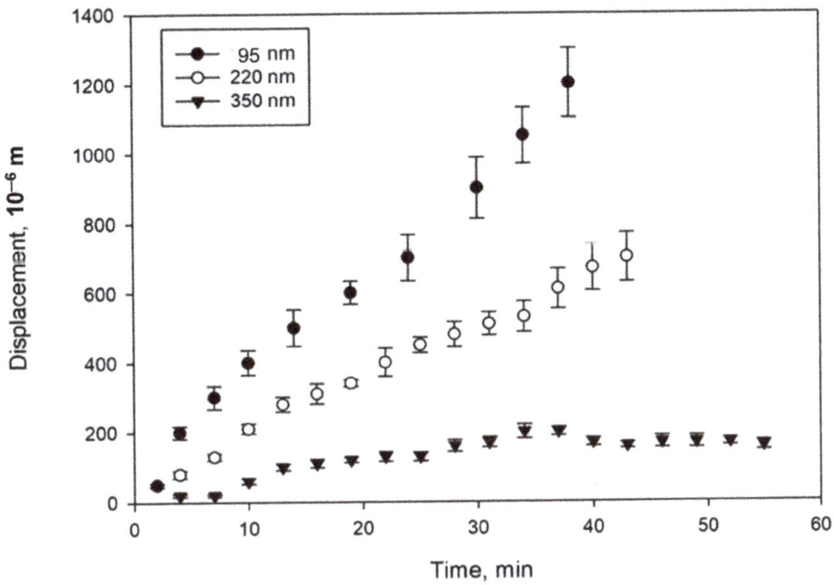

Figure 9. Relationships of magnetically driven displacement of CD-SPIONs of differing size versus time in agar porous medium with the mean pore size of 350–400 nm.

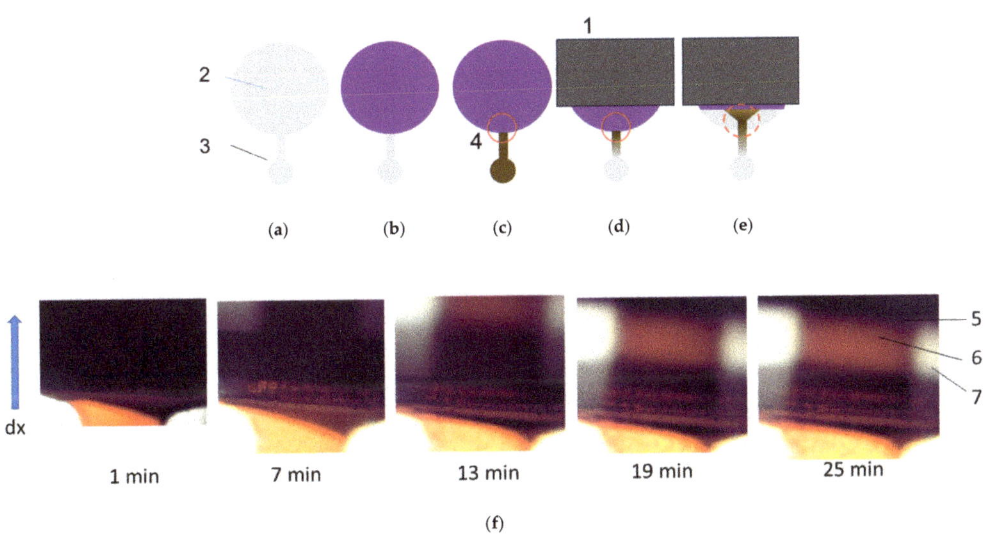

Figure 10. The experimental setup for CD-SPIONs transport in rat brain. (**a**) Geometry of the microfluidic system; (**b**) sample of the rat brain shown in blue; (**c**) CD-SPIONs suspension shown in brown; (**d**) application of magnetic field and migration of CD-SPIONs; (**e**) displacement of the brain tissue; (**f**) photographs of the stages (**c**–**e**) with CD-SPIONs of 350 nm size. 1—neodymium magnet, 2—basin for brain tissue, 3—basin and channel for CD-SPIONs suspension, 4—aperture of observation in the microscope marked with red line, 5—border of the displaced tissue with CD-SPIONs suspension underneath, 6—flow of remaining CD-SPIONs moving in MF, as in (**e**), 7—part of basin for tissue after (4).

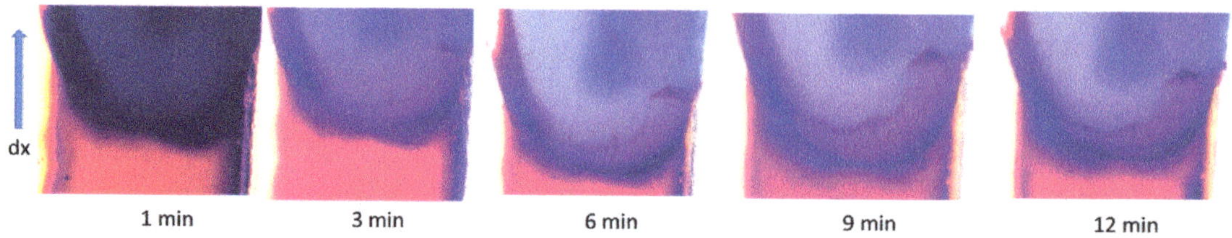

(**a**)

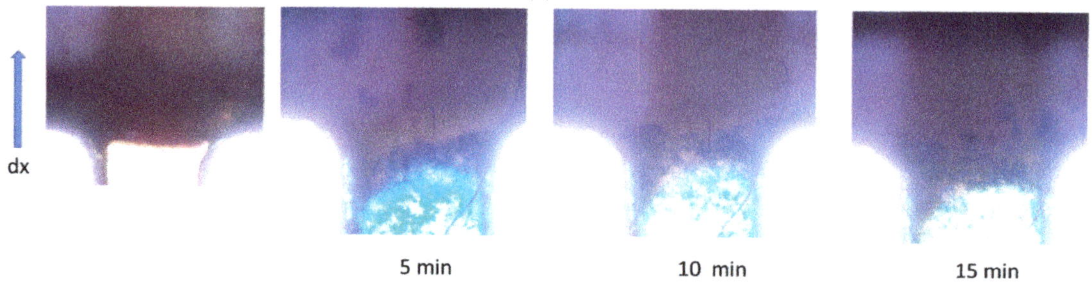

(**b**)

Figure 11. Magnetic transport of CD-SPIONs of the hydrodynamic size 220 nm (**a**) and 95 nm (**b**). Time of the pictures capture was shown below each photograph. In (**b**), the left image shows the brain sample border.

Figure 11a demonstrates that CD-SPIONs of the size 220 nm could not penetrate rat brain tissue, and at the same time, they are not capable of compressing the tissue. Figure 11a shows the initial stage of the experiment, when the CD-SPIONs suspension is filling compartment 3, while migrating in the magnetic field, positioned vertically in the picture. Phase (b) demonstrates the microchip in a 24 min period, when all the particles moved through the vasculature and slit along the walls of the basin, while the tissue remained intact. The conclusion was checked by the buffer purging the basin after the experiment when all the suspension was washed away.

The experiments with particles of 95 nm hydrodynamic size demonstrated similar patterns. Thus, Figure 11b shows the initial stage, when the CD-SPIONs sample was concentrated near the "tissue–solvent" interface, and the next frame shows the CD-SPIONs penetrating vasculature and the slits between the tissue and the casing.

4. Discussion

The study of the transport behavior of MNPs in porous media, including polymer gel (agar) of various mean pore size, as well the live brain tissue, makes it possible to evaluate the character of localization of nanoparticles under the action of a magnetic field. A significant increase in the concentration of MNPs near the surface of the tissue was observed: in several minutes, the concentration of CD-SPIONs increased by 5 to 10 times, depending on the particle size. Moreover, the active movement of CD-SPIONS was observed immediately after the MF was turned on. At the same time, the penetration of CD-SPIONs of hydrodynamic size down to 95 nm into the brain tissue was not observed. The SPIONs suspension penetrates the tissue via the vascular network and the slits near the walls of the microsystem. This means that the lower size particles are needed, which are at a higher price and less available. In this respect, the bacterial magnetosomes could be of interest; however, such materials are also not easily available.

At the same time, as regards the use of low cost and easily available CD-SPIONs, they could be of good use in express diagnostics systems (oncotheranostics) comprising hybrid materials, including polymer gels that are accessible for MNPs and captured cancer cells [39], as presented in Figure 12. In this case, the express testing of the fixed cells could be made by the targeted delivery of MNPs loaded with selected medications in the tumor-on-a-chip format microfluidic systems [41].

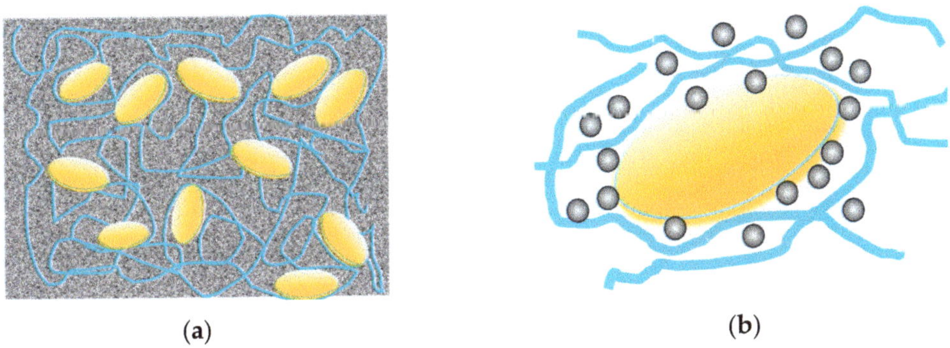

Figure 12. Oncotheranostic system, containing polymer gel of specific pore size (blue lines), fixing the cancer cells and infiltrated with suspension of MNPs (**a**) carrying the selected anticancer drug (**b**).

Another direction of MNPs target delivery development approaches is the angiogenesis methods, which will enable delivering drugs using larger size MNPs, which are more available, less expensive and more susceptible to magnetic guidance.

Microfluidic systems developed for modeling tumor tissues (tumor-on-a-chip) are an important tool for studying the transport characteristics of nanoparticles. These systems allow simulating physiological conditions in tumor tissues and vessels, which is especially important when testing new methods for drug delivery. In this paper, we describe systems based on polymer sandwich structures allowing the control and study of nanoparticles transport in capillary channels simulating the vascular system. The experimental results showed that when using a magnetic field, it is possible to achieve the significant localization of nanoparticles on the surface of tumor tissues, which facilitates their delivery to target cells. However, it should be noted that in real tumor tissues, the vascular system differs from normal tissues in its increased permeability, which requires careful calibration of the magnetic field strength and nanoparticle sizes.

To model tissues in the experiments, porous media based on agar gels with different mean pore sizes were used. These models allow the tissue characteristics such as density and porosity to be reproduced, which makes it possible to evaluate the transport characteristics of nanoparticles. In addition, such environments are easily integrated into microfluidic systems, which allows us to create complex research systems on a chip with their help.

Based on the obtained data on the transport of MNPs in tumor tissues, several promising areas for future research can be identified. These areas may focus on improving nanoparticle delivery methods, the in-depth study of biological interactions, and the optimization of microfluidic systems for oncotheranostics and other biomedical applications. Further improvement of such systems is necessary to more accurately model the tumor microenvironment, including the mechanical properties of tumor tissues, nutrient and oxygen concentration gradients, and imitation of dynamic interactions between tumor cells and healthy tissues. The creation of tumor-on-a-chip systems accurately mimicking the characteristics of individual patient tumors is the subject of research by many scientific groups and laboratories around the world. This may lead to significant progress in personalized medicine, providing the ability to select optimal dosages and combinations of drugs individually for a patient.

The microfluidic systems presented in this paper have several key advantages, making them a unique tool for research in the field of targeted delivery of nanoparticles. Firstly, they imitate the physiological conditions of tissues and vascular structures, which ensures reliable modeling of the processes of migration and accumulation of nanoparticles. Secondly, due to the use of porous media with an adjustable pore size, these systems allow varying experimental parameters and adapting them to different types of tissues. Finally, the technology of manufacturing microfluidic chips based on thermal compression and laser ablation ensures high accuracy and reproducibility at a relatively low cost, which makes them available for a wide range of studies. These advantages distinguish the systems presented from analogs and open new prospects for research in the field of nanomedicine and oncotheranostics.

5. Conclusions

The use of magnetic nanoparticles for the targeted delivery of drugs to tumor tissues is one of the promising areas in oncotheranostics. To achieve the advantages of this technique, it is important to precisely control the movement of nanoparticles using an external magnetic field, which allows them to be directed to specific areas of the tumor. Microfluidic systems developed for modeling tumor tissues (tumor-on-a-chip) are the perspective tool for studying the transport characteristics of magnetic nanoparticles.

In this work, models of tissues were prepared using planar sandwich-type polymer chips with basins into which polymer gels of various mean pore size and brain tissue of laboratory rats were deposited.

Two types of magnetic nanoparticles were prepared: magnetite nanoparticles (mMNPs) and carboxymethyl dextran coated SPIONS (CD-SPIONs).

Studies of the flow of an aqueous suspension of mMNPs in a capillary in the form of parallel flows with water showed a rapid agglomeration of mMNPs into microparticles within 10 min, then into large spatial structures causing the formation of laminar vortices within 20 min with subsequent destruction of the flow and, finally, the formation of a network covering all particles in the channel and consisting of magnetic fibers.

Studies of the magnetically driven migration of CD-SPIONs were carried out in polymer gel and brain tissue media. In polymer agar gels with mean pore sizes of 100–150 nm and 200–300 nm, no migration of CD-SPIONs was observed. In the gel with mean pore size of 350–400 nm, CD-SPIONs of 95–350 nm range were migrating in the field gradient of 40 T/m with differing velocities of 33, 16 and 5 μm/min correspondingly. No compression of the gel was observed.

CD-SPIONs of 350 nm size did not penetrate the brain tissue and compressed it at a speed of 100 μm/min. Particles of 220 nm were also compressing the brain tissue, but at a considerably lower rate, and particles of 95 nm were partially penetrating the edge of the tissue and did not compress it.

The agarose gels of sufficiently large mean pore size could be used in the tumor-on-a-chip format as a scaffold for tumor cells to which magnetic particles loaded with drugs could be operatively delivered in magnetic field for testing.

Further research could be aimed at improving the properties of nanoparticles to enhance their invasion into dense biological tissues as well as creating more complex tumor-on-a-chip models that consider not only the structural but also the biochemical properties of real tumors, making them more personalized.

Supplementary Materials: The following supporting information can be downloaded at: https://www.mdpi.com/article/10.3390/nano14242030/s1.

Author Contributions: Conceptualization, N.S. and M.S.; methodology, T.Z. and K.G.; validation, T.Z., N.S. and K.G.; formal analysis, T.Z. and M.S.; investigation, T.Z., N.S., K.B., D.T., V.F., K.S., S.C. and N.M.; data curation, N.S.; writing—original draft preparation, all authors; writing—review and editing, all authors; visualization, T.Z.; supervision, M.S.; project administration, M.S.; funding acquisition, K.S., S.C. and M.S. All authors have read and agreed to the published version of the manuscript.

Funding: This work was financially supported by the Ministry of Science and Higher Education of the Russian Federation (Agreement No. 075-15-2022-301).

Data Availability Statement: The original contributions presented in this study are included in the article/Supplementary Material. Further inquiries can be directed to the corresponding authors.

Conflicts of Interest: The authors declare no conflicts of interest.

References

1. WHO. Cancer. Available online: https://www.who.int/news-room/fact-sheets/detail/cancer (accessed on 28 August 2024).
2. Turner, J.H. An Introduction to the Clinical Practice of Theranostics in Oncology. *Br. J. Radiol.* **2018**, *91*, 20180440. [CrossRef]
3. Arnold, C. Theranostics Could Be Big Business in Precision Oncology. *Nat. Med.* **2022**, *28*, 606–608. [CrossRef] [PubMed]
4. Tietze, R.; Zaloga, J.; Unterweger, H.; Lyer, S.; Friedrich, R.P.; Janko, C.; Pöttler, M.; Dürr, S.; Alexiou, C. Magnetic Nanoparticle-Based Drug Delivery for Cancer Therapy. *Biochem. Biophys. Res. Commun.* **2015**, *468*, 463–470. [CrossRef] [PubMed]
5. Zimina, T.M.; Sitkov, N.O.; Gareev, K.G.; Fedorov, V.; Grouzdev, D.S.; Koziaeva, V.V.; Gao, H.; Combs, S.E.; Shevtsov, M. Biosensors and Drug Delivery in Oncotheranostics Using Inorganic Synthetic and Biogenic Magnetic Nanoparticles. *Biosensors* **2022**, *12*, 789. [CrossRef] [PubMed]
6. Nikolaev, B.; Yakovleva, L.; Fedorov, V.; Yudintceva, N.; Ryzhov, V.; Marchenko, Y.; Ischenko, A.; Zhakhov, A.; Dobrodumov, A.; Combs, S.E.; et al. Magnetic Relaxation Switching Assay Using IFNα-2b-Conjugated Superparamagnetic Nanoparticles for Anti-Interferon Antibody Detection. *Biosensors* **2023**, *13*, 624. [CrossRef] [PubMed]
7. Yudintceva, N.; Lomert, E.; Mikhailova, N.; Tolkunova, E.; Agadzhanian, N.; Samochernych, K.; Multhoff, G.; Timin, G.; Ryzhov, V.; Deriglazov, V.; et al. Targeting Brain Tumors with Mesenchymal Stem Cells in the Experimental Model of the Orthotopic Glioblastoma in Rats. *Biomedicines* **2021**, *9*, 1592. [CrossRef] [PubMed]
8. Li, J.; Wang, Q.; Xia, G.; Adilijiang, N.; Li, Y.; Hou, Z.; Fan, Z.; Li, J. Recent Advances in Targeted Drug Delivery Strategy for Enhancing Oncotherapy. *Pharmaceutics* **2023**, *15*, 2233. [CrossRef] [PubMed]
9. Shevtsov, M.; Stangl, S.; Nikolaev, B.; Yakovleva, L.; Marchenko, Y.; Tagaeva, R.; Sievert, W.; Pitkin, E.; Mazur, A.; Tolstoy, P.; et al. Granzyme B Functionalized Nanoparticles Targeting Membrane Hsp70-Positive Tumors for Multimodal Cancer Theranostics. *Small* **2019**, *15*, e1900205. [CrossRef] [PubMed]
10. Freeman, M.W.; Arrott, A.; Watson, J. Magnetism in Medicine. Available online: https://www.semanticscholar.org/paper/Magnetism-in-Medicine-Freeman-Arrott/488bd929580fc3650bc78f86bd434bcfca94ef7c (accessed on 2 December 2024).
11. Khan, U.S.; Khattak, N.K.; Rahman, A.; Khan, F. Historical Development of Magnetite Nanoparticles Synthesis. *J. Chem. Soc. Pak.* **2011**, *33*, 793–804.
12. Suwa, T.; Ozawa, S.; Ueda, M.; Ando, N.; Kitajima, M. Magnetic resonance imaging of esophageal squamous cell carcinoma using magnetite particles coated with anti-epidermal growth factor receptor antibody. *Int. J. Cancer* **1998**, *75*, 626–634. [CrossRef]
13. De Paoli, V.M.; De Paoli Lacerda, S.H.; Spinu, L.; Ingber, B.; Rosenzweig, Z.; Rosenzweig, N. Effect of an Oscillating Magnetic Field on the Release Properties of Magnetic Collagen Gels. *Langmuir* **2006**, *22*, 5894–5899. [CrossRef] [PubMed]
14. Yang, J.; Park, S.; Yoon, H.; Huh, Y.; Haam, S. Preparation of Poly ϵ-Caprolactone Nanoparticles Containing Magnetite for Magnetic Drug Carrier. *Int. J. Pharm.* **2006**, *324*, 185–190. [CrossRef] [PubMed]
15. Zhang, Y.; Kohler, N.; Zhang, M. Surface Modification of Superparamagnetic Magnetite Nanoparticles and Their Intracellular Uptake. *Biomaterials* **2002**, *23*, 1553–1561. [CrossRef] [PubMed]
16. Wilhelm, C.; Cebers, A.; Bacri, J.-C.; Gazeau, F. Gazeau Deformation of Intracellular Endosomes under a Magnetic Field. *Eur. Biophys. J.* **2003**, *32*, 655–660. [CrossRef]
17. Chen, Y.; Hou, S. Application of Magnetic Nanoparticles in Cell Therapy. *Stem Cell Res. Ther.* **2022**, *13*, 135. [CrossRef] [PubMed]
18. Aslam, H.; Shukrullah, S.; Naz, M.Y.; Fatima, H.; Hussain, H.; Ullah, S.; Assiri, M.A. Current and Future Perspectives of Multifunctional Magnetic Nanoparticles Based Controlled Drug Delivery Systems. *J. Drug Deliv. Sci. Technol.* **2022**, *67*, 102946. [CrossRef]
19. Chubarov, A.S. Serum Albumin for Magnetic Nanoparticles Coating. *Magnetochemistry* **2022**, *8*, 13. [CrossRef]
20. Koksharov, Y.A.; Gubin, S.P.; Taranov, I.V.; Khomutov, G.B.; Gulyaev, Y.V. Magnetic Nanoparticles in Medicine: Progress, Problems, and Advances. *J. Commun. Technol. Electron.* **2022**, *67*, 101–116. [CrossRef]
21. Hsieh, L.-C.; Le, T.-K.; Hu, F.-C.; Chen, Y.-T.; Hsieh, S.; Wu, C.-C.; Hsieh, S.-L. Targeted Colorectal Cancer Treatment: In Vitro Anti-Cancer Effects of Carnosine Nanoparticles Supported by Agar and Magnetic Iron Oxide. *Eur. J. Pharm. Biopharm.* **2024**, *203*, 114477. [CrossRef] [PubMed]
22. Chircov, C.; Ștefan, R.-E.; Dolete, G.; Andrei, A.; Holban, A.M.; Oprea, O.-C.; Vasile, B.S.; Neacșu, I.A.; Tihăuan, B. Dextran-Coated Iron Oxide Nanoparticles Loaded with Curcumin for Antimicrobial Therapies. *Pharmaceutics* **2022**, *14*, 1057. [CrossRef]
23. Alotaibi, I.; Alshammari, M.S.; Algessair, S.; Madkhali, N.; All, N.A.; Hjiri, M.; Alrub, S.A.; El Mir, L.; Lemine, O.M. Synthesis, Characterization and Heating Efficiency of Gd-Doped Maghemite (γ-Fe_2O_3) Nanoparticles for Hyperthermia Application. *Phys. B Condens. Matter* **2022**, *625*, 413510. [CrossRef]
24. Vangijzegem, T.; Lecomte, V.; Ternad, I.; Van Leuven, L.; Muller, R.N.; Stanicki, D.; Laurent, S. Superparamagnetic Iron Oxide Nanoparticles (SPION): From Fundamentals to State-of-The-Art Innovative Applications for Cancer Therapy. *Pharmaceutics* **2023**, *15*, 236. [CrossRef] [PubMed]

25. De Leo, V.; Maurelli, A.M.; Giotta, L.; Catucci, L. Liposomes Containing Nanoparticles: Preparation and Applications. *Colloids Surf. B Biointerfaces* **2022**, *218*, 112737. [CrossRef]
26. Kotakadi, S.M.; Borelli, D.P.R.; Nannepaga, J.S. Therapeutic Applications of Magnetotactic Bacteria and Magnetosomes: A Review Emphasizing on the Cancer Treatment. *Front. Bioeng. Biotechnol.* **2022**, *10*, 789016. [CrossRef] [PubMed]
27. Puri, R.; Arora, V.; Kabra, A.; Dureja, H.; Hamaal, S. Magnetosomes: A Tool for Targeted Drug Delivery in the Management of Cancer. *J. Nanomater.* **2022**, *2022*, 1–12. [CrossRef]
28. de Souza Cabral, A.; Verdan, M.; Presciliano, R.; Silveira, F.; Correa, T.; Abreu, F. Large-Scale Cultivation of Magnetotactic Bacteria and the Optimism for Sustainable and Cheap Approaches in Nanotechnology. *Mar. Drugs* **2023**, *21*, 60. [CrossRef] [PubMed]
29. Wan, L.; Neumann, C.A.; LeDuc, P.R. Tumor-on-a-Chip for Integrating a 3D Tumor Microenvironment: Chemical and Mechanical Factors. *Lab Chip* **2020**, *20*, 873–888. [CrossRef]
30. Monteiro, M.V.; Zhang, Y.S.; Gaspar, V.M.; Mano, J.F. 3D-Bioprinted Cancer-On-a-Chip: Level-up Organotypic in Vitro Models. *Trends Biotechnol.* **2021**, *40*, 4. [CrossRef] [PubMed]
31. Parihar, A.; Choudhary, N.K.; Parihar, D.S.; Khan, R. Tumor-On-a-Chip: Microfluidic Models of Hypoxic Tumor Microenvironment. In *Hypoxia in Cancer: Significance and Impact on Cancer Therapy*; Springer: Singapore, 2023; pp. 297–328. [CrossRef]
32. Anderson, S.R.; Stagner, E.J.; Sivakumar, H.; Skardal, A. Three-Dimensional Bioprinting of in Vitro Tumor Organoid and Organ-on-a-Chip Models. *MRS Bull.* **2023**, *48*, 643–656. [CrossRef]
33. Marino, A.; Battaglini, M.; Carmignani, A.; Pignatelli, F.; De Pasquale, D.; Tricinci, O.; Ciofani, G. Magnetic Self-Assembly of 3D Multicellular Microscaffolds: A Biomimetic Brain Tumor-On-a-Chip for Drug Delivery and Selectivity Testing. *APL Bioeng.* **2023**, *7*, 036103. [CrossRef] [PubMed]
34. Ustun, M.; Rahmani Dabbagh, S.; Ilci, I.S.; Bagci-Onder, T.; Tasoglu, S. Glioma-on-a-Chip Models. *Micromachines* **2021**, *12*, 490. [CrossRef] [PubMed]
35. Parlato, S.; Grisanti, G.; Sinibaldi, G.; Peruzzi, G.; Casciola, C.M.; Gabriele, L. Tumor-on-a-Chip Platforms to Study Cancer-Immune System Crosstalk in the Era of Immunotherapy. *Lab Chip* **2020**, *21*, 234–253. [CrossRef] [PubMed]
36. Iyer, A.K.; Khaled, G.; Fang, J.; Maeda, H. Exploiting the Enhanced Permeability and Retention Effect for Tumor Targeting. *Drug Discov. Today* **2006**, *11*, 812–818. [CrossRef] [PubMed]
37. Gawali, P.; Saraswat, A.; Bhide, S.; Gupta, S.; Patel, K. Human Solid Tumors and Clinical Relevance of the Enhanced Permeation and Retention Effect: A "Golden Gate" for Nanomedicine in Preclinical Studies? *Nanomedicine* **2023**, *18*, 169–190. [CrossRef] [PubMed]
38. Conte, M.; Volpe, A.; Gaudiuso, C.; Ancona, A. Bonding of PMMA to Silicon by Femtosecond Laser Pulses. *Sci. Rep.* **2023**, *13*, 5062. [CrossRef]
39. Madadi, M.; Madadi, A.; Zareifar, R.; Nikfarjam, A. A Simple Solvent-Assisted Method for Thermal Bonding of Large-Surface, Multilayer PMMA Microfluidic Devices. *Sens. Actuators A Phys.* **2023**, *349*, 114077. [CrossRef]
40. Nayak, N.C.; Yue, C.Y.; Lam, Y.C.; Tan, Y.L. Thermal Bonding of PMMA: Effect of Polymer Molecular Weight. *Microsyst. Technol.* **2009**, *16*, 487–491. [CrossRef]
41. Zimina, T.M.; Gareev, K.G.; Sitkov, N.O.; Brusina, K.E.; Shubina, M.A.; Ganiev, Z.; Bobkov, D.E.; Likhomanova, R.B.; Yudintseva, N.M.; Potrakhov, N.N.; et al. A Device for Studying Heterogeneous Tumor Cell Cultures Based on a Hybrid Microfluidic System. *Biomed. Eng.* **2024**, *58*, 153–157. [CrossRef]
42. Manshadi, M.K.D.; Saadat, M.; Mohammadi, M.; Shamsi, M.; Dejam, M.; Kamali, R.; Sanati-Nezhad, A. Delivery of Magnetic Micro/Nanoparticles and Magnetic-Based Drug/Cargo into Arterial Flow for Targeted Therapy. *Drug Deliv.* **2018**, *25*, 1963–1973. [CrossRef]
43. Thomsen, L.B.; Thomsen, M.S.; Moos, T. Targeted Drug Delivery to the Brain Using Magnetic Nanoparticles. *Ther. Deliv.* **2015**, *6*, 1145–1155. [CrossRef]
44. Hasan, M.; Choi, J.-G.; Lee, S.-S. External Magnet-Guided Targeted Delivery and Tissue Distribution Analysis of Anti-Cluster of Differentiation 3 Conjugated Magnetic Nanoparticles of Fe_3O_4 Using in Vivo Imaging System. *AIP Adv.* **2024**, *14*, 015128. [CrossRef]
45. Al-Madhagi, H.; Yazbik, V.; Abdelwahed, W.; Alchab, L. Magnetite Nanoparticle Co-Precipitation Synthesis, Characterization, and Applications: Mini Review. *BioNanoSci* **2023**, *13*, 853–859. [CrossRef]
46. Jiang, W.; Lai, K.-L.; Hu, H.; Zeng, X.-B.; Lan, F.; Liu, K.-X.; Wu, Y.; Gu, Z.-W. The Effect of $[Fe^{3+}]/[Fe^{2+}]$ Molar Ratio and Iron Salts Concentration on the Properties of Superparamagnetic Iron Oxide Nanoparticles in the Water/Ethanol/Toluene System. *J. Nanoparticle Res.* **2011**, *13*, 5135–5145. [CrossRef]
47. Wahajuddin, N.; Arora, S. Superparamagnetic Iron Oxide Nanoparticles: Magnetic Nanoplatforms as Drug Carriers. *Int. J. Nanomed.* **2012**, *7*, 3445–3471. [CrossRef] [PubMed]
48. Nikolaev, B.; Yakovleva, L.; Fedorov, V.; Li, H.; Gao, H.; Shevtsov, M. Nano- and Microemulsions in Biomedicine: From Theory to Practice. *Pharmaceutics* **2023**, *15*, 1989. [CrossRef] [PubMed]
49. Lee, J.W.; Kim, D.K. Carboxymethyl Group Activation of Dextran Cross-Linked Superparamagnetic Iron Oxide Nanoparticles. *J. Korean Ceram. Soc.* **2020**, *58*, 106–115. [CrossRef]
50. Guo, H.; Chen, W.; Sun, X.; Liu, Y.-N.; Li, J.; Wang, J. Theranostic Magnetoliposomes Coated by Carboxymethyl Dextran with Controlled Release by Low-Frequency Alternating Magnetic Field. *Carbohydr. Polym.* **2015**, *118*, 209–217. [CrossRef] [PubMed]

51. Vasić, K.; Knez, Ž.; Konstantinova, E.A.; Kokorin, A.I.; Gyergyek, S.; Leitgeb, M. Structural and Magnetic Characteristics of Carboxymethyl Dextran Coated Magnetic Nanoparticles: From Characterization to Immobilization Application. *React. Funct. Polym.* **2020**, *148*, 104481. [CrossRef]
52. Salvador, M.; Gutiérrez, G.; Noriega, S.; Moyano, A.; Blanco-López, M.C.; Matos, M. Microemulsion Synthesis of Superparamagnetic Nanoparticles for Bioapplications. *Int. J. Mol. Sci.* **2021**, *22*, 427. [CrossRef] [PubMed]
53. Okoli, C.; Sanchez-Dominguez, M.; Boutonnet, M.; Järås, S.; Civera, C.; Solans, C.; Kuttuva, G.R. Comparison and Functionalization Study of Microemulsion-Prepared Magnetic Iron Oxide Nanoparticles. *Langmuir* **2012**, *28*, 8479–8485. [CrossRef] [PubMed]
54. Li, J.M.; Liu, C.; Qiao, H.C.; Zhu, L.Y.; Chen, G.; Dai, X.D. Hot Embossing/Bonding of a Poly(Ethylene Terephthalate) (PET) Microfluidic Chip. *J. Micromech. Microeng.* **2007**, *18*, 015008. [CrossRef]
55. Keng, P.Y.; Esterby, M.; van Dam, R.M. Emerging technologies for decentralized production of PET tracers. In *Positron Emission Tomography-Current Clinical and Research Aspects*; IntechOpen: London, UK, 2012; pp. 153–182.
56. Yan, X.; Liu, W.; Yuan, Y.; Chen, C. Indium Tin Oxide Coated PET Film Contactless Conductivity Detector for Microchip Capillary Electrophoresis. *Anal. Methods* **2015**, *7*, 5295–5302. [CrossRef]
57. Chou, J.; Du, N.; Ou, T.; Floriano, P.N.; Christodoulides, N.; McDevitt, J.T. Hot Embossed Polyethylene Through-Hole Chips for Bead-Based Microfluidicdevices. *Biosens. Bioelectron.* **2013**, *42*, 653–660. [CrossRef]
58. Patil, S.M.; Rane, N.R.; Bankole, P.O.; Krishnaiah, P.; Ahn, Y.; Park, Y.-K.; Yadav, K.K.; Amin, M.A.; Jeon, B.-H. An Assessment of Micro- and Nanoplastics in the Biosphere: A Review of Detection, Monitoring, and Remediation Technology. *Chem. Eng. J.* **2022**, *430*, 132913. [CrossRef]
59. Wang, K.; He, L.; Manz, A.; Wu, W. User-Friendly Microfabrication Method for Complex Topological Structure and Three-Dimensional Microchannel with the Application Prospect in Polymerase Chain Reaction (PCR). *Anal. Chem.* **2020**, *93*, 1523–1528. [CrossRef] [PubMed]
60. Pekgor, M.; Nikzad, M.; Arablouei, R.; Masood, S. Design of a 3D-Printable UHF RFID Hybrid Liquid Antenna for Biosensing Applications. *Mater. Today Proc.* **2020**, *46*, 4619–4624. [CrossRef]
61. Matsuura, K.; Takata, K. Blood Cell Separation Using Polypropylene-Based Microfluidic Devices Based on Deterministic Lateral Displacement. *Micromachines* **2023**, *14*, 238. [CrossRef] [PubMed]
62. Seok, S.; Park, H.; Kim, J. Scotch-Tape Surface Wrinkling Based Thin-Film Material Properties Extraction. *J. Micromech. Microeng.* **2022**, *32*, 045002. [CrossRef]
63. Pradeep, A.; Raveendran, J.; Satheesh Babu, T.G. Design, Fabrication and Assembly of Lab-on-a-Chip and Its Uses. In *Progress in Molecular Biology and Translational Science*; Elsevier: Amsterdam, The Netherlands, 2022; pp. 121–162. [CrossRef]
64. Sun, H.; Chan, C.-W.; Wang, Y.; Yao, X.; Mu, X.; Lu, X.; Zhou, J.; Cai, Z.; Ren, K. Reliable and Reusable Whole Polypropylene Plastic Microfluidic Devices for a Rapid, Low-Cost Antimicrobial Susceptibility Test. *Lab Chip* **2019**, *19*, 2915–2924. [CrossRef] [PubMed]
65. Juang, Y.-J.; Chiu, Y.-J. Fabrication of Polymer Microfluidics: An Overview. *Polymers* **2022**, *14*, 2028. [CrossRef]
66. Rohr, T.; Ogletree, D.F.; Svec, F.; Fréchet, J.M.J. Surface Functionalization of Thermoplastic Polymers for the Fabrication of Microfluidic Devices by Photoinitiated Grafting. *Adv. Funct. Mater.* **2003**, *13*, 264–270. [CrossRef]
67. Wu, Y.; Zhang, X.; Zhao, Q.; Tan, B.; Chen, X.; Liao, J. Role of Hydrogels in Bone Tissue Engineering: How Properties Shape Regeneration. *J. Biomed. Nanotechnol.* **2020**, *16*, 1667–1686. [CrossRef]
68. Rai, A.; Singh, A.K.; Bleimling, N.; Posern, G.; Vetter, I.R.; Goody, R.S. Rep15 Interacts with Several Rab GTPases and Has a Distinct Fold for a Rab Effector. *Nat. Commun.* **2022**, *13*, 4262. [CrossRef] [PubMed]
69. Han, C.; Zhang, Z.; Guo, N.; Li, X.; Yang, M.; Peng, Y.; Ma, X.; Yu, K.; Wang, C. Effects of Sevoflurane Inhalation Anesthesia on the Intestinal Microbiome in Mice. *Front. Cell. Infect. Microbiol.* **2021**, *11*, 633527. [CrossRef] [PubMed]
70. Weir, G.J.; Chisholm, G.; Leveneur, J. The Magnetic Field about a Three-Dimensional Block Neodymium Magnet. *ANZIAM J.* **2021**, *62*, 386–405. [CrossRef]
71. Egea-Benavente, D.; Ovejero, J.G.; Morales, M.d.P.; Barber, D.F. Understanding MNPs Behaviour in Response to AMF in Biological Milieus and the Effects at the Cellular Level: Implications for a Rational Design That Drives Magnetic Hyperthermia Therapy toward Clinical Implementation. *Cancers* **2021**, *13*, 4583. [CrossRef]
72. Martínez-Sanz, M.; Ström, A.; Lopez-Sanchez, P.; Knutsen, S.H.; Ballance, S.; Zobel, H.K.; Sokolova, A.; Gilbert, E.P.; López-Rubio, A. Advanced Structural Characterisation of Agar-Based Hydrogels: Rheological and Small Angle Scattering Studies. *Carbohydr. Polym.* **2020**, *236*, 115655. [CrossRef] [PubMed]
73. Choi, M.; Choi, H.W.; Kim, H.E.; Hahn, J.; Choi, Y.J. Mimicking Animal Adipose Tissue Using a Hybrid Network-Based Solid-Emulsion Gel with Soy Protein Isolate, Agar, and Alginate. *Food Hydrocoll.* **2023**, *145*, 109043. [CrossRef]

nanomaterials

MDPI

Article

Enhancement of Triple-Negative Breast Cancer-Specific Induction of Cell Death by Silver Nanoparticles by Combined Treatment with Proteotoxic Stress Response Inhibitors

Christina M. Snyder [1], Beatriz Mateo [1], Khushbu Patel [1], Cale D. Fahrenholtz [1,2], Monica M. Rohde [1], Richard Carpenter [3] and Ravi N. Singh [1,4,*]

1 Department of Cancer Biology, Wake Forest University School of Medicine, Winston-Salem, NC 27157, USA; christinasnyder0@gmail.com (C.M.S.); bmateo@wakehealth.edu (B.M.); khpatel@wakehealth.edu (K.P.); cfahren@highpoint.edu (C.D.F.); monicarohde5@gmail.com (M.M.R.)

2 Fred Wilson School of Pharmacy, High Point University, High Point, NC 27268, USA

3 Department of Biochemistry and Molecular Biology, Indiana University School of Medicine, Bloomington, IN 47405, USA; richcarp@iu.edu

4 Atrium Health Wake Forest Baptist Comprehensive Cancer Center, Winston-Salem, NC 27157, USA

* Correspondence: rasingh@wakehealth.edu

Abstract: Metal nanoparticles have been tested for therapeutic and imaging applications in preclinical models of cancer, but fears of toxicity have limited their translation. An emerging concept in nanomedicine is to exploit the inherent drug-like properties of unmodified nanomaterials for cancer therapy. To be useful clinically, there must be a window between the toxicity of the nanomaterial to cancer and toxicity to normal cells. This necessitates identification of specific vulnerabilities in cancers that can be targeted using nanomaterials without inducing off-target toxicity. Previous studies point to proteotoxic stress as a driver of silver nanoparticle (AgNPs) toxicity. Two key cell stress responses involved in mitigating proteotoxicity are the heat shock response (HSR) and the integrated stress response (ISR). Here, we examine the role that these stress responses play in AgNP-induced cytotoxicity in triple-negative breast cancer (TNBC) and immortalized mammary epithelial cells. Furthermore, we investigate HSR and ISR inhibitors as potential drug partners to increase the anti-cancer efficacy of AgNPs without increasing off-target toxicity. We showed that AgNPs did not strongly induce the HSR at a transcriptional level, but instead decreased expression of heat shock proteins (HSPs) at the protein level, possibly due to degradation in AgNP-treated TNBC cells. We further showed that the HSR inhibitor, KRIBB11, synergized with AgNPs in TNBC cells, but also increased off-target toxicity in immortalized mammary epithelial cells. In contrast, we found that salubrinal, a drug that can sustain pro-death ISR signaling, enhanced AgNP-induced cell death in TNBC cells without increasing toxicity in immortalized mammary epithelial cells. Subsequent co-culture studies demonstrated that AgNPs in combination with salubrinal selectively eliminated TNBCs without affecting immortalized mammary epithelial cells grown in the same well. Our findings provide additional support for proteotoxic stress as a mechanism by which AgNPs selectively kill TNBCs and will help guide future efforts to identify drug partners that would be beneficial for use with AgNPs for cancer therapy.

Keywords: nanotoxicity; heat shock; proteotoxicity; integrated stress response; cancer therapy

1. Introduction

Proteotoxicity is a condition caused by the accumulation of misfolded and aggregated proteins in the cell and can lead to cell death if stress conditions are prolonged or unmitigated [1]. To alleviate proteotoxic stress, misfolded proteins are sequestered and delivered to proteasomes, lysosomes, and autophagosomes to be degraded [2]. Some cancer types exhibit high baseline levels of proteotoxic stress compared to non-cancer cells, and this may be an exploitable weakness [3]. For example, studies by Gupta et al. revealed that a subset of triple-negative breast cancer cells (TNBC) are vulnerable to small-molecule drugs, including tunicamycin and thapsigargin, which increased the accumulation of misfolded proteins [4,5]. These TNBCs are sensitive because they synthesized and secreted large amounts of extracellular matrix proteins, generating an enormous burden on the protein quality-control machinery. However, tunicamycin and thapsigargin have thus far failed to be effective clinically due to dose-limiting toxicities [6,7].

Nanotechnology offers the possibility to increase the efficacy and reduce off-target toxicity of small-molecule drugs used as proteotoxic stress inducers [8], but difficulty with formulation, characterization, and scale-up have hampered most efforts to translate drug-carrying nanoparticles for cancer therapy in humans [9]. The core technologies of clinically approved nanotherapeutics remain largely unchanged since the development of Doxil® [10]. Nanoparticles made of metals, including gold, silver, iron, and gadolinium, have been tested for therapeutic and imaging applications in pre-clinical models of cancer [11], but fears of toxicity have limited their translation. An emerging concept in nanomedicine is exploiting the inherent drug-like and cytotoxic properties of unmodified, metal nanomaterials for cancer therapy [12–16]. To be useful clinically, there must be a window between the toxicity of the nanomaterial to cancer and toxicity to normal cells. This necessitates identification of specific vulnerabilities in cancers that can be targeted using nanomaterials without inducing off-target toxicity. For example, TNBC cells, but not immortalized mammary epithelial cells, are sensitive to silver-nanoparticle (AgNP)-induced proteotoxic stress, as indicated by accumulation of protein aggregates, increased protein oxidation, and activation of proteotoxic stress-signaling responses [17–22]. Use of AgNPs (or other nanometals) themselves as the therapeutic agent simplifies formulation, characterization, and scale-up compared to conventional drug-loaded nanoparticles [23]. We hypothesize that the therapeutic efficacy of AgNPs could potentially be enhanced by co-delivery of drugs that modulate the proteotoxic stress response, but care must be taken to examine toxicities of therapeutics that will be administered together to ensure that off-target toxicity profiles are not increased.

The heat shock response (HSR) plays a key role in mitigating proteotoxicity. Heat shock proteins (HSPs) are a family of cytosolic, nuclear, and membrane-bound proteins that function as chaperones by assisting in protein folding, localization, and transport [24]. Activity of HSPs enables cells to endure proteotoxicity induced by stresses, such as heat, oxidative stress, nutrient deprivation, and impaired protein degradation machinery. HSPs include stress-inducible heat shock protein 70 (HSP70) and heat shock protein 27 (HSP27), as well as the ubiquitously expressed heat shock protein 90 (HSP90) [25]. Under normal conditions, heat shock factor 1 (HSF1), known as the 'master heat shock regulator', is present as an inactive monomer bound to HSP90 or HSP70 in the cytosol of cells [25]. When misfolded proteins are abundant, HSF1 dissociates from HSP70 or HSP90, trimerizes, autophosphorylates, and translocates to the nucleus [26]. HSF1 then binds heat shock elements (HSEs) on DNA to induce transcription of HSPs, which bind unfolded or misfolded proteins and act as chaperones to repair damaged proteins, when possible, or sequester damaged proteins to enable subsequent degradation [25]. When not bound to misfolded proteins, HSPs regulate HSF1 activity and can inhibit HSF1 activation or induce HSF1 degradation through multiple feedback inhibitory loops [27–29]. HSPs are frequently overexpressed in cancer, may promote malignant phenotypes, and HSR inhibitors have been investigated as cancer therapeutics [30]. Inhibition of the HSR increases sensitivity to proteotoxic stress, but the combination of AgNPs and HSR inhibitors has not been tested.

A second mechanism cells use to mitigate proteotoxicity is the integrated stress response (ISR). The ISR is initiated by stress-sensing kinases: double-stranded RNA-dependent protein kinase (PKR), PKR-like ER kinase (PERK), heme-regulated eIF2α kinase (HRI), and general control non-de-repressible 2 (GCN2) [31]. Each is activated by different types of stress, but all similarly initiate the ISR through phosphorylation of the alpha subunit of eukaryotic translation initiation factor 2 (eIF2α) on serine 51. Phosphorylation of eIF2α causes a decrease in global mRNA translation to reduce the protein synthesis burden, and simultaneously activates an alternative transcriptional program needed to boost protein folding and degradative capacity. Paradoxically, prolonged phosphorylation of eIF2α increases expression of C/EBP Homologous Protein (CHOP), which drives pro-apoptotic pathways. We and others observed phosphorylation of eIF2α in response to AgNPs in TNBC cells, but not in normal breast cells [17,20,22]. Because sustained ISR signaling initiates cell death programs [2,31], cells activate a signaling feedback that increases expression of Growth Arrest and DNA-damaged protein 34 (GADD34) and the serine/threonine protein phosphatase 1, which form a complex (GADD34:PP1) that dephosphorylates eIF2α to restore normal transcriptional programs [31]. Salubrinal is a drug that inhibits the phosphatase activity of the GADD34:PP1 complex, thereby sustaining ISR signaling. Because the ISR can produce both protective and lethal effects, use of salubrinal on cells under proteotoxic stress has the potential to be either protective or to increase cell death [32–35]. It is not known if salubrinal will protect against or enhance AgNP-induced cell death.

Here, we examine two types of proteotoxic stress responses, the HSR and the ISR, following AgNP exposure in TNBC and immortalized mammary epithelial cells. Furthermore, we employ drugs that modulate the HSR or ISR to identify potential pharmacologic partners that could increase the toxicity of AgNPs in TNBC without increasing toxicity in normal cells.

2. Materials and Methods

Silver nanoparticles: Here, 25 nm spherical AgNPs stabilized with polyvinylpyrrolidone (PVP) were purchased as dried powders from nanoComposix, Inc. (San Diego, CA, USA). The manufacturer specified a mass ratio of 17% Ag to 83% PVP for the nanoparticles. Nanoparticles were dispersed by bath sonication in phosphate-buffered saline (PBS), pH 7.4, without calcium or magnesium (Invitrogen, Carlsbad, CA, USA), at a concentration of 5 mg/mL based upon the mass of silver contained in the nanoparticles (i.e., excluding PVP), and were then diluted in cell culture medium to the final concentration listed in the figures prior to addition to wells containing cells. The physicochemical properties of this material were characterized previously [36]. Briefly, characteristics of the prepared stock of AgNPs were as follows: hydrodynamic diameter in PBS at pH 7.4: 36.5 $\pm$ 0.7 nm; ζ-potential in PBS at pH 7.4: -15.5 ± 1.6 mV; plasmon resonance peak in PBS at pH 7.4: 402 nm; soluble silver (Ag^+) present in the AgNP suspension: less than 0.001% Ag^+ by mass.

Cell culture: BT549, MDA-MB-231, and MCF-10A cells were purchased from the ATCC (Manassas, VA, USA). SUM159 cells were purchased from Asterand (now BioIVT, Westbury, NY, USA). iMEC cells were provided by Dr. Elizabeth Alli in the Department of Cancer Biology at Wake Forest University School of Medicine. Cell lines were expanded, and low-passage stocks were stored in liquid nitrogen and maintained by the Wake Forest Comprehensive Cancer Center Cell Engineering Shared Resource. Growth media for these cell lines are listed in Table 1. Cells were verified to be free from mycoplasma contamination by routine testing using the MycoAlert Mycoplasma Detection Kit (Lonza, Basel, Switzerland). Cells were passaged, and the medium was changed twice weekly. Cell monolayers were grown on tissue-culture-treated plastics purchased from Corning Life Sciences (Corning, NY, USA). Cells were maintained in culture for no longer than 4 months before new cultures were established from low-passage frozen stocks.

MTT assay: Cells were seeded on 96-well plates at a density of 5000 cells per well in 100 µL of complete media, recovered for 24 h, and then were exposed to combinations of AgNPs, KRIBB11 (Selleck Chemicals, Houston TX, USA), or salubrinal (MedChemExpress, Monmouth Junction, NJ, USA) in 100 µL of complete media containing doses of each drug, as listed in the figures. After 72 h, media containing treatments were replaced with 200 µL of media containing 0.5 mg/mL of 3-(4,5- dimethylthiazol-2-yl)-2,5-diphenyltetrazolium bromide (MTT; Sigma-Aldrich, St. Louis, MO, USA) and incubated for 1 h at 37 °C. After that, media were removed, cells were lysed in 200 µL of dimethyl sulfoxide (Thermo Fisher

Scientific, Pittsburgh, PA, USA), and absorbance was read using a Molecular Devices Emax Precision Microplate Reader (San Jose, CA, USA) at 560 nm. Absorbance measurements at 650 nm were subtracted from each reading to correct for any inconsistencies in optical properties between wells.

Table 1. Complete media formulations for cell culture.

Cell Line	Media Formulation
SUM-159	HAM's F12 supplemented with penicillin (250 units/mL), streptomycin (250 μg/mL), 2 mM L-glutamine, 5 μg/mL insulin, 1 μg/mL hydrocortisone, 10 μM HEPES, and 5% fetal bovine serum
BT-549	RPMI supplemented with penicillin (250 units/mL), streptomycin (250 μg/mL), and 10% fetal bovine serum
MDA-MB-231	DMEM/F12 supplemented with penicillin (250 units/mL), streptomycin (250 μg/mL), 2 mmol/L L-glutamine, and 10% fetal bovine serum
MCF-10A	DMEM/F12 supplemented with penicillin (250 units/mL), streptomycin (250 μg/mL), 2 mM L-glutamine, 10 μg/mL insulin, 20 ng/mL EGF, 0.5 μg/mL hydrocortisone, 100 ng/mL cholera toxin, and 5% heat-inactivated horse serum
iMEC	DMEM/F12 supplemented with 10 μg/mL insulin, 20 ng/mL hEGF, 0.5 μg/mL hydrocortisone, and 10% fetal bovine serum

Western Blotting: Cells were plated on 6 cm dishes at a density of 1,000,000 cells in 4 mL of complete media. Cells were allowed to recover for 48 h and then were exposed to AgNPs with or without salubrinal for 24 h at 37 °C. Medium was removed and cells were washed twice with ice-cold phosphate-buffered saline before lysis using Mammalian Protein Extraction Reagent supplemented with Halt Protease and Phosphatase Inhibitor Cocktail (both from Thermo Fisher Scientific). The protein concentration was determined for each sample using a Pierce bicinchoninic acid (BCA) protein assay kit (Thermo Fisher Scientific), according to the manufacturer's instructions. Proteins were size fractionated by gel electrophoresis and then transferred to a nitrocellulose membrane (Thermo Fisher Scientific). Non-specific binding was blocked by incubation for 30 min at room temperature with tris-buffered saline containing 0.1% Tween-20 (TBS-T; Bio-Rad) and either 5% blotting-grade blocker (Bio-Rad) or 5% bovine serum albumin (BSA; Sigma-Aldrich). Membranes were incubated overnight at 4 °C in dilutions containing TBS-T and either 5% blotting-grade blocker or 5% BSA and primary antibody. Primary antibodies used included the following: phospho(Ser51)-eIF2α (9721), eIF2α (5324), CHOP (2895), caspase-3 (9662), caspase-7 (9492), caspase-9 (9502), HSF1 (4356), HSP90 (4874), HSP70 (4876), phospho(Ser78)-HSP27 (2405), HSP27 (2402), GAPDH (2118), and β-actin (4970), purchased from Cell Signaling Technologies (Danvers, MA, USA), and phospho(Ser326)-HSF1 (ab76076) purchased from Abcam (Waltham, MA, USA). Membranes were washed and then incubated for 1 h at room temperature with anti-rabbit (Cell Signaling Technologies) or anti-mouse (Cell Signaling Technologies) horseradish peroxidase (HRP)-conjugated secondary antibodies diluted in 5% blotting-grade blocker in TBS-T. Immunoreactive products were visualized by chemiluminescence using SuperSignal Pico West reagent (Thermo Fisher Scientific).

Heat shock reporter assay: Cells were seeded on 24-well plates at a density of 100,000 cells per well in 1 mL media and allowed to recover for 2 days. Medium was replaced with 1 mL of fresh medium without penicillin or streptomycin. pGL4.41(luc/HSE/hygro) (Promega, Madison, WI, USA) was mixed with optiMEM (Thermo Fisher Scientific) and Xtreme Gene 9 (Sigma Aldrich) at a ratio of 33:3:1 (media (vol.):lipid (vol.):DNA (mass)). After 15 min, 0.2 μg DNA complexed to Xtreme Gene 9 was added to each well. After 24 h, cells were treated with AgNPs, then exposed to heat shock (43 °C) or physiologic temperature (37 °C) by incubation for 1 h in medium warmed to the respective temperature. After recovery for 3 h at 37 °C, luciferase activity was assessed using a commercial assay kit

(Promega Luciferase Assay System), as per the manufacturer's instructions. Briefly, cells were harvested in reporter lysis buffer and centrifuged for 5 min at 4 °C at 15,000 RPM in a microcentrifuge. Lysates were transferred to a white 96-well plate (Greiner Bio-One, Monroe, NC, USA), luciferin was added, and luminescence was evaluated using a FLUOstar Optima plate reader (BMG Labtech, Cary, NC, USA). Total protein in lysates was estimated using the Pierce BCA protein assay kit (Thermo Fisher Scientific), as per the manufacturer's instructions. Luminescence in each well was normalized based on the protein concentration.

Co-culture and flow cytometry: BT549 cells expressing mCherry ($BT549^{mCherry}$) and iMEC cells expressing GFP ($iMEC^{GFP}$) were produced by lentiviral transduction using concentrated lentivirus vectors produced by the Cell Engineering Shared Resource at Atrium Health Wake Forest Baptist Comprehensive Cancer Center. Briefly, pLenti-CMV-GFP-Puro-LV (a gift from Eric Campeau and Paul Kaufman; Addgene plasmid #17448; http://n2t.net/addgene:17448; RRID:Addgene_17448) [37] or pCDH-CMV-mCherry-T2A-Puro-LV (a gift from Kazuhiro Oka; Addgene plasmid #72264; http://n2t.net/addgene:72264; RRID:Addgene_72264) were co-transfected into HEK293T cells (ATCC) with plasmids from the pPACKH1 HIV Lentivector Packaging Kit (Systems Biosciences, Palo Alto, CA, USA), according to the manufacturer's instructions. Viral supernatants were concentrated using the Lenti-X Concentrator Kit (Takara Biosciences USA, San Jose, CA, USA) according to the manufacturer's instructions. Infectious lentivirus titers were determined in HEK293T cells, and cell lines were transduced at a multiplicity of infection of 10 and selected for 2 weeks in puromycin. For co-culture studies, cells were plated in 6-well plates at a density of 100,000 $BT549^{mCherry}$ and 100,000 $iMEC^{GFP}$ cells per well in 4 mL of complete iMEC media and allowed to adhere overnight. The next day, cells were treated with a range of AgNP doses with or without salubrinal. Medium was changed 3 days after treatments, and cells were grown in the absence of AgNPs or salubrinal drugs for 5 days. Cells were trypsinized, washed, diluted to a concentration of 10^6 cells per mL in PBS, and fixed in paraformaldehyde (1% final concentration). The percentage of mCherry (excitation: 561, emission: 610/20) and GFP (excitation: 488, emission: 530/30) expressing cells was quantified by flow cytometry using an LSRFortessa X-20 (BD Biosciences, San Jose, CA, USA), and data were analyzed using BD FACSDiva V6.0 software.

Statistical Analysis: Analysis was performed as described in the figure legends using Prism 9.1.3 software (GraphPad, San Diego, CA, USA). The number of technical and biological replicates for each experiment is included in the figure legends. Synergy was examined using the Chou–Talalay method [38]. Isobolograms at the 50% fraction affected dose (Fa = 0.5) were plotted using Excel.

3. Results

AgNPs do not strongly induce a protective HSR and decrease HSPs in TNBC cells to a greater degree than in mammary epithelial cells.

We verified our previous findings [17,22,36,39] of differences in the response of TNBCs (BT549, MDA-MB-231, and SUM159) and immortalized mammary epithelial cells (iMEC and MCF-10A) to AgNP-induced cytotoxicity. As shown in Figure 1A, the cytotoxic IC_{50} values of AgNPs were 5–10-fold less for TNBCs compared to immortalized mammary epithelial cells. We next investigated the effect AgNPs had on the heat shock response. Cells were treated with doses of AgNPs approximating the mean IC_{50} (18.75 µg/mL), twice the IC_{50} (37.5 µg/mL), and four times the IC_{50} (75 µg/mL) for AgNPs in TNBC cells. We noted dose-dependent decreases in phosphorylated (Serine 326) HSF1 and total HSP70 in both BT549 and MDA-MB-231 cells after AgNP exposure (Figure 1B). AgNP treatment also increased HSP27 phosphorylation in BT549 and MDA-MB-231 cells but decreased total HSP27 in BT549 cells. No change in pHSF1 was observed in AgNP-treated iMEC or MCF10A cells, but HSP70 decreased at high doses. No changes in pHSP27 or total HSP27 were seen in AgNP-treated iMEC or MCF10A cells. Neither HSP90 nor total HSF1 expression changed with AgNP treatment in any cell line.

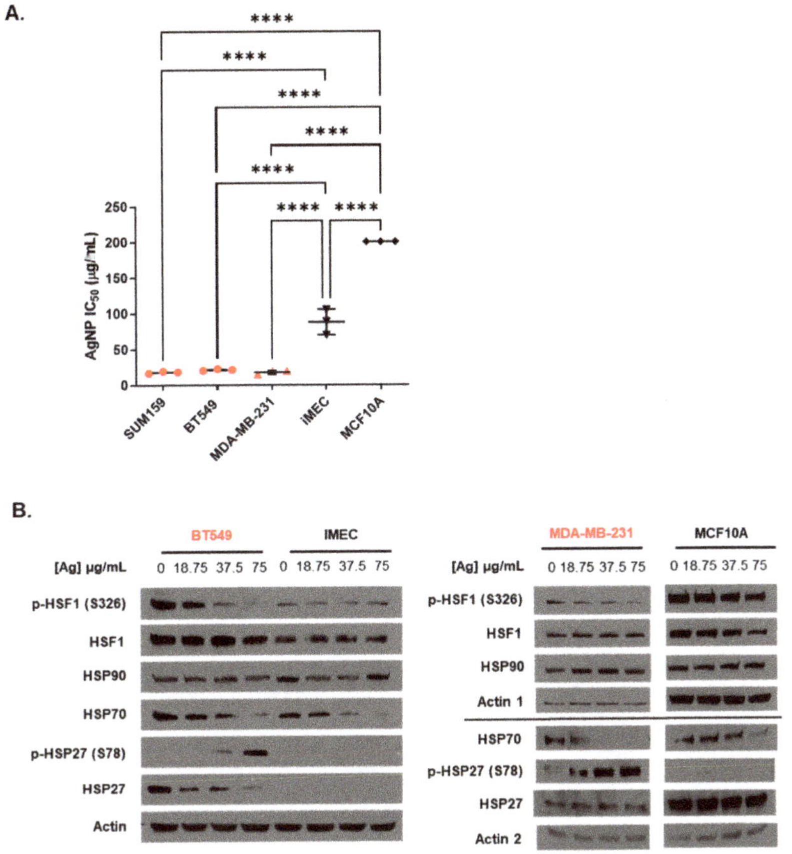

Figure 1. Low doses of AgNPs are more cytotoxic and modulate levels of heat shock proteins to a greater degree in TNBC cells than in immortalized mammary epithelial cells. (**A**) The cytotoxic IC_{50} values calculated from triplicate 10-point dose–response curves following 72 h exposure of each cell line to AgNPs are shown. Data points are in red for TNBC cell lines and in black for immortalized mammary epithelial cells. IC_{50} values for MCF10A cells were in excess of the dose range of AgNPs tested but were at least 200 µg/mL, as indicated in the figure. Statistical analysis was performed by one-way ANOVA followed by the post hoc Tukey test. Significant differences in IC_{50} values are indicated (**** $p < 0.0001$). (**B**) Heat shock protein expression was examined by Western blot 24 h after cells were exposed to AgNPs. Results are representative of duplicate independent experiments.

These studies indicated that AgNPs decreased the levels of multiple proteins involved in the HSR in TNBCs. In contrast, AgNPs had only modest effects on HSP expression in immortalized mammary epithelial cells. To determine if the effects of AgNPs on the heat shock response in TNBCs were caused at the transcriptional level, we transduced BT549 cells with a plasmid vector containing four copies of a heat shock response element (HSE), to which pHSF1 trimers bind and drive transcription of a luciferase reporter gene, indicative of a heat shock response (Figure 2A). Then, 24 h after transduction, cells were treated with AgNPs overnight, exposed to hyperthermic (43 °C) or physiologic temperature (37 °C) conditions, and luciferase activity was quantified 3 h later, as shown schematically in Figure 2B. The results of this experiment are shown in Figure 2C. AgNPs alone did not significantly increase HSE-driven luciferase activity at any of the doses tested. Exposure of cells to hyperthermia significantly increased luciferase activity, and this was reduced by co-treatment with KRIBB11, a selective inhibitor of HSF1, indicating that increased luciferase activity was dependent upon HSF1. Luciferase activity in AgNP-treated BT549 cells that were subsequently exposed to hyperthermia was similar to that of cells that were treated with hyperthermia alone. These data showed that AgNPs did not strongly induce a heat shock response in BT549 cells but also did not inhibit the capacity of cells to induce a heat shock response at a transcriptional level. Overall, our results suggest that AgNPs did not induce a protective HSR in TNBC cells. Decreased expression of HSPs in AgNP-treated cells is not due to transcriptional repression of the HSR and may be due to increased turnover or degradation of HSPs at the protein level.

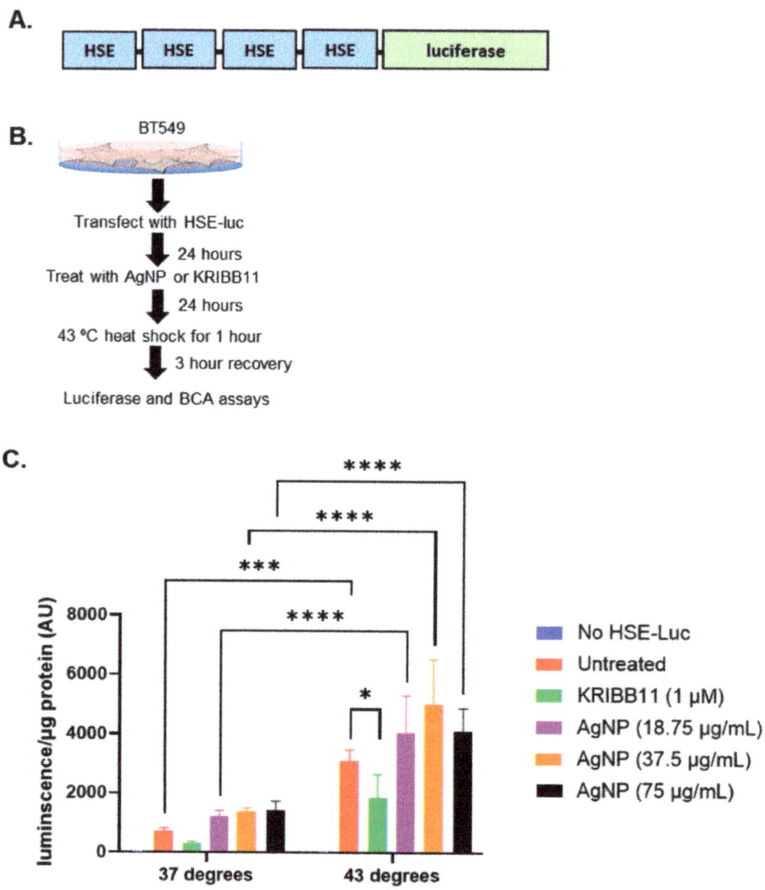

Figure 2. AgNP exposure does not induce a heat shock response at the transcriptional level or inhibit the heat shock response to hyperthermia. (**A**) A schematic is shown of the plasmid vector containing 4 copies of heat shock sequence elements (HSE), which drive transcription of a luciferase reporter that was used for these studies. (**B**) A schematic is shown for the experimental design of experiments to determine the impact of AgNPs on the transcriptional regulation of the heat shock response. (**C**) Luciferase expression indicative of a pHSF1-driven heat shock response is shown for cells treated as described in (**B**). Results are representative of at least four biological replicates per condition and duplicate independent experiments. Statistical analysis was performed by two-way ANOVA followed by the post hoc Tukey test. Significant differences are indicated (**** $p < 0.0001$, *** $p < 0.001$, * $p < 0.05$).

Heat shock inhibitors synergistically increase the cytotoxicity of AgNPs in both TNBC and immortalized mammary epithelial cells.

Next, we asked if the combination of AgNPs and KRIBB11 resulted in a synergistic interaction in treatment efficacy compared to the individual agents. To perform this analysis, cells must be treated with fractions or multiples of a constant dose ratio of the two agents fixed at a 1:1 ratio of IC_{50} for the individual agents, as required by the Chou–Talalay method for drug combination analysis [38]. The IC_{50} for AgNPs for these two cell lines was determined in Figure 1A, and we performed triplicate experiments using similar approaches to determine the IC_{50} of KRIBB11 in BT549 and iMEC cells (Figure 3A). Based upon these experiments, the concentration ratio needed to achieve a 1:1 ratio of the IC_{50} of AgNP (μg/mL):KRIBB11 (μM) was defined for each cell line as 3.5:1 for BT549 cells or 4.15:1 for iMEC cells. We treated cells with the individual agents or the fixed ratio combination of the two agents and plotted the survival curves for each treatment for BT549 (Figure 3B) and iMEC cells (Figure 3C). To the right of the survival curves, we plotted the combination dose of AgNPs and KRIBB11 that achieved a fraction affected (Fa) of 0.5 (indicating 50% loss of cell viability), together with the predicted additive isobole for AgNPs combined with KRIBB11. The data indicated that the combination of AgNPs and KRIBB11 was synergistic because the point representing Fa = 0.5 for the concentrations of concurrently administered AgNP and was below the additive isobole line connecting the IC_{50} of AgNPs and KRIBB11 used as single agents. We further calculated the combination index (CI) relative to Fa = 0.5. CI < 1 denotes a synergistic interaction, CI > 1 represents antagonism, and CI = 1 represents an additive interaction. The CIs for AgNPs combined with KRIBB11 for BT549 cells and iMEC cells were 0.38 and 0.32, respectively, indicating synergistic dose enhancement in both cases. This indicated that inhibition of the HSR can potentially increase both on- and off-target toxicity of AgNPs.

Salubrinal potentiates AgNP toxicity in mesenchymal TNBC without affecting immortalized mammary epithelial cells.

We wanted to know if inhibition of eIF2α dephosphorylation using salubrinal would protect against or enhance AgNP-induced cell death. Initially, we performed studies to determine the cytotoxic IC_{50} of salubrinal alone in TNBC cells (BT549 and SUM159) or immortalized mammary epithelial cells (iMEC and MCF10A). Due to the limited solubility of salubrinal in aqueous media, we were not able to treat cells at a high enough dose to obtain an accurate cytotoxic IC_{50}. Across the dose range tested, we observed no significant differences in the cytotoxicity of salubrinal among the cell lines (Figure 4A). Because we were unable to determine the cytotoxic IC_{50} of salubrinal, we could not perform synergy analysis for the combination of salubrinal and AgNPs. Instead, we combined increasing doses of AgNPs with fixed doses of salubrinal (5 or 10 μM) to determine if this combination affected AgNP cytotoxicity. We found that fixed doses of salubrinal significantly increased the cytotoxicity of AgNPs in BT549 and SUM159 cells but did not substantially increase toxicity of AgNPs in iMEC or MCF10A cells (Figure 4B).

We investigated ISR and apoptosis signaling in BT549 cells treated with combinations of AgNPs and salubrinal for 24 h. We observed increased phosphorylation of eIF2α in BT549 cells exposed to the combination of AgNPs and salubrinal compared to the individual treatments (Figure 5). This is consistent with previous studies showing that salubrinal prevents dephosphorylation of eIF2α, resulting in increased peIF2α when cells are exposed to proteotoxic stressors [40]. Additionally, we found that the combination of salubrinal and AgNPs induced increased expression of downstream effectors of peIF2α-induced cell death, including pro-apoptosis protein CHOP and decreased expression of full-length caspase-3, caspase-7, and caspase-9, which is indicative of caspase cleavage.

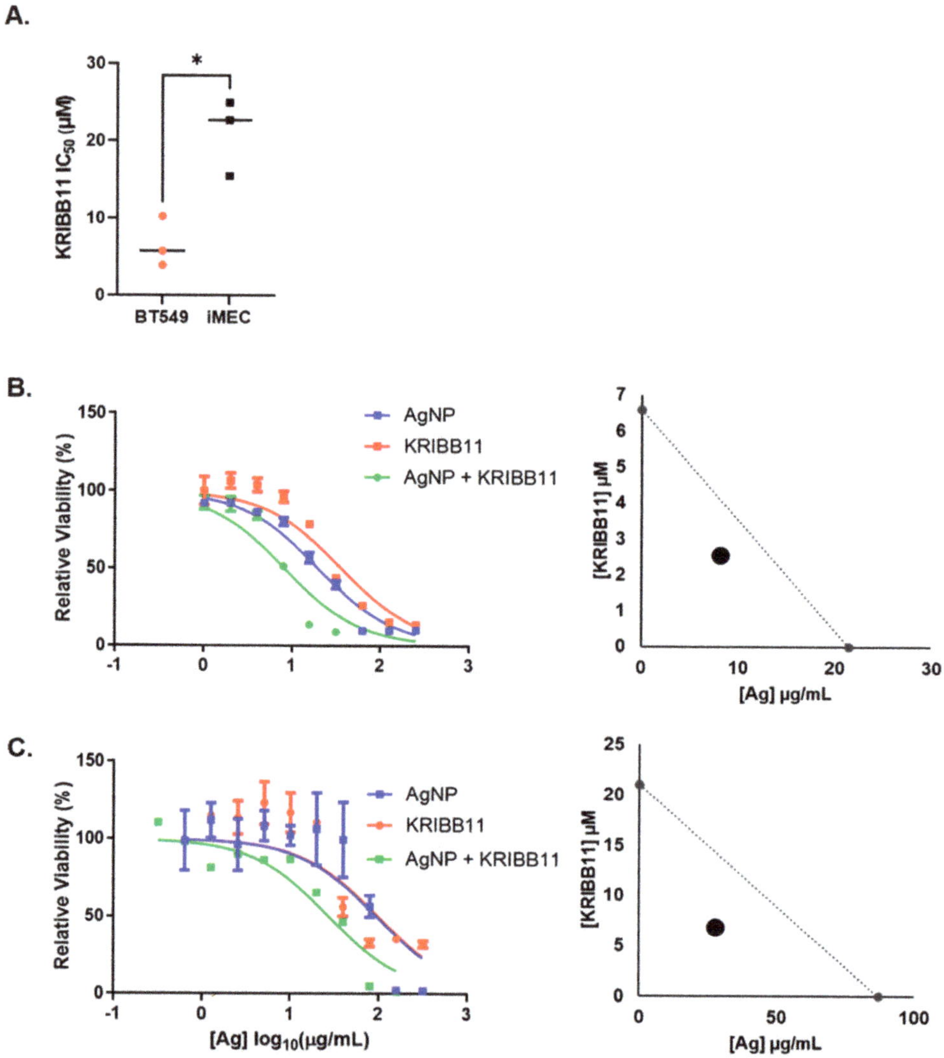

Figure 3. KRIBB11, an HSF1 inhibitor, synergistically increases the cytotoxicity of AgNPs in both TNBC and immortalized mammary epithelial cells. (**A**) The cytotoxic IC_{50} values calculated from triplicate 10-point dose–response curves following 72 h exposure of each cell line to KRIBB11 are shown. Statistical analysis was performed by Student's *t*-test. A significant difference is shown (* $p < 0.05$). (**B**) BT549 or (**C**) iMEC cells were treated with AgNPs, KRIBB11, or AgNPs + KRIBB11 for 72 h and viability was assessed by the MTT assay, as shown. Isobolograms were plotted showing the expected additive effects of AgNPs and KRIBB11, and the dose of the combination that yielded Fa = 0.5 is plotted for each cell line. Results are representative of at least four biological replicates per treatment and two independent experiments per cell line.

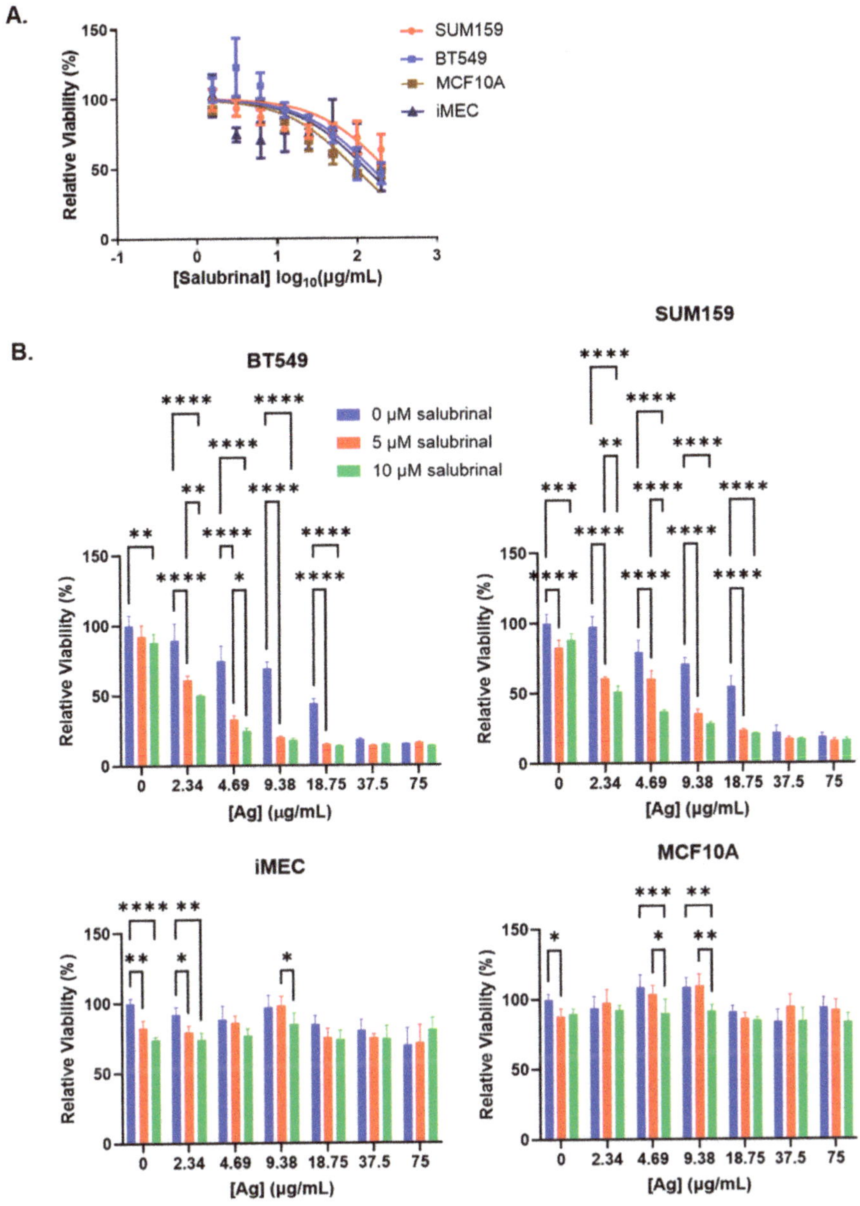

Figure 4. Salubrinal increases the cytotoxicity of AgNPs in TNBC but not in immortalized mammary epithelial cells. (**A**) Cells were treated with increasing doses of salubrinal for 72 h, and viability was assessed by the MTT assay. (**B**) Cells were treated with AgNPs with or without salubrinal for 72 h, and viability was assessed by the MTT assay. Results are representative of at least four biological replicates per condition and duplicate independent experiments. Statistical analysis was performed by two-way ANOVA followed by the post hoc Tukey test. Significant differences are indicated (**** $p < 0.0001$, *** $p < 0.001$, ** $p < 0.01$, and * $p < 0.05$).

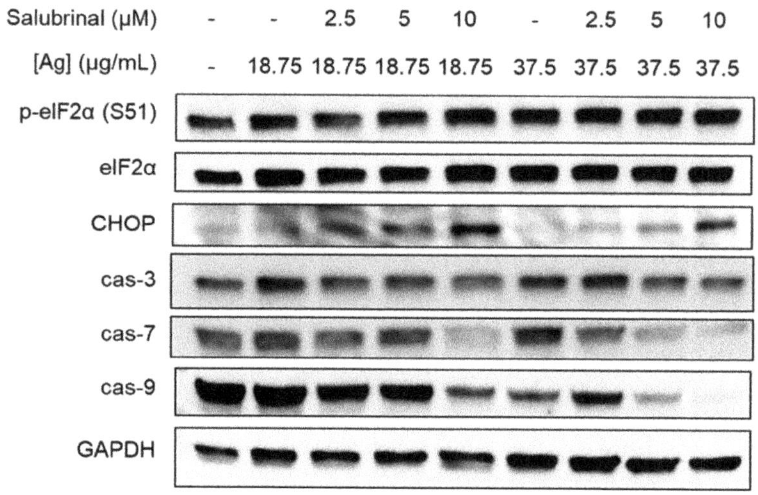

Figure 5. Salubrinal increases AgNP-induced ISR and apoptosis in TNBC cells. BT549 cells were treated with combinations of AgNPs and salubrinal for 24 h, and ISR and apoptosis markers were examined by Western blot. Results are representative of duplicate independent experiments.

These results implied that the combination of AgNPs and salubrinal could be effective for eliminating TNBCs without damaging normal breast cells. To assess this directly, we performed a co-culture experiment in which BT549 and iMEC cells were grown together in a single well and treated with AgNPs, salubrinal, or a combination of the two. To distinguish between cell types, BT549 cells were transduced using a lentivirus to express mCherry, and iMEC cells were transduced to express eGFP. Cells were grown in co-culture at a 1:1 ratio, as shown in Figure 6A. The cells were treated with AgNPs, salubrinal, or a combination of AgNPs and salubrinal. The proportion of $iMEC^{GFP}$ cells and $BT549^{mCherry}$ cells that survived treatment was determined by flow cytometry, as shown in Figure 6B. In agreement with data obtained from monoculture studies, AgNPs alone decreased the percentage of $BT549^{mCherry}$ cells in the co-culture, resulting in a concomitant increase in the percent of $iMEC^{GFP}$ cells present. Salubrinal on its own had little effect, but the combination of AgNPs and salubrinal almost completely eliminated $BT549^{mCherry}$ from the co-culture, while preserving $iMEC^{GFP}$ cells. Quantification of results from triplicate experiments indicated that the surviving population was significantly depleted of $BT549^{mCherry}$ cells (resulting in an increased fraction of $iMEC^{GFP}$ cells) following AgNP exposure, and this effect was further enhanced when AgNPs were combined with salubrinal, as shown in Figure 6C. A portion (approximately 15%) of each cell line had low-level expression or did not express either fluorophore (seen in the lower left quadrant of each panel in the graphs in Figure 6B) and, therefore, the total percentage of cells quantified did not add to 100%. The results indicate that AgNPs alone selectively killed TNBC cells without affecting the growth of immortalized mammary epithelial cells grown together in co-culture, and this selectivity was enhanced when AgNPs were combined with salubrinal.

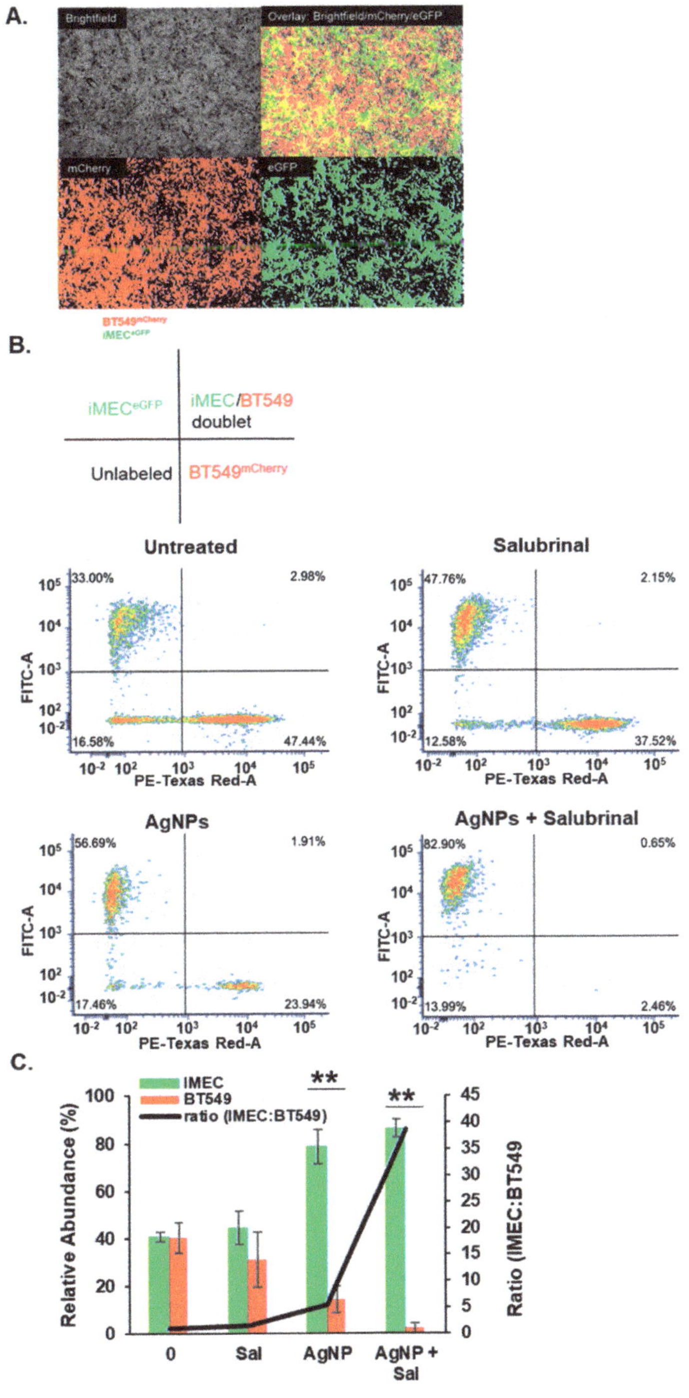

Figure 6. The combination of AgNPs and salubrinal selectively eliminates TNBC cells grown in co-culture with immortalized mammary epithelial cells. (**A**) BT549^{mCherry} (red) and iMEC^{eGFP} cells

(green) were grown in co-culture and imaged using fluorescent microscopy. Representative images from the individual visible, red, and green light channels, and an overlay of the three channels, are shown for untreated cells. (**B**) BT549^{mCherry} and iMEC^{eGFP} cells grown in co-culture were treated with salubrinal alone (10 μM), AgNPs alone (37.5 μg/mL), or a combination of 10 μM salubrinal and 37.5 μg/mL for 8 days. Flow cytometry was used to quantify the relative fraction of each cell type remaining at the end of treatment. A schematic is shown to indicate the quadrants of the scattergrams below. The results are representative of triplicate independent experiments. (**C**) Data from triplicate experiments performed, as shown in (**B**), were quantified to show the mean percentage of iMEC^{eGFP} or BT549^{mCherry} cells (green and red bars) and the ratio of iMEC^{eGFP}:BT549^{mCherry} cells (black line) detected after each treatment. Statistical analysis was performed by two-way ANOVA followed by the post hoc Tukey test. Significant differences are indicated (** $p < 0.01$).

4. Discussion

AgNPs have enormous potential for clinical applications [11,17,20,22,36,39], but fears of toxicity have limited translation [41–47]. Our previous studies showed that AgNPs induced proteotoxicity and cell death in TNBC cells at doses that did not affect the homeostasis of normal breast epithelial cells in vitro [17,36] or in vivo [22]. Two key cell stress responses involved in mitigating proteotoxicity are the HSR and the ISR, but few studies have examined the roles of the HSR and ISR in mitigating or enhancing toxicities due to AgNP exposure. Here, we showed that the use of drugs that modulate these pathways can enhance the proteotoxic effects of AgNPs, and we identified combinations that are lethal to TNBC cells without increasing off-target toxicity on normal mammary epithelial cells.

Phosphorylation of HSF1 is indicative of HSR activation. Surprisingly, we observed decreased levels of pHSF1 in TNBC cells exposed to AgNPs, but no change in pHSF1 in immortalized mammary epithelial cells. We investigated if AgNPs affected the ability of HSF1 to transcriptionally activate the HSR. AgNPs alone did not significantly induce HSF1-dependent transcription, and AgNP-treated cells were capable of generating a robust HSF1-dependent transcriptional response after exposure to hyperthermia (43 °C). This indicated that AgNPs did not inhibit heat-induced transcriptional activity driven by pHSF1. Thus, our results suggested that decreased pHSF1 protein following AgNP exposure was due to increased turnover or degradation, possibly by a negative feedback inhibition loop [25,29]. We also observed decreased expression of HSP70 in both TNBC and immortalized mammary epithelial cells treated with AgNPs. This is in contrast to literature findings in other cell lines, indicating increased HSP70 following AgNP exposure [41,44,48–50]. It is unclear what factors drive these differences, but we have shown that Ag^+ contamination in AgNP solutions can change physiological responses [36]. As discussed below, studies that fail to characterize Ag^+ content or use AgNP solutions that contain Ag^+ ions may be unreliable. HSP70 plays a key role in sequestering misfolded proteins, aids in delivering them to proteasomes, autophagosomes, and lysosomes, and may be degraded along with damaged proteins as cells attempt to recover from proteotoxic stress [51]. We also observed decreased total HSP27 in TNBCs treated with AgNPs. Similar to HSP70, HSP27 can increase the degradation of damaged proteins in stressed cells through the proteosomal pathway and is degraded in the process [52]. In contrast to total HSP27, we observed increased pHSP27 in AgNP-treated TNBC cells. Non-phosphorylated HSP27 forms long oligomers that mediate most of its chaperone functions. HSP27 phosphorylation leads to a significant decrease of the oligomeric size and reduces chaperone action. The smaller size of pHSP27 enables nuclear translocation of the cytosolic pool of HSP27, and this provides a mechanism for HSP27 to leave the cytosol and enter the nucleus, where it can be dephosphorylated and form oligomers with full chaperone capacity [53]. Therefore, a plausible hypothesis accounting for why AgNPs decrease total HSP27 and concomitantly increase pHSP27 expression is that pHSP27 increases in stressed cells but does not participate in chaperone functions and, therefore, does not enter protein degradation pathways.

In an effort to discover drug partners for AgNPs to enhance on-target and lower potential off-target toxic effects, we examined combinations of AgNPs with drugs that interfere with proteotoxic stress responses. We found that the HSF1 inhibitor, KRIBB11, synergistically increased the cytotoxicity of AgNPs in both TNBCs and immortalized mammary epithelial cells. These data indicated that the HSPs play a key role in mitigating AgNP toxicity, and inhibition of the HSR may increase off-target toxicity of AgNPs used for cancer therapy. We next tested salubrinal, an inhibitor of the eIF2α phosphatase complex [54], for its interaction with AgNPs. Inhibition of dephosphorylation of eIF2α using salubrinal enhanced pro-death ISR signaling and increased toxicity following AgNP treatment in TNBCs, but it had no effect on AgNP toxicity in immortalized mammary epithelial cells. When TNBC and immortalized mammary epithelial cells were treated in co-culture with a combination of salubrinal and AgNPs, the combination was substantially more toxic to TNBC cells than either AgNPs or salubrinal alone, but it had little effect on immortalized mammary epithelial cells. This is consistent with previous studies demonstrating that AgNP exposure increases peIF2α in TNBC cells but not in immortalized mammary epithelial cells [17,20,22,36]. Because ISR signaling is not activated by AgNPs in immortalized mammary epithelial cells, salubrinal cannot prolong ISR activation and drive cell death. Salubrinal has not been tested in humans, and further research will be needed to determine its translational potential. Despite this limitation, our results provide support to investigate the use of a combination of AgNPs and drugs that can sustain ISR signaling for treatment of TNBC.

Four distinct stress-sensing kinases (PKR, HRI, GCN2, and PERK) can initiate the ISR through phosphorylation of eIF2α [31]. In our study, we did not specifically determine which of these kinases drives the ISR in response to AgNPs. Previous studies linked AgNP exposure to activation of endoplasmic reticulum (ER) stress, as indicated by the unfolded protein response (UPR). Briefly summarized, the UPR is initiated by three ER-bound stress-sensing proteins, PERK, activating transcription factor 6α/β (ATF6), and inositol requiring enzyme 1α/β (IRE1). In unstressed states, the protein chaperone GRP78 binds and inhibits PERK, IRE1, and ATF6 on the ER membrane [55]. If the load of unfolded proteins is greater than the capacity of ER protein chaperones to sequester and fold, GRP78 dissociates from PERK, IRE1, and ATF6 to increase the folding capacity. This enables PERK and IRE1 to form homodimers and trans-autophosphorylate to initiate downstream signaling cascades, and frees ATF6 to translocate to the Golgi, where it is cleaved to its active form [56–58]. Although some studies showed that all three arms of the UPR are activated in retinal pigment cells [59] or in Chang liver and human lung fibroblasts [60] following AgNP exposure, other studies, including our own previous work [17,22,36], did not demonstrate activation of all arms of the UPR. In one study, AgNPs increased CHOP expression in drug-resistant breast cancer cells, but PERK dependence was not examined, nor was activation of the other arms of the UPR [19]. In another study, AgNP exposure increased CHOP and total IREα, but no evidence of phosphorylation of PERK, phosphorylation of IREα, cleavage of ATF6, or activation of other downstream signaling cascades was shown [61]. Other studies offer conflicting evidence with regard to activation of the UPR by AgNPs. For example, Simmard et al. showed that AgNPs increased pPERK; however, treatments that induced the highest levels of peIF2a and its downstream effectors ATF4 and CHOP did not correlate with treatments that induced the highest levels of pPERK, possibly indicating other drivers of eIF2a phosphorylation [20]. In the same report, the effects of AgNPs on pIRE were heterogeneous and were not dose-dependent. ATF6 cleavage was not examined, but in subsequent studies, they observed that AgNPs increased degradation of ATF6 [21]. We previously performed proteomic analysis in lung cancer cells with differing sensitivities to AgNPs [45]. We found that in AgNP-sensitive lung cancer cells, but not in AgNP-insensitive ones, there was a substantial decrease in overall protein synthesis following AgNP exposure. This is consistent with phosphorylation of eIF2a, which blocks translation of most mRNAs, leading to a global decrease in protein synthesis. The most affected signaling pathways in AgNP-sensitive cells were protein synthesis (including decreased activity of eIF2 and eIF4 pathways, both of which are critical for protein synthesis), protein stability, ER signaling, cell cycle, and autophagy/mitophagy. These responses are associated with mitigating proteotoxicity and, therefore, these data provide support for proteotoxic stress as a key feature of AgNP toxicity. In our current study, we could not definitively state that all three arms of the UPR were active, nor did previous studies clearly demonstrate dependence of AgNP-induced cell death on UPR signaling. Given the heterogeneity of the responses, the only consistently identified pathway is increased peIF2A and downstream effectors, including CHOP. Therefore, we conclude that the ISR, with some degree of overlap with the UPR, is a key proteotoxic stress response to AgNP exposure.

Although proteotoxic stress increased following AgNP exposure, the precise biological features that render some cells more sensitive to AgNPs than others remain to be determined. One hypothesis is that high baseline stress on the proteostasis machinery due to synthesis, folding, posttranslational modification, and secretion of ECM proteins in TNBC cells that contributes to their increased sensitivity to small molecules that induce ER stress [4,5] could also drive sensitivity to AgNPs. Alternatively, differences in the vesicle acidification rate leading to increased intracellular release of Ag^+ could play a role [62]. Differences in AgNP uptake may be a contributing factor, but our previous studies and those of others showed that the relative sensitivity of a specific cell line to AgNPs does not necessarily correlate with the mass of AgNPs taken up by the cells, indicating that other factors dominate [17,47].

There is considerable debate regarding the potential toxicity of AgNPs used for clinical applications. Failure to separate dissolved silver cations (Ag^+) from AgNPs before toxicity testing likely contributes to the lack of a definitive answer. Ag^+ is highly toxic and has a distinct cytotoxic mechanism of action compared to AgNPs; specifically, AgNPs induce proteotoxicity by a mechanism that does not depend on oxidative stress, while Ag^+ cytotoxicity is dependent upon oxidative stress [36]. Many studies that attempted to characterize the toxicity of AgNPs failed to account for Ag^+, making it difficult to generalize their findings [47]. Estimates from various sources indicate that Ag^+ levels in AgNP dispersions vary from 2.6 to 5.9% to greater than 70% in the majority of lab-made and commercially produced AgNPs tested [41,63]. Not all cells are equally sensitive to AgNPs or Ag^+, and sensitivity to one does not necessarily correlate with sensitivity to the other [17,36]. Contamination of AgNP suspensions with as little as 1% mass fraction of Ag^+ alters the toxicity profile of AgNPs [36]. Importantly, we previously showed that the AgNPs used for the experiments presented here contained less than 0.001% Ag^+ by mass, even after 1 month of storage [36]. Another issue clouding the AgNP toxicity debate is that in vivo testing may use doses that are either sub-therapeutic (too low) or excessively high. Studies using PVP-coated AgNPs for treatment of cancer [22] or viral diseases [64] showed that doses in the range of 4–6 mg/kg delivered systemically for multiple weeks induced therapeutic responses, indicating that this range is reasonable for evaluation of the potential toxic effects of AgNPs. Direct comparison of the toxicity of citrate-coated AuNPs (10 mg/kg intravenous (IV) weekly for 8 weeks) and citrate-coated AgNPs (5 mg/kg IV weekly for 8 weeks), which roughly corresponds to a 1:1 ratio of silver and gold atoms, indicated that neither AuNPs nor AgNPs elicited any overt toxicity, though tissue discoloration and enlarged spleens were apparent [65]. Rats receiving intravenous injections of citrate-coated AgNPs (6 mg/kg for 28 consecutive days) exhibited no dose-limiting toxicities, though transient effects on liver and immune cell function were noted [66]. Metabolomic studies in mice injected with polyethylene-glycol-coated AgNPs (8 mg/kg IV) found no evidence of dose-limiting toxicity, but modest effects on liver function were observed [67]. These effects were transient and not indicative of persistent liver injury. Studies in humans receiving intravenous AgNPs are limited. However, a recent clinical trial described the use of AgNPs for prevention of severe illness and death due to COVID-19 [68]. Here, 40 patients were injected intravenously with 1.8 mg of AgNPs for 3 consecutive days (combined with standard COVID-19 treatments), and the group receiving AgNPs had significantly greater survival rates and spent fewer days on supplemental oxygen than those who did not receive AgNPs. Importantly, all of the in vivo studies described above used AgNPs with minimal contamination of Ag^+. Taken together, these studies indicate that within the therapeutic dose range, AgNPs exhibit low off-target toxicity in vivo.

Overall, our findings provide insight into the mechanism by which AgNPs selectively kill TNBCS. Our data implied that HSPs and ISR are essential for mitigating AgNP-induced damage, pointing to accumulation of misfolded proteins as a driver of AgNP-induced cell death. At present, AgNPs, KRIBB11, and salubrinal are not used clinically, and additional studies are needed to determine their safety and utility for cancer therapy. Research defining the drivers of sensitivity to AgNPs will enable selection of cancer patients who will benefit most from AgNP exposure and will help guide future efforts to identify potential synergistic drug partners that would be beneficial for use with AgNPs for cancer therapy.

Author Contributions: All authors contributed substantially to the work, reviewed and edited the manuscript, and approved the final manuscript. Specific contributions include the following: conceptualization and design, C.M.S., B.M., K.P., C.D.F., M.M.R. and R.N.S.; formal analysis, C.M.S., B.M., K.P., C.D.F., M.M.R. and R.N.S.; resources, R.C. and R.N.S.; initial drafting, C.M.S., B.M., K.P. and R.N.S. All authors have read and agreed to the published version of the manuscript.

Funding: The research in this work was supported by NIH/NCI R01CA207222. Support for MMR was provided in part by NIH/NIGMS T32GM127261. Assistance provided by the Atrium Health Wake Forest Baptist Comprehensive Cancer Center (AHWFUCCC) Cell Engineering and Flow Cytometry Shared Resources was supported in part by NIH/NCI CCSG P30CA012197. The content is solely the responsibility of the authors and does not necessarily represent the official views of the NIH or NCI.

Data Availability Statement: Raw data is available upon request to the senior author, R.N.S.

Conflicts of Interest: The authors declare no conflicts of interest.

References

1. Brancolini, C.; Iuliano, L. Proteotoxic Stress and Cell Death in Cancer Cells. *Cancers* **2020**, *12*, 2385. [CrossRef] [PubMed]
2. Ho Zhi Guang, M.; Kavanagh, E.L.; Dunne, L.P.; Dowling, P.; Zhang, L.; Lindsay, S.; Bazou, D.; Goh, C.Y.; Hanley, C.; Bianchi, G.; et al. Targeting Proteotoxic Stress in Cancer: A Review of the Role that Protein Quality Control Pathways Play in Oncogenesis. *Cancers* **2019**, *11*, 66. [CrossRef] [PubMed]
3. Sannino, S.; Brodsky, J.L. Targeting protein quality control pathways in breast cancer. *BMC Biol.* **2017**, *15*, 109. [CrossRef]
4. Feng, Y.X.; Sokol, E.S.; Del Vecchio, C.A.; Sanduja, S.; Claessen, J.H.; Proia, T.A.; Jin, D.X.; Reinhardt, F.; Ploegh, H.L.; Wang, Q.; et al. Epithelial-to-Mesenchymal Transition Activates PERK-eIF2 alpha and Sensitizes Cells to Endoplasmic Reticulum Stress. *Cancer Discov.* **2014**, *4*, 702–715. [CrossRef]
5. Feng, Y.X.; Jin, D.X.; Sokol, E.S.; Reinhardt, F.; Miller, D.H.; Gupta, P.B. Cancer-specific PERK signaling drives invasion and metastasis through CREB3L1. *Nat. Commun.* **2017**, *8*, 1079. [CrossRef] [PubMed]
6. Jaskulska, A.; Janecka, A.E.; Gach-Janczak, K. Thapsigargin—From Traditional Medicine to Anticancer Drug. *Int. J. Mol. Sci.* **2020**, *22*, 4. [CrossRef]
7. Banerjee, S.; Ansari, A.A.; Upadhyay, S.P.; Mettman, D.J.; Hibdon, J.R.; Quadir, M.; Ghosh, P.; Kambhampati, A.; Banerjee, S.K. Benefits and Pitfalls of a Glycosylation Inhibitor Tunicamycin in the Therapeutic Implication of Cancers. *Cells* **2024**, *13*, 395. [CrossRef]
8. Shen, S.; Du, X.J.; Liu, J.; Sun, R.; Zhu, Y.H.; Wang, J. Delivery of bortezomib with nanoparticles for basal-like triple-negative breast cancer therapy. *J. Control. Release* **2015**, *208*, 14–24. [CrossRef]
9. Anchordoquy, T.; Artzi, N.; Balyasnikova, I.V.; Barenholz, Y.; La-Beck, N.M.; Brenner, J.S.; Chan, W.C.W.; Decuzzi, P.; Exner, A.A.; Gabizon, A.; et al. Mechanisms and Barriers in Nanomedicine: Progress in the Field and Future Directions. *ACS Nano* **2024**, *18*, 13983–13999. [CrossRef]
10. Barenholz, Y. Doxil(R)—The first FDA-approved nano-drug: Lessons learned. *J. Control. Release* **2012**, *160*, 117–134. [CrossRef]
11. Gawel, A.M.; Singh, R.; Debinski, W. Metal-Based Nanostructured Therapeutic Strategies for Glioblastoma Treatment—An Update. *Biomedicines* **2022**, *10*, 1598. [CrossRef] [PubMed]
12. Sen, P.; Saha, M.; Ghosh, S.S. Nanoparticle mediated alteration of EMT dynamics: An approach to modulate cancer therapeutics. *Mater. Adv.* **2020**, *1*, 2614–2630. [CrossRef]
13. Arvizo, R.R.; Saha, S.; Wang, E.; Robertson, J.D.; Bhattacharya, R.; Mukherjee, P. Inhibition of tumor growth and metastasis by a self-therapeutic nanoparticle. *Proc. Natl. Acad. Sci. USA* **2013**, *110*, 6700–6705. [CrossRef] [PubMed]
14. Li, X.J.; Song, L.; Hu, X.; Liu, C.; Shi, J.; Wang, H.; Zhan, L.; Song, H. Inhibition of Epithelial-Mesenchymal Transition and Tissue Regeneration by Waterborne Titanium Dioxide Nanoparticles. *ACS Appl. Mater. Interfaces* **2018**, *10*, 3449–3458. [CrossRef]

15. Liu, Y.; Chen, C.; Qian, P.; Lu, X.; Sun, B.; Zhang, X.; Wang, L.; Gao, X.; Li, H.; Chen, Z. Gd-metallofullerenol nanomaterial as non-toxic breast cancer stem cell-specific inhibitor. *Nat. Commun.* **2015**, *6*, 5988. [CrossRef]
16. Seaberg, J.; Clegg, J.R.; Bhattacharya, R.; Mukherjee, P. Self-Therapeutic Nanomaterials: Applications in Biology and Medicine. *Mater. Today* **2023**, *62*, 190–224. [CrossRef] [PubMed]
17. Snyder, C.M.; Rohde, M.M.; Fahrenholtz, C.D.; Swanner, J.; Sloop, J.; Donati, G.L.; Furdui, C.M.; Singh, R. Low Doses of Silver Nanoparticles Selectively Induce Lipid Peroxidation and Proteotoxic Stress in Mesenchymal Subtypes of Triple-Negative Breast Cancer. *Cancers* **2021**, *13*, 4217. [CrossRef]
18. Fageria, L.; Bambroo, V.; Mathew, A.; Mukherjee, S.; Chowdhury, R.; Pande, S. Functional Autophagic Flux Regulates AgNP Uptake and the Internalized Nanoparticles Determine Tumor Cell Fate by Temporally Regulating Flux. *Int. J. Nanomed.* **2019**, *14*, 9063–9076. [CrossRef]
19. Gopisetty, M.K.; Kovács, D.; Igaz, N.; Rónavári, A.; Bélteky, P.; Rázga, Z.; Venglovecz, V.; Csoboz, B.; Boros, I.M.; Kónya, Z.; et al. Endoplasmic reticulum stress: Major player in size-dependent inhibition of P-glycoprotein by silver nanoparticles in multidrug-resistant breast cancer cells. *J. Nanobiotechnol.* **2019**, *17*, 9. [CrossRef]
20. Simard, J.C.; Durocher, I.; Girard, D. Silver nanoparticles induce irremediable endoplasmic reticulum stress leading to unfolded protein response dependent apoptosis in breast cancer cells. *Apoptosis* **2016**, *21*, 1279–1290. [CrossRef]
21. Simard, J.C.; Vallieres, F.; de Liz, R.; Lavastre, V.; Girard, D. Silver nanoparticles induce degradation of the endoplasmic reticulum stress sensor activating transcription factor-6 leading to activation of the NLRP-3 inflammasome. *J. Biol. Chem.* **2015**, *290*, 5926–5939. [CrossRef] [PubMed]
22. Swanner, J.; Fahrenholtz, C.D.; Tenvooren, I.; Bernish, B.W.; Sears, J.J.; Hooker, A.; Furdui, C.M.; Alli, E.; Li, W.; Donati, G.L.; et al. Silver nanoparticles selectively treat triple-negative breast cancer cells without affecting non-malignant breast epithelial cells in vitro and in vivo. *FASEB Bioadv.* **2019**, *1*, 639–660. [CrossRef] [PubMed]
23. Arvizo, R.R.; Bhattacharyya, S.; Kudgus, R.A.; Giri, K.; Bhattacharya, R.; Mukherjee, P. Intrinsic therapeutic applications of noble metal nanoparticles: Past, present and future. *Chem. Soc. Rev.* **2012**, *41*, 2943–2970. [CrossRef] [PubMed]
24. Shevtsov, M.; Balogi, Z.; Khachatryan, W.; Gao, H.; Vígh, L.; Multhoff, G. Membrane-Associated Heat Shock Proteins in Oncology: From Basic Research to New Theranostic Targets. *Cells* **2020**, *9*, 1263. [CrossRef]
25. Wang, X.; Chen, M.; Zhou, J.; Zhang, X. HSP27, 70 and 90, anti-apoptotic proteins, in clinical cancer therapy (Review). *Int. J. Oncol.* **2014**, *45*, 18–30. [CrossRef]
26. Green, M.; Schuetz, T.J.; Sullivan, E.K.; Kingston, R.E. A heat shock-responsive domain of human HSF1 that regulates transcription activation domain function. *Mol. Cell. Biol.* **1995**, *15*, 3354–3362. [CrossRef] [PubMed]
27. Ali, A.; Bharadwaj, S.; O'Carroll, R.; Ovsenek, N. HSP90 interacts with and regulates the activity of heat shock factor 1 in Xenopus oocytes. *Mol. Cell. Biol.* **1998**, *18*, 4949–4960. [CrossRef]
28. Krakowiak, J.; Zheng, X.; Patel, N.; Feder, Z.A.; Anandhakumar, J.; Valerius, K.; Gross, D.S.; Khalil, A.S.; Pincus, D. Hsf1 and Hsp70 constitute a two-component feedback loop that regulates the yeast heat shock response. *eLife* **2018**, *7*, e31668. [CrossRef]
29. Brunet Simioni, M.; De Thonel, A.; Hammann, A.; Joly, A.L.; Bossis, G.; Fourmaux, E.; Bouchot, A.; Landry, J.; Piechaczyk, M.; Garrido, C. Heat shock protein 27 is involved in SUMO-2/3 modification of heat shock factor 1 and thereby modulates the transcription factor activity. *Oncogene* **2009**, *28*, 3332–3344. [CrossRef]
30. Ciocca, D.R.; Arrigo, A.P.; Calderwood, S.K. Heat shock proteins and heat shock factor 1 in carcinogenesis and tumor development: An update. *Arch. Toxicol.* **2013**, *87*, 19–48. [CrossRef]
31. Pakos-Zebrucka, K.; Koryga, I.; Mnich, K.; Ljujic, M.; Samali, A.; Gorman, A.M. The integrated stress response. *EMBO Rep.* **2016**, *17*, 1374–1395. [CrossRef] [PubMed]
32. Gao, B.; Zhang, X.Y.; Han, R.; Zhang, T.T.; Chen, C.; Qin, Z.H.; Sheng, R. The endoplasmic reticulum stress inhibitor salubrinal inhibits the activation of autophagy and neuroprotection induced by brain ischemic preconditioning. *Acta Pharmacol. Sin.* **2013**, *34*, 657–666. [CrossRef] [PubMed]
33. Li, R.J.; He, K.L.; Li, X.; Wang, L.L.; Liu, C.L.; He, Y.Y. Salubrinal protects cardiomyocytes against apoptosis in a rat myocardial infarction model via suppressing the dephosphorylation of eukaryotic translation initiation factor 2α. *Mol. Med. Rep.* **2015**, *12*, 1043–1049. [CrossRef]
34. Matsuoka, M.; Komoike, Y. Experimental Evidence Shows Salubrinal, an eIF2α Dephosphorylation Inhibitor, Reduces Xenotoxicant-Induced Cellular Damage. *Int. J. Mol. Sci.* **2015**, *16*, 16275–16287. [CrossRef]
35. Wu, C.T.; Sheu, M.L.; Tsai, K.S.; Chiang, C.K.; Liu, S.H. Salubrinal, an eIF2α dephosphorylation inhibitor, enhances cisplatin-induced oxidative stress and nephrotoxicity in a mouse model. *Free Radic. Biol. Med.* **2011**, *51*, 671–680. [CrossRef]
36. Rohde, M.M.; Snyder, C.M.; Sloop, J.; Solst, S.R.; Donati, G.L.; Spitz, D.R.; Furdui, C.M.; Singh, R. The mechanism of cell death induced by silver nanoparticles is distinct from silver cations. *Part. Fibre Toxicol.* **2021**, *18*, 37. [CrossRef]
37. Campeau, E.; Ruhl, V.E.; Rodier, F.; Smith, C.L.; Rahmberg, B.L.; Fuss, J.O.; Campisi, J.; Yaswen, P.; Cooper, P.K.; Kaufman, P.D. A versatile viral system for expression and depletion of proteins in mammalian cells. *PLoS ONE* **2009**, *4*, e6529. [CrossRef] [PubMed]
38. Chou, T.C. Drug combination studies and their synergy quantification using the Chou-Talalay method. *Cancer Res.* **2010**, *70*, 440–446. [CrossRef]
39. Swanner, J.; Mims, J.; Carroll, D.L.; Akman, S.A.; Furdui, C.M.; Torti, S.V.; Singh, R.N. Differential cytotoxic and radiosensitizing effects of silver nanoparticles on triple-negative breast cancer and non-triple-negative breast cells. *Int. J. Nanomed.* **2015**, *10*, 3937–3953.

40. Tsaytler, P.; Bertolotti, A. Exploiting the selectivity of protein phosphatase 1 for pharmacological intervention. *FEBS J.* **2013**, *280*, 766–770. [CrossRef] [PubMed]
41. Beer, C.; Foldbjerg, R.; Hayashi, Y.; Sutherland, D.S.; Autrup, H. Toxicity of silver nanoparticles—Nanoparticle or silver ion? *Toxicol. Lett.* **2012**, *208*, 286–292. [CrossRef] [PubMed]
42. Falconer, J.L.; Grainger, D.W. In vivo comparisons of silver nanoparticle and silver ion transport after intranasal delivery in mice. *J. Control. Release* **2018**, *269*, 1–9. [CrossRef] [PubMed]
43. Gao, X.; Topping, V.D.; Keltner, Z.; Sprando, R.L.; Yourick, J.J. Toxicity of nano- and ionic silver to embryonic stem cells: A comparative toxicogenomic study. *J. Nanobiotechnol.* **2017**, *15*, 31. [CrossRef] [PubMed]
44. Ho, C.-C.; Huang, L.-J.; Yang, R.-C. Silver nanoparticles induces heat shock response and provides an anti-inflammatory effect in Clone 9 cells. *FASEB J.* **2013**, *27*, lb708. [CrossRef]
45. Holmila, R.; Wu, H.; Lee, J.; Tsang, A.W.; Singh, R.; Furdui, C.M. Integrated Redox Proteomic Analysis Highlights New Mechanisms of Sensitivity to Silver Nanoparticles. *Molelular Cell. Proteom.* **2021**, 100073. [CrossRef]
46. Holmila, R.J.; Vance, S.A.; King, S.B.; Tsang, A.W.; Singh, R.; Furdui, C.M. Silver Nanoparticles Induce Mitochondrial Protein Oxidation in Lung Cells Impacting Cell Cycle and Proliferation. *Antioxidants* **2019**, *8*, 552. [CrossRef]
47. Smith, J.N.; Thomas, D.G.; Jolley, H.; Kodali, V.K.; Littke, M.H.; Munusamy, P.; Baer, D.R.; Gaffrey, M.J.; Thrall, B.D.; Teeguarden, J.G. All that is silver is not toxic: Silver ion and particle kinetics reveals the role of silver ion aging and dosimetry on the toxicity of silver nanoparticles. *Part. Fibre Toxicol.* **2018**, *15*, 47. [CrossRef]
48. Ahamed, M.; Posgai, R.; Gorey, T.J.; Nielsen, M.; Hussain, S.M.; Rowe, J.J. Silver nanoparticles induced heat shock protein 70, oxidative stress and apoptosis in Drosophila melanogaster. *Toxicol. Appl. Pharmacol.* **2010**, *242*, 263–269. [CrossRef]
49. Tsai, T.N.; Lee, T.Y.; Liu, M.S.; Ho, J.J.; Huang, L.J.; Liu, C.J.; Chen, T.J.; Yang, R.C. Nonlethal dose of silver nanoparticles attenuates TNF-alpha-induced hepatic epithelial cell death through HSP70 overexpression. *Am. J. Physiol.-Cell Physiol.* **2015**, *308*, C959–C963. [CrossRef]
50. Xin, L.; Wang, J.; Wu, Y.; Guo, S.; Tong, J. Increased oxidative stress and activated heat shock proteins in human cell lines by silver nanoparticles. *Hum. Exp. Toxicol.* **2015**, *34*, 315–323. [CrossRef]
51. Alagar Boopathy, L.R.; Jacob-Tomas, S.; Alecki, C.; Vera, M. Mechanisms tailoring the expression of heat shock proteins to proteostasis challenges. *J. Biol. Chem.* **2022**, *298*, 101796. [CrossRef] [PubMed]
52. Parcellier, A.; Brunet, M.; Schmitt, E.; Col, E.; Didelot, C.; Hammann, A.; Nakayama, K.; Nakayama, K.I.; Khochbin, S.; Solary, E.; et al. HSP27 favors ubiquitination and proteasomal degradation of p27Kip1 and helps S-phase re-entry in stressed cells. *FASEB J.* **2006**, *20*, 1179–1181. [CrossRef] [PubMed]
53. Bryantsev, A.L.; Kurchashova, S.Y.; Golyshev, S.A.; Polyakov, V.Y.; Wunderink, H.F.; Kanon, B.; Budagova, K.R.; Kabakov, A.E.; Kampinga, H.H. Regulation of stress-induced intracellular sorting and chaperone function of Hsp27 (HspB1) in mammalian cells. *Biochem. J.* **2007**, *407*, 407–417. [CrossRef]
54. Boyce, M.; Bryant, K.F.; Jousse, C.; Long, K.; Harding, H.P.; Scheuner, D.; Kaufman, R.J.; Ma, D.; Coen, D.M.; Ron, D.; et al. A selective inhibitor of eIF2alpha dephosphorylation protects cells from ER stress. *Science* **2005**, *307*, 935–939. [CrossRef] [PubMed]
55. Bertolotti, A.; Zhang, Y.; Hendershot, L.M.; Harding, H.P.; Ron, D. Dynamic interaction of BiP and ER stress transducers in the unfolded-protein response. *Nat. Cell Biol.* **2000**, *2*, 326–332. [CrossRef]
56. Harding, H.P.; Zhang, Y.; Ron, D. Protein translation and folding are coupled by an endoplasmic-reticulum-resident kinase. *Nature* **1999**, *397*, 271–274. [CrossRef]
57. Korennykh, A.V.; Egea, P.F.; Korostelev, A.A.; Finer-Moore, J.; Zhang, C.; Shokat, K.M.; Stroud, R.M.; Walter, P. The unfolded protein response signals through high-order assembly of Ire1. *Nature* **2009**, *457*, 687–693. [CrossRef]
58. Cox, J.S.; Walter, P. A novel mechanism for regulating activity of a transcription factor that controls the unfolded protein response. *Cell* **1996**, *87*, 391–404. [CrossRef]
59. Quan, J.H.; Gao, F.F.; Lee, M.; Yuk, J.M.; Cha, G.H.; Chu, J.Q.; Wang, H.; Lee, Y.H. Involvement of endoplasmic reticulum stress response and IRE1-mediated ASK1/JNK/Mcl-1 pathways in silver nanoparticle-induced apoptosis of human retinal pigment epithelial cells. *Toxicology* **2020**, *442*, 152540. [CrossRef]
60. Zhang, R.; Piao, M.J.; Kim, K.C.; Kim, A.D.; Choi, J.Y.; Choi, J.; Hyun, J.W. Endoplasmic reticulum stress signaling is involved in silver nanoparticles-induced apoptosis. *Int. J. Biochem. Cell Biol.* **2012**, *44*, 224–232. [CrossRef]
61. Dey, S.; Fageria, L.; Sharma, A.; Mukherjee, S.; Pande, S.; Chowdhury, R.; Chowdhury, S. Silver nanoparticle-induced alteration of mitochondrial and ER homeostasis affects human breast cancer cell fate. *Toxicol. Rep.* **2022**, *9*, 1977–1984. [CrossRef] [PubMed]
62. Wang, X.; Wang, W.X. Tracking the Cellular Degradation of Silver Nanoparticles: Development of a Generic Kinetic Model. *ACS Nano* **2024**, *18*, 13308–13321. [CrossRef] [PubMed]
63. Kumar, A.; Goia, D.V. Comparative Analysis of Commercial Colloidal Silver Products. *Int. J. Nanomed.* **2020**, *15*, 10425–10434. [CrossRef] [PubMed]
64. Morris, D.; Ansar, M.; Speshock, J.; Ivanciuc, T.; Qu, Y.; Casola, A.; Garofalo, R.P. Antiviral and Immunomodulatory Activity of Silver Nanoparticles in Experimental RSV Infection. *Viruses* **2019**, *11*, 732. [CrossRef]
65. Weaver, J.L.; Tobin, G.A.; Ingle, T.; Bancos, S.; Stevens, D.; Rouse, R.; Howard, K.E.; Goodwin, D.; Knapton, A.; Li, X.; et al. Evaluating the potential of gold, silver, and silica nanoparticles to saturate mononuclear phagocytic system tissues under repeat dosing conditions. *Part. Fibre Toxicol.* **2017**, *14*, 25. [CrossRef]

66. De Jong, W.H.; Van Der Ven, L.T.; Sleijffers, A.; Park, M.V.; Jansen, E.H.; Van Loveren, H.; Vandebriel, R.J. Systemic and immunotoxicity of silver nanoparticles in an intravenous 28 days repeated dose toxicity study in rats. *Biomaterials* **2013**, *34*, 8333–8343. [CrossRef]
67. Jarak, I.; Carrola, J.; Barros, A.S.; Gil, A.M.; Pereira, M.L.; Corvo, M.L.; Duarte, I.F. From the Cover: Metabolism Modulation in Different Organs by Silver Nanoparticles: An NMR Metabolomics Study of a Mouse Model. *Toxicol. Sci.* **2017**, *159*, 422–435. [CrossRef]
68. Wieler, L.; Vittos, O.; Mukherjee, N.; Sarkar, S. Reduction in the COVID-19 pneumonia case fatality rate by silver nanoparticles: A randomized case study. *Heliyon* **2023**, *9*, e14419. [CrossRef]

nanomaterials

MDPI

Article

Activity of Hydrophilic, Biocompatible, Fluorescent, Organic Nanoparticles Functionalized with Purpurin-18 in Photodynamic Therapy for Colorectal Cancer

Rayan Chkair [1], Justine Couvez [2], Frédérique Brégier [1], Mona Diab-Assaf [3], Vincent Sol [1], Mireille Blanchard-Desce [2,*], Bertrand Liagre [1] and Guillaume Chemin [1,*]

1 University Limoges, LABCiS, UR 22722, 87000 Limoges, France; rayan.chkair@unilim.fr (R.C.); frederique.bregier@unilim.fr (F.B.); vincent.sol@unilim.fr (V.S.); bertrand.liagre@unilim.fr (B.L.)
2 University Bordeaux, CNRS, Bordeaux INP, ISM (UMR5255), Bat A12, 351 Cours de la Libération, 33405 Talence, France; justine.couvez@u-bordeaux.fr
3 Doctoral School of Sciences and Technology, Lebanese University, Hadath, Beirut 21219, Lebanon; mdiabassaf@ul.edu.lb
* Correspondence: mireille.blanchard-desce@u-bordeaux.fr (M.B.-D.); guillaume.chemin@unilim.fr (G.C.)

Abstract: Photodynamic therapy (PDT) is a clinically approved, non-invasive therapy currently used for several solid tumors, triggering cell death through the generation of reactive oxygen species (ROS). However, the hydrophobic nature of most of the photosensitizers used, such as chlorins, limits the overall effectiveness of PDT. To address this limitation, the use of nanocarriers seems to be a powerful approach. From this perspective, we have recently developed water-soluble and biocompatible, fluorescent, organic nanoparticles (FONPs) functionalized with purpurin-18 and its derivative, chlorin p6 (Cp6), as new PDT agents. In this study, we aimed to investigate the induced cell death mechanism mediated by these functionalized nanoparticles after PDT photoactivation. Our results show strong phototoxic effects of the FONPs[Cp6], mediated by intracellular ROS generation, and subcellular localization in HCT116 and HT-29 human colorectal cancer (CRC) cells. Additionally, we proved that, post-PDT, the FONPs[Cp6] induce apoptosis via the intrinsic mitochondrial pathway, as shown by the significant upregulation of the Bax/Bcl-2 ratio, the activation of caspases 9, 3, and 7, leading poly-ADP-ribose polymerase (PARP-1) cleavage, and DNA fragmentation. Our work demonstrates the photodynamic activity of these nanoparticles, making them promising candidates for the PDT treatment of CRC.

Keywords: colorectal cancer; photodynamic therapy; purpurin-18; nanocarriers; FONPs; apoptosis

1. Introduction

Colorectal cancer (CRC) is one of the most prevalent malignancies worldwide, ranking second in terms of cancer-related deaths and third in terms of incidence in 2022, according to the World Health Organization (WHO). By 2035, it is estimated that CRC incidence will increase by 80%, resulting in approximately 2.4 million cases and 1.3 million deaths globally [1,2]. Once diagnosed, cancer patients receive several treatment modalities, alone or in combination, depending on the stage and location of the tumor. Conventional therapies that are commonly used include surgery, radiotherapy, chemotherapy, and immunotherapy. These therapies are very effective in the early stages, but their efficacy becomes very limited in the later stages. Besides this limitation, conventional treatment is often associated with severe side effects, cancer recurrence, and resistance [3]. Hence, it has become vital to develop new therapeutic strategies that selectively target tumor sites while improving patients' quality of life.

Photodynamic therapy (PDT) has emerged as a potential localized treatment, showing promising outcomes with solid tumors (such as gastrointestinal, prostate, and bladder cancers) [4,5], with improved tumor selectivity and minimal adverse effects compared to conventional cancer therapies [6]. Moreover, phase I and II clinical trials showed the clinical efficacy of PDT in CRC as a successful approach [7]. The mechanism of action relies on the dynamic interaction between three non-toxic components: a photosensitizer (PS), light with a suitable wavelength that activates the PS, and molecular oxygen. Once administered, the PS preferentially accumulates within the tumor and converts from its ground state (PS_0) to an excited singlet state (PS_1), under tumor illumination. The latter is unstable: it can return to the ground state by fluorescence emission or generate heat via internal conversion. However, this excited state can generate a more stable excited triplet state (T_1) through an intersystem crossing. By this triplet state, the PS can undergo two kinds of photochemical reactions, leading to the production of reactive oxygen species (ROS), which are highly reactive and can damage biomolecules, triggering PDT-induced cancer cell death. In the case of type I reactions, the triplet state reacts with the cellular compounds by electron transfer, producing free radicals that, in the presence of molecular oxygen, generate hydroxyl radical (•OH), superoxide anion radical ($O_2^{\bullet -}$), and hydrogen peroxide (H_2O_2). On the other hand, the type II reactions occur when the triplet state interacts directly with the molecular oxygen, resulting in singlet oxygen ($^1O_2$) generation [8].

Like any therapeutic option, PDT presents some limitations, such as the hydrophobicity of most of the traditional PSs used in the literature, PS aggregation occurring in biological environments, and selective cellular uptake [6]. These factors reduce the PDT's clinical efficiency, limiting its overall therapeutic potential. To address these limitations, and thanks to the development of nanotechnology, especially in the context of CRC treatment, PS delivery efficiency may be improved by the encapsulation of the PS within nanoparticles (NPs), providing enhanced water solubility, tumor selectivity through passive or active approaches, and the reduction of toxic effects [7,9]. For this purpose, a wide range of nanocarriers were used in the literature, mainly categorized into organic and inorganic NPs [10,11]. The main advantage of organic NPs is their high biocompatibility, as they can be easily modified to control their physical and chemical characteristics [12]. Overall, the preclinical studies demonstrate the benefits of using nano-encapsulated PSs in the CRC model [7,13,14].

Among chlorins, purpurin-18 (Pp-18), is one of the most commonly used PSs in the literature. With an absorption maximum of 700 nm and a high singlet oxygen quantum yield, Pp-18 has been shown as a strong inducer of PDT-mediated cell death [15–17]. However, its clinical application is limited due to its high hydrophobicity. Thanks to its chemical structure that offers the possibility of different structural modifications, several approaches have been reported to improve its physicochemical properties, such as conjugation with peptides [18], gold nanoparticles [19], polyethylene glycol (PEG) [20], and sugars [21]. In addition, targeted nanoparticles as drug delivery systems has become one of the most promising strategies for CRC cancer therapy [22].

In this context, we have recently developed a new nano-formulation, based on dedicated soft fluorescent organic nanoparticles (FONPs) as nanocarriers of Pp-18 and its derivative chlorin p6 (Cp6). The functionalized FONPs, hereafter referred to as FONPs[Cp6], represent potential biocompatible nanoparticles for PDT treatment in CRC [23]. We stress that targeted nanoparticles The FONPs were prepared from an equimolar mixture of citric acid and diethylenetriamine (DETA) via polycondensation reactions using precise synthesis protocols (Scheme S1). The dehydration can be achieved either by heating a water solution under microwave activation [24] or by the dry heating of the diethlylenetriamonium citrate salt at 200 °C [23]. Importantly, controlled reaction conditions need to be implemented to avoid the carbonization and formation of the graphitic region. As a result, the FONPs had dry diameters of approximately 10–15 nm (Figure S1) and showed high water-solubility (>200 g/L) and biocompatibility [23]. Aqueous FONP solutions strongly absorb in the near UV-region ($\lambda_{abs}{}^{max}$ = 360 nm) and generate bright blue fluorescence under excitation due to an endogenous chromophore formed during the condensation process [23,24]. The FONPs showed a large number of CO_2H and NH_2 surface groups, which were responsible for their high-water solubility. These groups are of major interest for further functionalization using standard conjugation reactions. In order to implement such conjugations, the FONPs were activated either (i) with succinic acid [24], leading to organic nanoparticles enriched in CO_2H surface groups, or (ii) with ethylene diamine [23] leading to organic nanoparticles enriched NH_2 surface groups. In particular, the reaction of the latter with Pp18 resulted in the FONPs[Cp6] having a total PS loading (i.e., Cp6+ Pp18) of approximately 35 μmol.g^{-1}. These nanoparticles retained good water solubility and showed the characteristic absorption bands of Cp6 and Pp18 in the visible region (Figure S2), with peaks at 420 nm (Soret Band), 500 nm, 560 nm, 660 nm, and 710 nm (Q bands). Interestingly, the FONPs[Cp6] showed deep-red emissions upon excitation in the visible region, as well as weak NIR emissions when excited at 720 nm (Figure S3). Moreover, the FONPs[Cp6] exhibited a high singlet oxygen generation quantum yield in aqueous media (0.72 upon excitation at 405 nm) [23].

The FONPs[Cp6] were found to show a strong in vitro phototoxicity against two CRC cell lines, HCT116 and HT-29, after illumination at 650 nm. Indeed, our previous results demonstrated the PDT potential of these new NPs with half-maximal inhibitory concentrations (IC_{50}) of 0.04 and 0.13 nmol/mL for the HCT116 and HT-29 CRC cell lines, respectively [23].

Based on these promising results, the present study aimed to investigate the cell death mechanism mediated by these FONPs. First, we showed an excellent safety profile of the FONPs against human CRC cell lines, HCT116 and HT-29. Then, we demonstrated that the intracellular accumulation and the subcellular localization of the FONPs[Cp6] induced a strong anticancer efficacy through ROS production. Consistent with other in vitro PDT studies, we validated that once photoactivated, the FONPs[Cp6] induced apoptosis via the intrinsic pathway.

2. Results

2.1. In Vitro Safety of FONPs

Our previous study demonstrated the in vitro phototoxicity of FONPs functionalized with Pp18 and its derivative Cp6 (FONPs[Cp6]) [23], so we aimed first to investigate the FONPs safety as drug delivery nanocarriers in human CRC cell lines, HCT116 and HT-29. For this aim, the cells were treated with the FONPs at 1 to 100 µg/mL concentrations. Then, the cells were exposed to red light (650 nm), and the phototoxic effects were monitored for 48 h post-illumination using an MTT assay. Our results showed no significant reduction in cell viability when the FONPs were exposed to light or kept in the dark, even at a concentration of 100 µg/mL in both cell lines (Figure 1). These results further complement the previous findings of the absence of the dark cytotoxicity of the FONPs[Cp6], suggesting the good safety of the FONPs in human CRC cell lines.

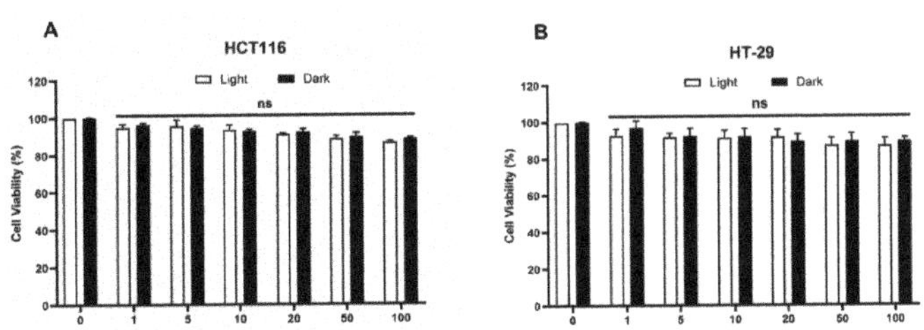

Figure 1. Phototoxic effects of FONPs on human CRC cell lines. HCT116 (**A**), and HT-29 (**B**) cells were seeded in 96-well plates for 24 h, then cells were treated or not with FONPs at different concentrations for 24 h. Cells were illuminated or not with red light (650 nm, 38 mW/cm^2). The cytotoxic effects were then monitored 48 h following illumination by MTT assay. Data are shown as mean ± SEM (n = 3); ns: not significant relative to the control group.

2.2. Cellular Internalization and Localization of FONPs[Cp6]

For confocal microscopy analyses, the HCT116 and HT-29 CRC cell lines were treated with the FONPs[Cp6] at IC_{50} concentrations already calculated: 1.40 and 3.86 μg/mL for the HCT116 and HT-29 cells, respectively, (corresponding to 0.04 nmol/mL and 0.13 nmol/mL of active PSs, respectively). The FONPs[Cp6] exhibited a deep-red or infrared NIR fluorescence [23], so their cellular uptake could be monitored by confocal microscopy when excited typically at 520 nm. The red fluorescence observed in the cytoplasm of the HCT116 and HT-29 CRC cells indicated the cellular internalization of the NPs in both of the cell lines (Figure 2 and Figure S5). Interestingly, compared to the noncancerous cell line HEK-293, we observed higher cellular accumulation within the HCT116 and HT-29 CRC cells, with a negligible uptake in HEK-293 cells, even at 10 μg/mL (Figure 2C). These results show that our NPs were preferentially internalized into CRC cells.

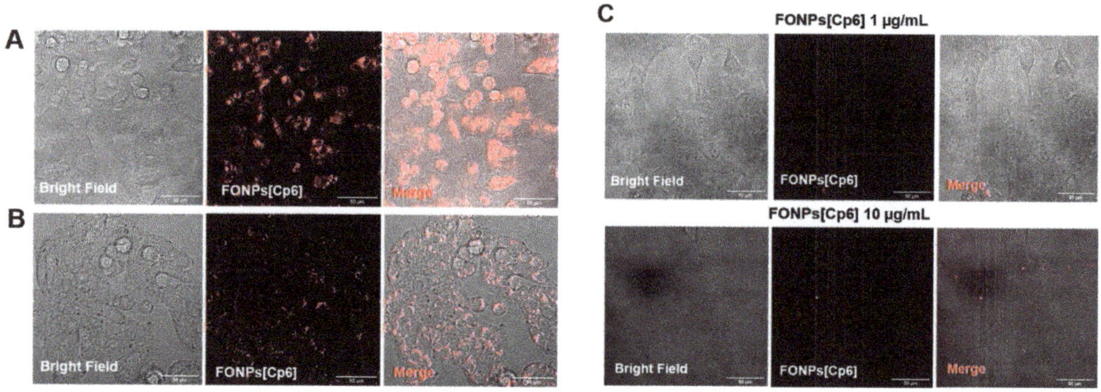

Figure 2. Cellular internalization of FONPs[Cp6] on HCT116 (**A**), HT-29 (**B**), and HEK-293 (**C**) cell lines. Cells were grown for 24 h in chamber slides coated with a type I collagen (3 mg/mL) with acetic acid (20 mM) gel. Cells were then treated with FONPs[Cp6] at IC_{50} concentrations for HCT116 and HT-29 and 1 or 10 μg/mL for HEK-293 cells. After 24 h, red fluorescence was assessed by confocal microscopy (Zeiss LSM880 confocal microscopeJena, Germany). Co-localization was analyzed using the ImageJ software. White scale bars represent 50 μm.

PDT-induced cell death depends on the intracellular localization of the PS. Thus, after confirming the cellular uptake, we looked to determine the subcellular localization of the FONPs[Cp6] in the CRC cells. For this aim, the cells were co-treated with MitoTracker, LysoTracker, or Endoplasmic Reticulum-Tracker (ER-Tracker) organelle probes. The confocal fluorescence images in Figure 3 showed that the FONPs[Cp6] were distributed in all organelles, including the mitochondria, lysosomes, and ER (as indicated by yellow fluorescence in merged images), demonstrating the co-localization of the FONPs[Cp6] in both of the cell lines.

2.3. ROS Generation

The PDT-induced cell death mechanism involves intracellular ROS generation. Therefore, ROS levels were quantified using DCFDA staining after PDT illumination. Flow cytometry analyses revealed that the photoactivation of the FONPs[Cp6] enhanced cellular ROS generation in both cell lines (Figure 4A,B). Moreover, our results showed a greater ROS production, with 85.7% of positive gated in the HCT116 CRC cells compared to 72.6% in the HT-29 CRC cells (Figure 4C). In both the cell lines, the FONPs[Cp6] induced a limited ROS generation in the dark, with 9% and 6.4% of positive gated in the HCT116 and the HT-29 CRC cells, respectively.

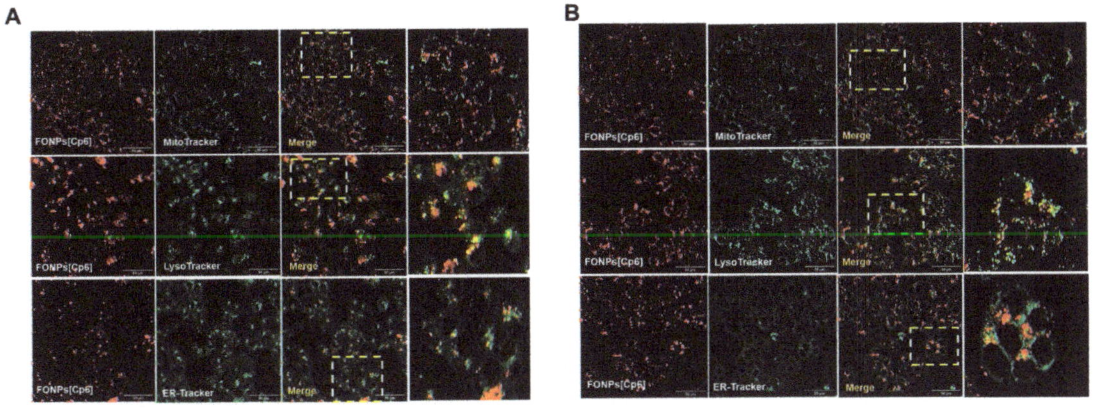

Figure 3. Cellular localization of FONPs[Cp6] on HCT116 (**A**) and HT-29 (**B**) CRC cell lines. Cells were seeded in chamber slides for 24 h prior to exposure to FONPs[Cp6] at IC_{50} concentrations. After 24 h, cells were co-treated with MitoTracker, LysoTracker, or ER-Tracker organelle probes. The images on the right of each panel correspond to the zoomed-in merged images (yellow box). Localization was determined by confocal microscopy, and co-localization was assessed using the ImageJ software. Scale bar = 50 μm.

These results were further confirmed by confocal microscopy, using the DCFDA as an ROS indicator. The confocal fluorescence images indicated a strong green fluorescence within the cells after the photoactivation of the FONPs[Cp6] compared to the control in both the cell lines (Figure 4D,E). Similarly, the HCT116 CRC cells showed an increased green fluorescence compared to the HT-29 CRC cells, demonstrating an enhanced ROS generation in the HCT116 CRC cells. Together, these data demonstrate that the photoactivation of the FONPs[Cp6] induces intracellular ROS generation in both the CRC cell lines.

2.4. Apoptotic Cell Death Induction

In order to investigate the cell death mechanism induced by the FONPs[Cp6], Annexin V-FITC and PI dual staining were performed to assess the level of apoptosis 24 and 48 h post-illumination by flow cytometry. The early apoptotic cells could only retain Annexin V-FITC and were placed in the lower right quadrant. However, due to increased cell permeability, the late apoptotic cells could retain Annexin V-FITC and PI and were placed in the upper right quadrant. As shown in Figure 5A,C, in the HCT116 CRC cells, the control, light control, and the FONPs[Cp6] exhibited high cellular viabilities of 86.97%, 86.47%, and 85.00%, respectively, at 24 h. Similarly, 85.60%, 85.46%, and 84.43% of the population of the HCT116 CRC cells were viable after exposure to the control, light control, and the FONPs[Cp6], respectively, at 48 h. However, 24 h after the FONPs[Cp6] photoactivation, the rate of early and late apoptosis both increased significantly to 46.23%, while the percentage of viable cells decreased to 47.83%. Importantly, 48 h post-PDT, the rate of apoptosis continued to rise to 69.94%, with only 25.50% being viable cells. Similar results were observed with the HT-29 CRC cells: 85.20%, 85.68%, and 83.49% of the cell population were viable after exposure to the control, light control, and the FONPs[Cp6], respectively, at 24 h. After 48 h, 86.97%, 85.18%, and 84.75% of the control, light control, and the FONPs[Cp6], respectively, remained mostly viable. Twenty-four hours after illumination, the FONPs[Cp6] induced a significant increase in the rate of apoptosis to 28.01%, with 64.45% being viable cells. Moreover, after 48 h of photoactivation, the rate of apoptosis increased further to reach 60.19%, while the percentage of viable cells decreased to 36.48% (Figure 5B,D). Together, these findings provide evidence that the photoactivation of the FONPs[Cp6] induces a significant increase in the rate of early and late apoptosis, thus confirming that apoptosis was one of the major induced cell death mechanisms.

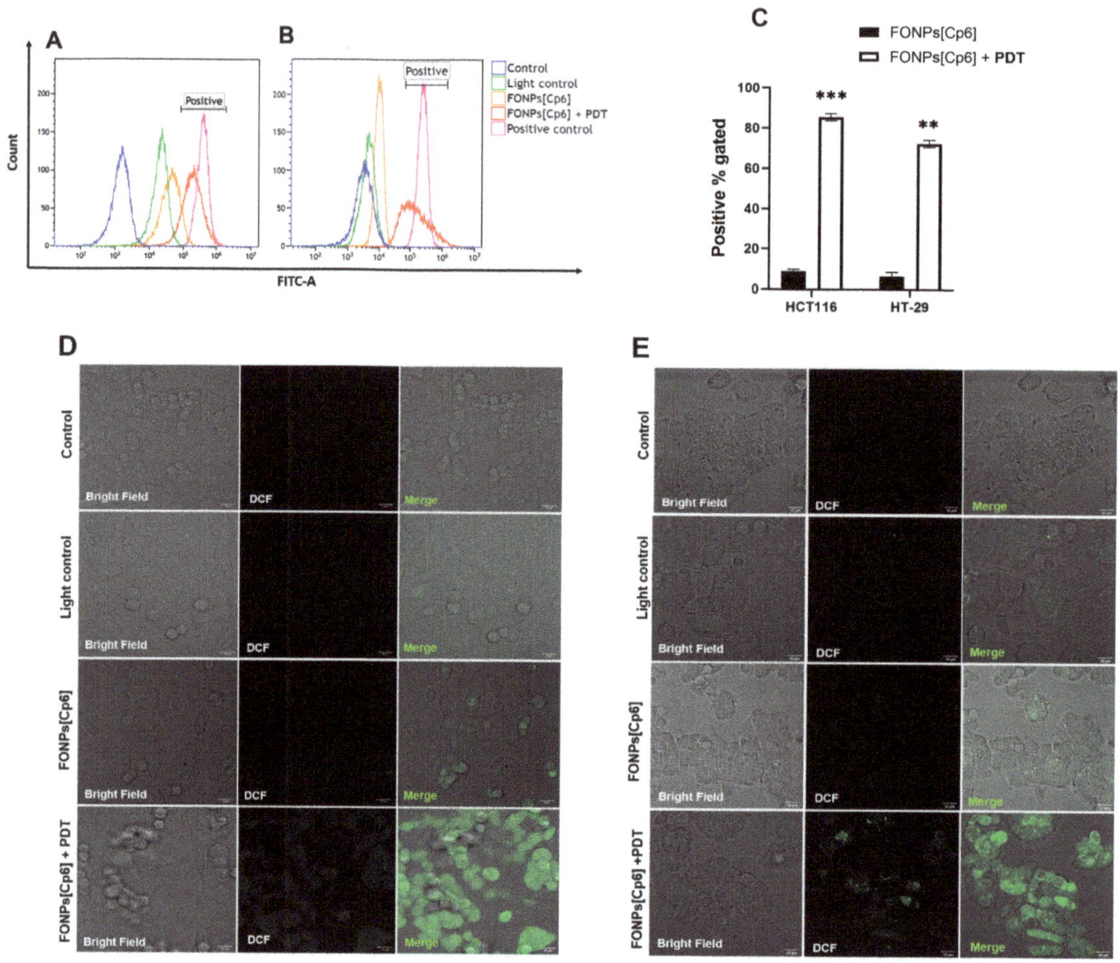

Figure 4. ROS generation by the FONPs[Cp6] in HCT116 (**A**,**C**,**D**) and HT-29 (**B**,**C**,**E**) CRC cell lines. For all experiments, cells were grown for 24 h. Then, cells were treated with the FONPs[Cp6] at IC_{50} concentrations and illuminated or not. Intracellular ROS generation was quantified by flow cytometry using DCFDA staining directly after PDT. An increased fluorescence intensity, resulting from increased 2′,7′-dichlorofluorescein (DCF) formation, causes a shift to the right. Hydrogen peroxide (H_2O_2) was used as a positive control (**A**,**B**). Results of flow cytometry analyses are shown in (**C**). The confocal microscopy images of DCFDA-labeled HCT116 (**D**) and HT-29 (**E**) CRC cells were captured immediately after PDT illumination. Scale bar = 20 μm. Data are represented as mean ± SEM (n = 3). ** $p < 0.01$ and *** $p < 0.001$, relative to the control.

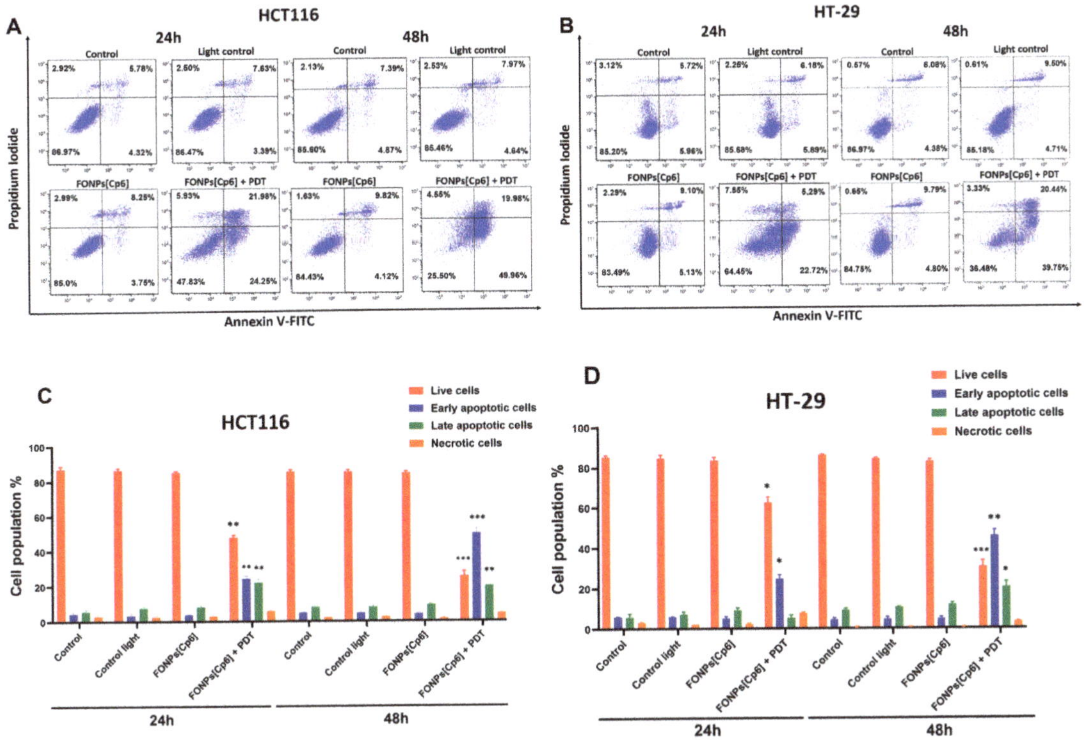

Figure 5. Apoptosis effects of the FONPs[Cp6] on HCT116 (**A**,**C**) and HT-29 (**B**,**D**) CRC cells. Cells were grown 24 h before exposure or not to the FONPs[Cp6] at IC_{50} concentrations. Then, cells were illuminated or not. Twenty-four and forty-eight hours after PDT illumination, cells were stained with Annexin V-FITC and PI, and cell apoptosis was measured by flow cytometry. (**A**,**B**) The results are represented in a scatter plot as four quadrants: living (lower left), early apoptotic (lower right), late apoptotic (upper right), and necrotic (upper left) cells. (**C**,**D**) The percentage of each cell population is expressed as mean ± SEM (n = 3). A t-test was used to compare each population in the treated group to its corresponding population in the control. * $p < 0.05$; ** $p < 0.01$, and *** $p < 0.001$, relative to the control.

2.5. PDT-Induced Apoptosis Mechanism

Caspases are known for their crucial role in the initiation and execution of apoptosis. Hence, to confirm the apoptotic cell death mechanism observed after the photoactivation of the FONPs[Cp6], the level of the caspases-3/7 activity, the main apoptosis executioner, was measured using the Incucyte S3 live-cell imaging system. For this purpose, the cells were labeled with caspases-3/7 green reagent immediately after PDT, and their fluorescence was monitored for 48 h. As observed in Figure 6A, the photoactivation of the FONPs[Cp6] in the HCT116 CRC cells induced a significant increase in the percentage of activated caspases-3/7 in a time-dependent manner, with 86.6% compared to 6.1% in the light control, at 48 h post-illumination. Similar results were obtained with the HT-29 CR+C cells, with a significant increase in caspase-3/7 activity level to 68.5%, compared to 9.6% in the light control (Figure 6B).

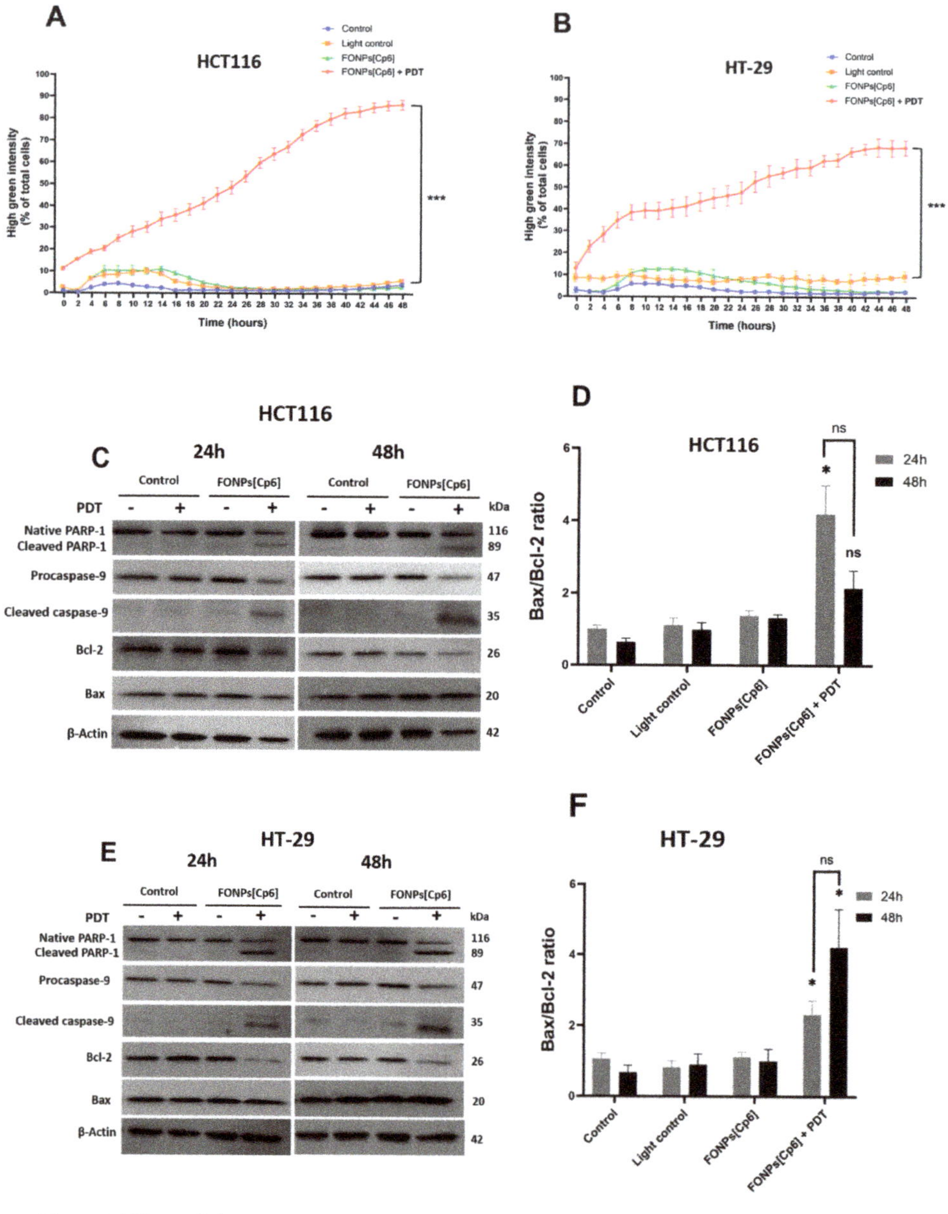

Figure 6. Effects of photoactivation of the FONPs[Cp6] on caspases-3/7 activity and protein expression in HCT116 (**A**,**C**,**D**) and HT-29 (**B**,**E**,**F**) CRC cells. Cells were cultured for 24 h, then treated with the FONPs[Cp6] at IC_{50} concentrations for 24 h, then illuminated. (**A**,**B**) Quantification of caspases-3/7 activity. Directly after PDT of cells co-treated with caspases-3/7 green reagent, fluorescence over time was monitored by Incucyte S3 live-cell imaging system. The ratio of green fluorescent cells to total cells was quantified using IncuCytesoftware version 2023A Rev1). (**C**–**F**) The expression of the key apoptotic proteins was assessed by western blotting 24 and 48 h after illumination. β-actin served as a loading control. Bands were quantified by densitometry using ImageJ software, and the Bax/Bcl-2 ratio was then calculated relative to β-actin. * $p < 0.05$ and *** $p < 0.001$, relative to the control; ns: not significant.

To further confirm the apoptotic cell death, we conducted western blotting to assess the protein expression of key apoptotic-related markers. We focused mainly on poly-ADP-ribose polymerase (PARP-1), which is a substrate of caspase-3, caspase-9 (initiator caspase), Bax (pro-apoptotic), and Bcl-2 (anti-apoptotic) protein expression, 24 and 48 h following photoactivation. In the HCT116 CRC cells, our results show that the photoactivation of the FONPs[Cp6] induced a significant increase in the Bax/Bcl-2 ratio, by 4.19-fold compared to the control. After 48 h, this ratio decreased to 2.1-fold. These observations were correlated with caspase-9 activation, leading to its cleavage, when the FONPs[Cp6] were photoactivated. Subsequently, the activation of caspase-3 (Figure 6C,D) resulted in PARP-1 cleavage (inactivation), associated with a decrease in the expression of the native PARP-1, 24 and 48 h after the FONPs[Cp6] illumination. Similar results were observed in the HT-29 CRC cells: the Bax/Bcl-2 ratio was significantly increased by 2.31- and 4.1-fold at 24 and 48 h, respectively, compared to the control. The increase in the Bax/Bcl-2 ratio was correlated with the activation of caspase-9, and its cleavage 24 and 48 h following PDT. Furthermore, the PARP-1 cleavage was mainly observed 24 and 48 h after the FONPs'[Cp6] photoactivation (Figure 6E,F).

In the absence of PARP-1 repair, DNA is cleaved by endonucleases. To evaluate the apoptotic process at the later stages and examine the nuclear changes induced by the FONPs[Cp6], DNA fragmentation was performed 24 and 48 h after PDT using an ELISA assay. In the HCT116 CRC cells, the results indicate that the FONPs[Cp6], after PDT, induced a strong increase in DNA fragmentation by 5.2- and 12.3-fold, at 24 and 48 h, respectively, compared to the control (Figure 7A). As expected, the FONPs[Cp6] in the dark induced very limited DNA fragmentation of 1.3- and 1.5-fold at 24 and 48 h, respectively, compared to the control. The same results were observed in the HT-29 CRC cells: the FONPs[Cp6] photoactivation induced a significant increase in DNA fragmentation by 3.3- and 6.3-fold at 24 and 48 h, respectively, when compared to the control. The FONPs[Cp6] in the dark showed no significant DNA fragmentation, with 1.3- and 1.5-fold changes at 24 and 48 h, respectively (Figure 7B).

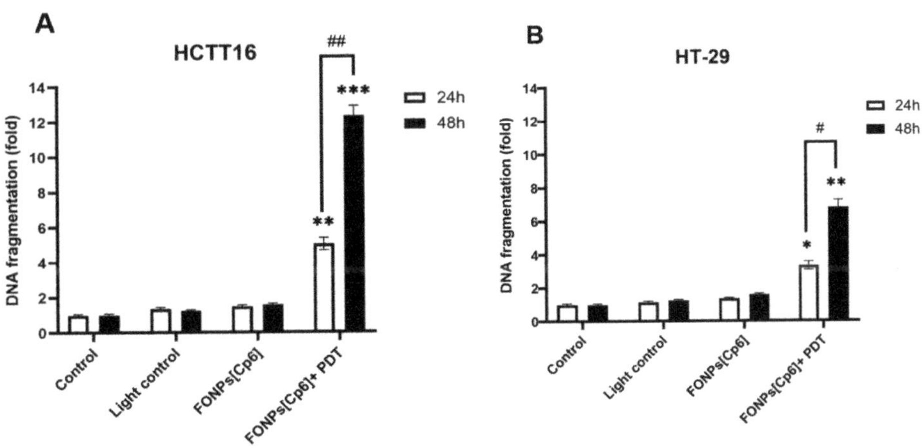

Figure 7. Effects of photoactivation of the FONPs[Cp6] on DNA fragmentation in HCT116 (**A**) and HT-29 (**B**) CRC cells. Cells were grown for 24 h before being treated or not with the FONPs[Cp6] at IC_{50} concentrations. After 24 h of incubation, cells were then illuminated or not. DNA fragmentation 24 and 48 h post-illumination was quantified from cytosol extracts by ELISA assay. The results are expressed as n-fold compared to the control. Data are expressed as mean ± SEM (n = 3). * $p < 0.05$, ** $p < 0.01$, and *** $p < 0.001$, compared to the control, or # $p < 0.05$ and ## $p < 0.01$, compared to 24 h data.

Taken together, these results demonstrate that the FONPs[Cp6] photoactivation induced apoptosis in the human CRC cell lines via the intrinsic pathway.

3. Discussion

Presently, extensive efforts have been made to overcome the conventional PDT limitations, particularly through nanocarrier platforms. The use of NPs seems to be an attractive strategy for improving traditional PSs' hydrophilicity and accumulation within tumor sites. Interestingly, targeted NPs are a novel class of NPs that were developed by conjugating with ligands or other molecules to facilitate cancer cell recognition. These NPs are widely used in the literature to treat CRC [22]. In a previous study [23], we investigated the anticancer efficiency of a new nano-formulation (FONPs[Cp6]) of hydrophobic PS, Pp-18, and its derivative, Cp6, on human CRC cell lines (HCT116 and HT-29). Our previous work demonstrated that the FONPs[Cp6] combine several attractive characteristics, including high water solubility and stability, a high singlet oxygen quantum yield, and deep-red or NIR fluorescence [23]. So, after confirming the photodynamic activity of these functionalized NPs, the main objective of the present study was to understand the induced cell death mechanism that contributes to anticancer effects. First, we evaluated the in vitro safety of the FONPs, and we showed that these nanocarriers were extremely safe on the human CRC cell lines and were found to selectively accumulate within the cancer cells, compared to noncancerous HEK-293 cells. These results confirmed a previous study using the FONPs on glioblastoma cells, in which they were shown to preferentially accumulate in glioblastoma cells, mainly by caveolin and lipid rafts endocytosis [24].

On the other hand, the subcellular localization of PSs plays a critical role in determination of the extent of the cellular damage and the mechanisms involved in cell death. Our nanocarriers were found to accumulate in the cytoplasm of both the cell lines, but not in the nucleus. These results correlate with a previous study, in which Pp-18 was loaded into solid lipid nanoparticles (SLNs) and evaluated against HeLa and A549 lines [25]. Moreover, our confocal fluorescence images showed that the FONPs[Cp6] seemed to localize in lysosomes, mitochondria, and ER organelles. Similarly, a research group found that the conjugation of Pp-18 with a polyethylene glycol (PEG) spacer led to improved accumulation in the mitochondria, lysosomes, and ER in various human cancer cells [20].

Apoptosis is the most commonly described cell death after PDT, mainly due to ROS generation [26]. The intracellular ROS generation results showed that the FONPs[Cp6] exhibited high photoactivity, with 85.7% and 72.6% of ROS production in the HCT116 and the HT-29 CRC cell lines, respectively. When compared to Liu et al.'s work [27], in which they obtained water dispersible organic nanoparticles (WSONs) from the nanoprecipitation in water of hydrophobic chlorin derivatives, the FONPs[Cp6] in the present study seem to have a higher photodynamic efficiency. Liu et al.'s study highlighted that WSONs obtained from purpurin18 methyl ester (pu18ME) induce mitochondrial apoptotic cell death, with singlet oxygen generation of less than 65% in HeLa cell lines [27]. The use of a carbon monoxide (CO) controllable release system has emerged as an efficient anticancer therapy [28]. An interesting approach, reported by Wu et al., is the PDT-driven CO release system, based on Cp6, using nanogels. They showed that, when exposed to light irradiation, this system induces synergistic anticancer effects shown by the high CO and singlet oxygen ($^1O_2$) generation, demonstrating a strong photodynamic efficiency in both in vitro and in vivo breast cancer models [29].

Regarding the apoptotic cell death mechanism, we validated that the FONPs[Cp6] induce apoptosis via the intrinsic pathway. As localized in the mitochondria, the photoactivation of the FONPs[Cp6] induces ROS generation that induces damage to the mitochondrial membrane, followed by an increase in the Bax/Bcl-2 ratio, leading to the subsequent activation of caspase-9. Once activated, caspase-9 cleaves and activates the executioner caspases 3 and 7, leading to PARP-1 cleavage and DNA fragmentation. Consistent with our results, a study reported that the photoactivation of Cp6 induces apoptosis and autophagy via ROS generation in human CRC SW620 cells [17]. Furthermore, Bharathiraja et al.

demonstrated that Cp6 conjugated with silica nanoparticles enhanced cellular uptake and induced apoptosis by intracellular ROS generation in breast cancer cells [30]. An in vivo study on hepatocellular carcinoma, using Pp-18 conjugated with gold NPs, showed that PDT-irradiation induced caspase-3 activation and DNA fragmentation [31].

Accumulating evidence suggests that autophagy is triggered after PDT due to ROS generation. In fact, autophagy seems to play a controversial role in the PDT context. Several studies have demonstrated that applying an autophagy inhibitor can enhance PDT-induced cell death [17,32]. However, other studies showed that autophagy inhibition significantly decreased PDT-induced cell death in breast cancer cells [33]. In our work, we proved that the FONPs[Cp6] were localized in the lysosomes and ER, so further investigations of the role of autophagy in the induced apoptotic cell death could be useful. Indeed, the use of 3D cell spheroid models of CRC that imitate the tumor microenvironment seems to be crucial to assessing the PDT potential of the FONPs[Cp6]. Additionally, the use of in vivo animal models (such as the subcutaneous xenograft mouse model of CRC) [34] to investigate these novel nanocarriers would be the ultimate goal of the present work.

4. Materials and Methods

4.1. Materials

The RPMI 1640 and RPMI 1640 red-phenol-free mediums were purchased from Gibco BRL (Cergy-Pontoise, France). The 3-(4,5-dimethylthiazol-2-yl)-2,5-diphenyltetrazoliumbromide (MTT), 2′,7′-dichlorofluorescein diacetate (DCFDA) kit, cell death detection enzyme-linked immunosorbent assay^{PLUS} (ELISA), and human anti-actin antibodies were acquired from Sigma-Aldrich (Burlington, MA, USA). The Poly-ADP-ribose polymerase (PARP), caspase-9, Bcl-2, and Bax antibodies, goat anti-rabbit IgG H&L horseradish peroxidase (HRP) secondary antibodies, and LysoTracker Green DND-26 were obtained from Cell Signaling Technology-Ozyme (Danvers, MA, USA). The MitoTracker Green FM, ER-Tracker™ Green, annexin V-FITC, propidium iodide (PI), and rabbit anti-mouse IgG-IgM H&L HRP secondary antibodies were obtained from Invitrogen-Thermo Fisher Scientific (Van Allen Way, Carlsbad, CA, USA).

4.2. Synthesis of the FONPs[Cp6]

The FONPs and the FONPs[Cp6] were prepared as described in ref [23]. Briefly, the ammonium citrate salt was first prepared by mixing equimolar amounts of diethylenetriamine (DETA) and citric triacid in pure water, then evaporating the water under vacuum to obtain a dry solid. The salt was subsequently heated at 200 °C for 30 min, leading to a brownish–yellowish crude material, which was washed with ethanol and then dried under vacuum. The bare FONPs were then enriched in NH_2 surface groups by heating in ethylene diamine at 120 °C for 16 h [35]. Finally, the FONPs^{NH_2} were reacted with Pp18 (Scheme S1, step 3) in a biphasic medium (H_2O/toluene) under vigorous stirring and heating at 100 C to yield the FONPs[Cp6] after the separation of the aqueous phase, freeze-drying, and washing with CH_2Cl_2, then drying under a vacuum. The FONPs[Cp6] showed a size distribution ranging between 10 nm and 20 nm, giving a mean diameter of 12.9 nm, as indicated by transmission electron microscopy (Figure S1). A small number of larger nanoparticles (approximately 25 and 35 nm in diameter) were also detected [23]. Gel electrophoresis (Figure S4) indicated that both the FONPs^{NH2} and the FONPs[Cp6] were mainly positively charged.

4.3. Cell Lines Culture and Treatment

The human CRC HCT116 and HT-29 adherent cell lines were purchased from the American Type Culture Collection (ATCC—LGC Standards, Molsheim, France). We chose these human CRC cell lines because they are of different stages, enabling us to evaluate the possible resistance to our treatments: The HCT116 CRC line was isolated from a stage I colorectal carcinoma from an adult male. The HT-29 CRC line was derived from a stage II colorectal adenocarcinoma from a 44-year-old woman.

The cells were grown in an RPMI medium and cultured as previously mentioned [23], and the HEK cells were seeded at 2×10^4 cells/cm^2. The FONPs[Cp6] and FONP platform stock solutions were prepared in ultrapure water and diluted in a culture medium to prepare the working concentrations. The in vitro PDT protocol was performed as previously explained [23].

4.4. In Vitro Cytotoxicity Assay of FONPs

The cytotoxicity of the FONPs was assessed using the MTT assay, as described in [23], which reveals dehydrogenase activity in living cells. The cells were seeded in 96-well culture plates for 24 h before exposure or not to the compounds. After 24 h of incubation, the cells were illuminated with a 650 nm lamp (PhotoCure ASA, Oslo, Norway) as a light source at 38 mW/cm^2. The cell viability percentage was then measured 48 h after illumination, by normalizing control cells.

4.5. Intracellular ROS Generation by the FONPs[Cp6]

ROS production was quantified using a cell-permeable fluorescent probe, 2′,7′-dichloro fluorescein diacetate (DCFDA), to determine the intracellular oxidative stress. To this end, the cells were seeded in 25 cm^2 tissue culture flasks for 24 h before being treated or not with the FONPs[Cp6] at IC_{50} concentrations. After 24 h, the cells were stained with DCFDA (20 µM) at 37 °C. After 30 min of incubation, the cells were illuminated or not. ROS generation was assessed by flow cytometry (BD Biosciences, San Jose, CA, USA) directly after PDT. Hydrogen peroxide (H_2O_2) was used as a positive control (800 µM).

For confocal microscopy images, the cells were plated for 24 h in chamber slides (ibidi µ-Slide 8 well; Clinisciences, Gräfelfing, Germany) and coated with a type I collagen (3 mg/mL) and acetic acid (20 mM) gel. After this, the cells were treated with the FONPs[Cp6] at IC_{50} values. Then, after 24 h, DCFDA (20 µM) was added for 30 min, and the fluorescence images were taken under excitation of 488 nm at 40× magnification by confocal microscopy (Zeiss LSM880 confocal microscope).

4.6. Cellular Uptake and Localization of the FONPs[Cp6]

The cells were seeded for 24 h before treatment with the FONPs[Cp6] at IC_{50} concentrations in chamber slides coated as mentioned above to determine the cellular localization. Then, after 24 h, the cells were treated with MitoTracker Green FM (50 nM), ER-Tracker™ Green (500 nM), and LysoTracker Green DND-26 (50 nM) for 45 min, 30 min, and extemporaneously at 37 °C, respectively. The FONPs′[Cp6] localization was examined by confocal microscopy, using the FONPs[Cp6] fluorescence (excitation/emission: 520/665 nm) with MitoTracker fluorescence (490/516 nm), LysoTracker fluorescence (504/511 nm), and ER-Tracker fluorescence (504/511 nm). The co-localization was analyzed using the ImageJ software (version 1.52p).

4.7. Apoptotic Cell Death Induction by the FONPs[Cp6]

4.7.1. Dual Staining Assay with Annexin V-FITC and PI

Annexin V-FITC/PI dual staining was performed to evaluate the apoptosis induction. The cells were treated with the FONPs[Cp6] for each cell line at IC_{50} concentrations and were illuminated or not. Twenty-four and forty-eight hours post-PDT, a total of 2.5×10^5 cells/conditions were collected, then washed with PBS, centrifuged at 1500 rpm for 5 min, and incubated with 100 µL binding buffer (1X) with 5 µL annexin V-FITC and 1 µL of PI at 100 µg/mL for 15 min in the dark. After that, 200 µL of the binding buffer was added to the samples, and the percentage of apoptotic cells was analyzed by flow cytometry. Our data are represented in a scatter plot as four quadrants: living, early apoptosis, late apoptosis, and necrotic cells.

4.7.2. DNA Fragmentation

A further analysis for cell death was performed using a Cell Death Detection ELISA^{PLUS} to detect DNA fragmentation. For this, the cells were treated or not with the FONPs[Cp6] at IC_{50} concentrations for 24 h, and then illuminated or not. Twenty-four and forty-eight hours after PDT, cytosolic extracts of 2×10^5 cells/condition were obtained, and the amounts of mono- and oligonucleosomes produced from apoptotic cells were quantified, following the manufacturer's protocol. The absorbance was measured at 405 nm, using a microplate reader (Thermoscientific MULTISKAN FC). The results are expressed as n-fold compared to the control.

4.7.3. Quantification of Caspase-3/7 Activity

The amounts of activated caspases-3/7 were monitored using an IncucyteS3 system (Sartorius, Göttingen, Germany), a live-cell imaging device that allows fluorescence tracking over time. The cells were seeded in 96-well plates for 24 h before being treated or not with the FONPs[Cp6] at IC_{50} concentrations and illuminated. Then, the cells were labeled with caspases-3/7 green reagent (Sartorius) at 5 µM. The photos were taken every 2 h at ×20 magnification in phase contrast/green fluorescence. The Green Object Count/Phase Object Count ratio was quantified using Incucyte software (version 2023A Rev1) to determine the percentage of green fluorescent cells.

4.7.4. Protein Expression by Western Blotting

Each cell line was treated with IC_{50} concentrations and illuminated. Twenty-four and forty-eight hours after PDT, the total cells of each condition were collected, and whole-cell lysate proteins were extracted using an RIPA lysis buffer (containing 50 mM HEPES, pH 7.5, 150 mM NaCl, 1% sodium deoxycholate, 1% NP-40, 0.1% SDS, and 20 mg/mL of aprotinin). The protein levels were quantified using the Bradford method. Briefly, the proteins (60–80 µg) were separated by electrophoresis on SDS-PAGE gels and then transferred to polyvinylidene difluoride (PVDF) membranes (Amersham Pharmacia Biotech, Saclay, France). After that, the membranes were incubated with primary antibodies against human apoptosis-related proteins, including PARP-1 (1:1000), caspase-9 (1:1000), Bcl-2 (1:1000), and Bax (1:1000). β-actin (1:5000) was used as a loading control. The membranes were then incubated with secondary antibodies, and bands were revealed using the Immobilon Western HRP Substrate (Merck, Darmstadt, Germany) and a G: BOX system (Syngene, Cambridge, UK). The bands were quantified by densitometry, using ImageJ software.

4.8. Statistical Analyses

All the data are presented as the mean ± standard error of the mean (SEM) of three independent experiments. The statistical analyses were performed using GraphPad Prism 9.0 by two-tailed unpaired Student's *t*-tests. *p* values of * $p < 0.05$, ** $p < 0.01$, and *** $p < 0.001$ or # $p < 0.05$ and ## $p < 0.01$ were considered statistically significant.

5. Conclusions

In this study, we evaluated the anticancer efficiency of novel biocompatible FONPs that were functionalized with a hydrophobic PS (Pp-18 and its derivative Cp6) on human CRC cell lines (HCT116 and HT-29). We validated the in vitro safety of the FONPs as delivery nanocarriers for the hydrophobic PS. Interestingly, the FONPs[Cp6] showed a preferential internalization in the CRC cell lines. Furthermore, the confocal microscopy analysis confirmed that the FONPs[Cp6] were localized in the mitochondria, lysosomes, and ER, which are the desired targets for PDT. We can conclude that the FONPs[Cp6] exhibit strong photodynamic activity through ROS generation, resulting in induced apoptosis via the intrinsic pathway. Together, the in vitro results demonstrate the interesting anticancer efficiency of this nano-formulation against CRC. To further this study, the use of 3D cancer models could be the first step, followed by in vivo experiments to validate our findings.

Supplementary Materials: The following supporting information can be downloaded at: https://www.mdpi.com/article/10.3390/nano14191557/s1. Synthesis and characterizations of FONPs[Cp6]. DAPI staining. Western blot analysis of the autophagy markers.

Author Contributions: Conceptualization, G.C., M.D.-A. and B.L.; methodology, R.C.; software, R.C.; validation, G.C., B.L. and M.B.-D.; formal analysis, R.C.; investigation, R.C., J.C., F.B. and V.S.; resources, J.C., F.B., V.S. and M.B.-D.; data curation, G.C. and B.L.; writing—original draft preparation, R.C.; writing—review and editing, B.L., M.B.-D. and G.C.; visualization, R.C. and G.C.; supervision, B.L. and G.C.; project administration, B.L., M.D.-A. and G.C.; funding acquisition, G.C., M.D.-A., V.S. and M.B.-D. All authors have read and agreed to the published version of the manuscript.

Funding: This study was funded by the Ligue contre le Cancer Haute-Vienne, 23 Av. des Bénédictins, 87000 Limoges, PFI grant number R2107610 and the Université de Limoges, 33 Rue François Mitterrand, 87000 Limoges, PFI grant number RET23840. This work was supported by Région Nouvelle Aquitaine (project NANOPHOT, convention AAPR2020-2019-8225810). This study received financial support from the French government in the framework of the University of Bordeaux's France 2030 program/RRI "IMPACT".

Data Availability Statement: The data are contained within the article and Supplementary Materials.

Acknowledgments: We are grateful to the BISCEm platform of OmegaHealth Institut of Limoges University, especially Catherine OUK and Claire CARRION for the flow cytometry and the confocal microscopy analyses. We thank Olivier Sandre for his contribution to funding acquisition (grant to J.C.). We thank Aline Pinon for her technical assistance during the requested revisions.

Conflicts of Interest: The authors declare no conflicts of interest.

Abbreviations

ATCC: American Type Culture Collection; Cp6: chlorin p6; CRC: colorectal cancer; DCFDA: 2′,7′-dichlorofluorescein diacetate; DNA: deoxyribonucleic acid; ELISA: enzyme-linked immunosorbent assay; ER-Tracker: endoplasmic reticulum tracker; FONPs: fluorescent organic nanoparticles; HEPES: 4-(2-hydroxyethyl)-1-piperazine ethane sulfonic acid; HEK-293 cells: human embryonic kidney-293 cells; H_2O_2: hydrogen peroxide; HRP: horseradish peroxidase; IC_{50}: half-maximal inhibitory concentration; MTT: 3-(4,5-dimethylthiazol-2-yl)-2,5-diphenyltetrazolium bromide; NPs: nanoparticles; $^1O_2$: singlet oxygen; •OH: hydroxyl radical; $O_2^{\bullet-}$: superoxide anion; PARP-1: poly-ADP-ribose polymerase; PBS: phosphate-buffered saline; PDT: photodynamic therapy; PI: propidium Iodide; PS: photosensitizer; PVDF membrane: polyvinylidene fluoride membrane; Pp-18: purpurin-18; RIPA: radio immuno precipitation assay; ROS: reactive oxygen species; RPMI: Roswell Park Memorial Institute medium; SDSPAGE: electrophoresis in polyacrylamide gel containing sodium dodecyl sulfate; SEM: standard error of the mean.

References

1. Douaiher, J.; Ravipati, A.; Grams, B.; Chowdhury, S.; Alatise, O.; Are, C. Colorectal Cancer—Global Burden, Trends, and Geographical Variations. *J. Surg. Oncol.* **2017**, *115*, 619–630. [CrossRef] [PubMed]
2. Sawicki, T.; Ruszkowska, M.; Danielewicz, A.; Niedźwiedzka, E.; Arłukowicz, T.; Przybyłowicz, K.E. A Review of Colorectal Cancer in Terms of Epidemiology, Risk Factors, Development, Symptoms and Diagnosis. *Cancers* **2021**, *13*, 2025. [CrossRef] [PubMed]
3. Nompumelelo Simelane, N.W.; Kruger, C.A.; Abrahamse, H. Photodynamic Diagnosis and Photodynamic Therapy of Colorectal Cancer: In Vitro and In Vivo. *RSC Adv.* **2020**, *10*, 41560–41576. [CrossRef] [PubMed]
4. Sharifkazemi, H.; Amini, S.M.; Koohi Ortakand, R.; Narouie, B. A Review of Photodynamic Therapy in Different Types of Tumors. *Transl. Res. Urol.* **2022**, *4*, 61–70. [CrossRef]
5. Azzouzi, A.R.; Vincendeau, S.; Barret, E.; Cicco, A.; Kleinclauss, F.; van der Poel, H.G.; Stief, C.G.; Rassweiler, J.; Salomon, G.; Solsona, E.; et al. Padeliporfin Vascular-Targeted Photodynamic Therapy versus Active Surveillance in Men with Low-Risk Prostate Cancer (CLIN1001 PCM301): An Open-Label, Phase 3, Randomised Controlled Trial. *Lancet Oncol.* **2017**, *18*, 181–191. [CrossRef]
6. Rodrigues, J.A.; Correia, J.H. Photodynamic Therapy for Colorectal Cancer: An Update and a Look to the Future. *Int. J. Mol. Sci.* **2023**, *24*, 12204. [CrossRef]

7. Simelane, N.W.N.; Abrahamse, H. Nanoparticle-Mediated Delivery Systems in Photodynamic Therapy of Colorectal Cancer. *Int. J. Mol. Sci.* **2021**, *22*, 12405. [CrossRef]
8. Correia, J.H.; Rodrigues, J.A.; Pimenta, S.; Dong, T.; Yang, Z. Photodynamic Therapy Review: Principles, Photosensitizers, Applications, and Future Directions. *Pharmaceutics* **2021**, *13*, 1332. [CrossRef]
9. Lee, D.; Kwon, S.; Jang, S.Y.; Park, E.; Lee, Y.; Koo, H. Overcoming the Obstacles of Current Photodynamic Therapy in Tumors Using Nanoparticles. *Bioact. Mater.* **2022**, *8*, 20–34. [CrossRef]
10. Dos Santos, A.F.; Arini, G.S.; de Almeida, D.R.Q.; Labriola, L. Nanophotosensitizers for Cancer Therapy: A Promising Technology? *J. Phys. Mater.* **2021**, *4*, 032006. [CrossRef]
11. Bhattacharya, D.; Mukhopadhyay, M.; Shivam, K.; Tripathy, S.; Patra, R.; Pramanik, A. Recent developments in photodynamic therapy and its application against multidrug resistant cancers. *Biomed. Mater.* **2023**, *18*. [CrossRef] [PubMed]
12. Debele, T.A.; Peng, S.; Tsai, H.-C. Drug Carrier for Photodynamic Cancer Therapy. *Int. J. Mol. Sci.* **2015**, *16*, 22094–22136. [CrossRef] [PubMed]
13. Rajan, S.S.; Chandran, R.; Abrahamse, H. Overcoming Challenges in Cancer Treatment: Nano-Enabled Photodynamic Therapy as a Viable Solution. *Wiley Interdiscip. Rev. Nanomed. Nanobiotechnol.* **2024**, *16*, e1942. [CrossRef] [PubMed]
14. Nkune, N.W.; Kruger, C.A.; Abrahamse, H. Possible Enhancement of Photodynamic Therapy (PDT) Colorectal Cancer Treatment When Combined with Cannabidiol. *Anticancer. Agents Med. Chem.* **2020**, *21*, 137–148. [CrossRef] [PubMed]
15. Pavlíčková, V.; Škubník, J.; Jurášek, M.; Rimpelová, S. Advances in Purpurin 18 Research: On Cancer Therapy. *Appl. Sci.* **2021**, *11*, 2254. [CrossRef]
16. Huang, P.; Zhang, B.; Yuan, Q.; Zhang, X.; Leung, W.; Xu, C. Photodynamic Treatment with Purpurin 18 Effectively Inhibits Triple Negative Breast Cancer by Inducing Cell Apoptosis. *Lasers Med. Sci.* **2021**, *36*, 339–347. [CrossRef]
17. Xue, Q.; Wang, X.; Wang, P.; Zhang, K.; Liu, Q. Role of P38 MAPK in Apoptosis and Autophagy Responses to Photodynamic Therapy with Chlorin E6. *Photodiagnosis Photodyn. Ther.* **2015**, *12*, 84–91. [CrossRef]
18. Srivatsan, A.; Ethirajan, M.; Pandey, S.K.; Dubey, S.; Zheng, X.; Liu, T.H.; Shibata, M.; Missert, J.; Morgan, J.; Pandey, R.K. Conjugation of CRGD Peptide to Chlorophyll a Based Photosensitizer (HPPH) Alters Its Pharmacokinetics with Enhanced Tumor-Imaging and Photosensitizing (PDT) Efficacy. *Mol. Pharm.* **2011**, *8*, 1186–1197. [CrossRef]
19. Oluwole, D.O.; Manoto, S.L.; Malabi, R.; Maphanga, C.; Ombinda-Lemboumba, S.; Mthunzi-Kufa, P.; Nyokong, T. Evaluation of the Photophysicochemical Properties and Photodynamic Therapy Activity of Nanoconjugates of Zinc Phthalocyanine Linked to Glutathione Capped Au and Au3Ag1 Nanoparticles. *Dye. Pigment.* **2018**, *150*, 139–150. [CrossRef]
20. Pavlíčková, V.; Rimpelová, S.; Jurášek, M.; Záruba, K.; Fähnrich, J.; Křížová, I.; Bejček, J.; Rottnerová, Z.; Spiwok, V.; Drašar, P.; et al. PEGylated Purpurin 18 with Improved Solubility: Potent Compounds for Photodynamic Therapy of Cancer. *Molecules* **2019**, *24*, 4477. [CrossRef]
21. Kato, A.; Kataoka, H.; Yano, S.; Hayashi, K.; Hayashi, N.; Tanaka, M.; Naitoh, I.; Ban, T.; Miyabe, K.; Kondo, H.; et al. Maltotriose Conjugation to a Chlorin Derivative Enhances the Antitumor Effects of Photodynamic Therapy in Peritoneal Dissemination of Pancreatic Cancer. *Mol. Cancer Ther.* **2017**, *16*, 1124–1132. [CrossRef] [PubMed]
22. Cisterna, B.A.; Kamaly, N.; Choi, W.I.; Tavakkoli, A.; Farokhzad, O.C.; Vilos, C. Targeted nanoparticles for colorectal cancer. *Nano-Medicine* **2016**, *11*, 2443–2456. [CrossRef] [PubMed]
23. Sasaki, I.; Brégier, F.; Chemin, G.; Daniel, J.; Couvez, J.; Chkair, R.; Vaultier, M.; Sol, V.; Blanchard-Desce, M. Hydrophilic Biocompatible Fluorescent Organic Nanoparticles as Nanocarriers for Biosourced Photosensitizers for Photodynamic Therapy. *Nanomaterials* **2024**, *14*, 216. [CrossRef] [PubMed]
24. Daniel, J.; Montaleytang, M.; Nagarajan, S.; Picard, S.; Clermont, G.; Lazar, A.N.; Dumas, N.; Correard, F.; Braguer, D.; Blanchard-Desce, M.; et al. Hydrophilic Fluorescent Nanoprodrug of Paclitaxel for Glioblastoma Chemotherapy. *ACS Omega* **2019**, *4*, 18342–18354. [CrossRef]
25. Yeo, S.; Song, H.H.; Kim, M.J.; Hong, S.; Yoon, I.; Lee, W.K. Synthesis and Design of Purpurin-18-Loaded Solid Lipid Nanoparticles for Improved Anticancer Efficiency of Photodynamic Therapy. *Pharmaceutics* **2022**, *14*, 1064. [CrossRef]
26. Gunaydin, G.; Gedik, M.E.; Ayan, S. Photodynamic Therapy—Current Limitations and Novel Approaches. *Front. Chem.* **2021**, *10*, 691697. [CrossRef]
27. Liu, Y.; Lee, T.H.; Lee, S.H.; Li, J.; Lee, W.K.; Yoon, I. Mitochondria-Targeted Water-Soluble Organic Nanoparticles of Chlorin Derivatives for Biocompatible Photodynamic Therapy. *ChemNanoMat* **2020**, *6*, 610–617. [CrossRef]
28. Wegiel, B.; Gallo, D.; Csizmadia, E.; Harris, C.; Belcher, J.; Vercellotti, G.M.; Penacho, N.; Seth, P.; Sukhatme, V.; Ahmed, A.; et al. Carbon monoxide expedites metabolic exhaustion to inhibit tumor growth. *Cancer Res.* **2013**, *73*, 7009–7021. [CrossRef] [PubMed]
29. Wu, L.; Cai, X.; Zhu, H.; Li, J.; Shi, D.; Su, D.; Yue, D.; Gu, Z. PDT-Driven Highly Efficient Intracellular Delivery and Controlled Release of CO in Combination with Sufficient Singlet Oxygen Production for Synergistic Anticancer Therapy. *Adv. Funct. Mater.* **2018**, *28*, 1804324. [CrossRef]
30. Bharathiraja, S.; Moorthy, M.S.; Manivasagan, P.; Seo, H.; Lee, K.D.; Oh, J. Chlorin E6 Conjugated Silica Nanoparticles for Targeted and Effective Photodynamic Therapy. *Photodiagnosis Photodyn. Ther.* **2017**, *19*, 212–220. [CrossRef]
31. Kwon, J.-G.; Song, I.-S.; Kim, M.-S.; Lee, B.H.; Kim, J.H.; Yoon, I.; Shim, Y.K.; Kim, N.; Han, J.; Youm, J.B. Pu-18-N-Butylimide-NMGA-GNP Conjugate Is Effective against Hepatocellular Carcinoma. *Integr. Med. Res.* **2013**, *2*, 106–111. [CrossRef] [PubMed]
32. Xiong, L.; Liu, Z.; Ouyang, G.; Lin, L.; Huang, H.; Kang, H.; Chen, W.; Miao, X.; Wen, Y. Autophagy Inhibition Enhances Photocytotoxicity of Photosan-II in Human Colorectal Cancer Cells. *Oncotarget* **2017**, *8*, 6419–6432. [CrossRef] [PubMed]

33. Zhu, J.; Tian, S.; Li, K.T.; Chen, Q.; Jiang, Y.; Lin, H.D.; Yu, L.H.; Bai, D.Q. Inhibition of Breast Cancer Cell Growth by Methyl Pyropheophenylchlorin Photodynamic Therapy Is Mediated Though Endoplasmic Reticulum Stress-Induced Autophagy in vitro and vivo. *Cancer Med.* **2018**, *7*, 1908–1920. [CrossRef]
34. Bretin, L.; Pinon, A.; Bouramtane, S.; Ouk, C.; Richard, L.; Perrin, M.-L.; Chaunavel, A.; Carrion, C.; Bregier, F.; Sol, V.; et al. Photodynamic Therapy Activity of New Porphyrin-Xylan-Coated Silica Nanoparticles in Human Colorectal Cancer. *Cancers* **2019**, *11*, 1474. [CrossRef] [PubMed]
35. Sasaki, I.; Daniel, J.; Marais, S.; Clermont, G.; Verlhac, J.B.; Vaultier, M.; Blanchard-Desce, M. Soft fluorescent organic nanodots as nanocarriers for porphyrins. *J. Porphyr. Phthalocyanines* **2019**, *23*, 1463–1469. [CrossRef]

nanomaterials

MDPI

Review

Monte Carlo Simulations in Nanomedicine: Advancing Cancer Imaging and Therapy

James C. L. Chow [1,2]

[1] Radiation Medicine Program, Princess Margaret Cancer Centre, University Health Network, Toronto, ON M5G 1X6, Canada; james.chow@uhn.ca; Tel.: +1-416-946-4501
[2] Department of Radiation Oncology, University of Toronto, Toronto, ON M5T 1P5, Canada

Abstract: Monte Carlo (MC) simulations have become important in advancing nanoparticle (NP)-based applications for cancer imaging and therapy. This review explores the critical role of MC simulations in modeling complex biological interactions, optimizing NP designs, and enhancing the precision of therapeutic and diagnostic strategies. Key findings highlight the ability of MC simulations to predict NP bio-distribution, radiation dosimetry, and treatment efficacy, providing a robust framework for addressing the stochastic nature of biological systems. Despite their contributions, MC simulations face challenges such as modeling biological complexity, computational demands, and the scarcity of reliable nanoscale data. However, emerging technologies, including hybrid modeling approaches, high-performance computing, and quantum simulation, are poised to overcome these limitations. Furthermore, novel advancements such as FLASH radiotherapy, multifunctional NPs, and patient-specific data integration are expanding the capabilities and clinical relevance of MC simulations. This topical review underscores the transformative potential of MC simulations in bridging fundamental research and clinical translation. By facilitating personalized nanomedicine and streamlining regulatory and clinical trial processes, MC simulations offer a pathway toward more effective, tailored, and accessible cancer treatments. The continued evolution of simulation techniques, driven by interdisciplinary collaboration and technological innovation, ensures that MC simulations will remain at the forefront of nanomedicine's progress.

Keywords: nanomaterials; cancer therapy; medical imaging; Monte Carlo simulation; nanoparticle-enhanced radiotherapy; dose enhancement; DNA damage; DNA dosimetry; nanodosimetry; nanoparticle; theranostics

1. Introduction

Nanoparticles (NPs) have transformed the biomedical sciences, especially in the areas of cancer imaging and therapy. Their unique physicochemical properties—such as small size, high surface area-to-volume ratio [1], and the ability to be functionalized [2]—allow for enhanced targeting, controlled drug release, and improved imaging contrast. These features make NPs ideal candidates for overcoming the limitations of conventional cancer therapies, such as poor specificity and systemic toxicity, and for advancing precision medicine [3]. In medical imaging, NP-based agents have facilitated the development of highly sensitive diagnostic tools, enabling earlier detection and better monitoring of cancer progression [4]. Similarly, in cancer therapy, they provide innovative platforms for delivering chemotherapeutic drugs, gene therapies, and even enabling modalities like photothermal and photodynamic therapy [5]. Figure 1 shows the contribution of NPs in nanomedicine, highlighting their roles in cancer imaging and therapy. It illustrates how NP design (e.g., size, shape, and surface functionalization) enables targeted drug delivery, enhances imaging (e.g., Magnetic Resonance Imaging (MRI), Computed Tomography (CT), and Positron Emission Tomography (PET) contrast agents, and fluorescent labels), and supports therapeutic applications like photothermal therapy, photodynamic therapy, radiotherapy, chemotherapy, and gene therapy.

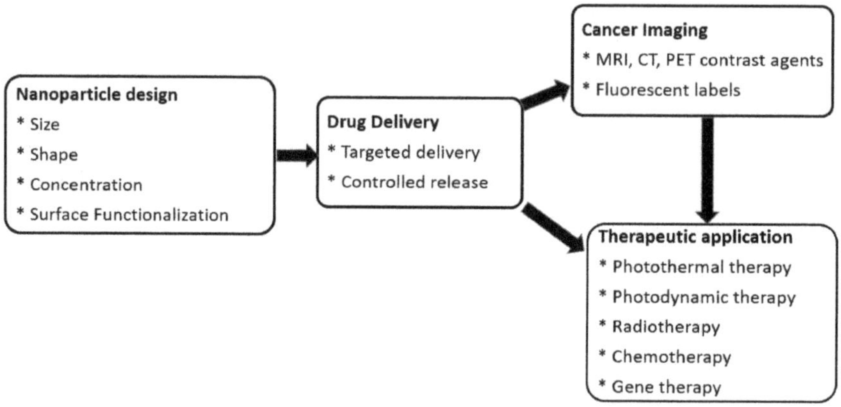

Figure 1. Contribution of NPs in nanomedicine for cancer imaging and therapy.

Despite the promise of NPs, their design, optimization, and translation into clinical applications remain challenging. The interactions of NPs with biological systems are highly complex and dynamic, influenced by numerous factors including size, shape, surface charge, and the biological environment [6]. Understanding these interactions at both the cellular and systemic levels is crucial for developing effective nanomedicine strategies. However, experimental approaches alone often fall short due to time, cost, and the complexity of biological systems. This is where computational techniques, particularly Monte Carlo (MC) simulations, play a pivotal role [7].

MC simulations have emerged as a powerful computational tool in biomedical research [8], providing stochastic methods to model complex systems and predict the behavior of NPs under various conditions. These simulations are particularly valuable in studying the interactions of NPs with tissues and cells, optimizing dosimetry in cancer therapy, and improving medical imaging techniques [9]. By simulating the probabilistic nature of particle interactions, MC methods enable researchers to explore a vast parameter space, optimize NP designs, and predict outcomes that would be difficult or impossible to achieve through experimental means alone [10].

The objective of this topical review is to provide a comprehensive overview of the applications of MC simulations in the field of nanomedicine, with a focus on their role in advancing cancer imaging and therapy. We will examine the principles of MC methods, discuss their use in modeling NP behavior, and highlight studies that demonstrate their impact on improving cancer diagnostics and treatment. Beyond summarizing current advancements, this review aims to serve as a roadmap for researchers, outlining the future directions and emerging opportunities in NP research using MC simulations. By offering a prospective view, we hope to guide the development of more effective, personalized, and clinically translatable nanomedicine strategies through the continued evolution of these powerful computational tools.

2. MC Simulations: Fundamentals and Applications

2.1. Principles of MC Methods

MC methods are a class of computational algorithms that rely on repeated random sampling to obtain numerical results. These methods are particularly effective for solving problems that are deterministic in nature but analytically intractable due to their complexity [11]. MC simulations are widely used in scenarios where the solution involves a large number of variables or uncertain inputs, making direct computation impractical.

At its core, the MC method involves three fundamental steps: (1) Problem definition: The problem is framed in terms of a probabilistic model. This typically involves defining the input variables, their probability distributions, and the relationships among them; (2) Random sampling: Random samples are generated from the defined probability distributions. These samples simulate the stochastic behavior of the system under study; (3) Computation and aggregation: For each set of sampled inputs, the system's response or outcome is calculated [12]. Repeated sampling produces a distribution of outcomes, from which statistical measures like the mean, variance, or probabilities of specific events can be estimated. MC methods excel in approximating solutions to high-dimensional integrals, optimizing complex systems, and modeling random processes [13]. The accuracy of these simulations improves with the number of samples, following the law of large numbers. However, the computational cost also scales with the number of simulations, highlighting the need for efficient sampling techniques and parallel computation [14,15].

MC methods were developed during the 1940s, largely in response to problems encountered in nuclear physics [16]. The term "Monte Carlo" was coined by physicists Stanislaw Ulam and John von Neumann while working on the Manhattan Project [17,18], referencing the famous Monte Carlo Casino due to the method's reliance on randomness and chance [19]. Their work was instrumental in solving neutron transport problems, which involve tracking the random interactions of particles in a medium. In physics, MC simulations have since become indispensable for studying systems with many degrees of freedom, such as molecular dynamics, statistical mechanics, and quantum field theory [20]. They provide critical insights into phenomena like phase transitions, particle behavior, and radiation transport. In medicine, MC methods are extensively used for modeling the behavior of radiation in biological tissues, which is crucial in fields such as radiation therapy [21] and medical imaging [22]. For example, in radiation therapy, MC simulations help optimize treatment plans by accurately predicting the dose distribution within the patient's body, ensuring maximum damage to cancerous tissues while sparing healthy ones [23]. Similarly, in diagnostic imaging techniques like PET and CT scans, these simulations improve image reconstruction and enhance diagnostic accuracy [24]. Figure 2 illustrates the MC models for a cone-beam CT and PET detector system, along with their respective phantoms used for dosimetric measurements.

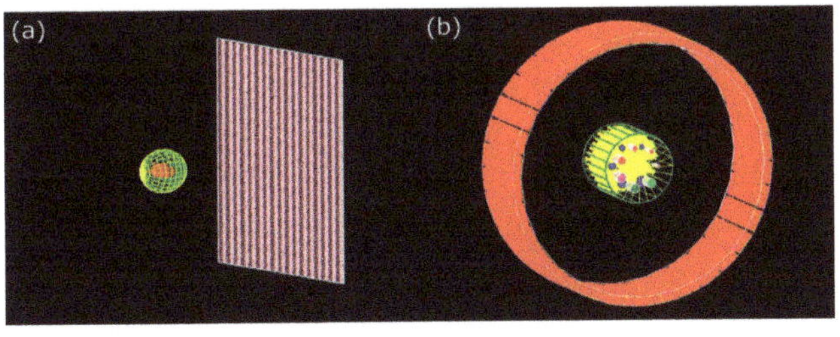

Figure 2. MC models of (**a**) cone-beam CT detector and Shepp–Logan phantom and (**b**) PET detection system and ACR-type phantom. Reproduced from reference [9] under the Creative Commons Attribution 4.0 International License (https://creativecommons.org/licenses/by/4.0/ (accessed on 9 December 2024)).

2.2. Role in NP Research

The unique properties of NPs, such as their small size, large surface area-to-volume ratio, and tunable surface chemistry, have expanded their applications across fields like medicine, environmental science, and materials engineering [25]. However, understanding their behavior in complex biological environments is challenging. MC simulations provide a valuable computational approach to model and predict NP interactions, transport, dosimetry, and distribution in these systems [26]. In biological environments, NPs interact dynamically with various entities, including cells, proteins, and other biomolecules. These interactions are influenced by particle size, shape, surface charge, and the biological milieu [27]. MC methods are particularly useful for modeling such stochastic interactions. For example, they can simulate the formation of the "protein corona"—a layer of proteins that adsorbs onto the NP surface upon entering a biological system [28,29]. The corona alters the NP's biological identity, affecting its interactions with cells and biomolecules. MC simulations model this process by simulating protein adsorption and desorption based on binding affinities, concentrations, and NP surface properties. These models predict corona composition and dynamics, providing insights into how it influences cellular interactions and uptake. MC simulations also help model cellular uptake processes, such as receptor-mediated endocytosis, by predicting the likelihood and rates of NP internalization [30]. This information is crucial for designing NPs with optimized biological functionality.

Another critical area of NP research involves predicting their transport, distribution, and accumulation in biological systems. These processes are inherently complex, involving passive diffusion, active transport, and interactions with biological barriers such as blood vessels and cell membranes [31,32]. MC simulations enable researchers to model the random motion of NPs, including their diffusion and transport within the bloodstream. By simulating these dynamics, researchers can predict how NPs distribute across different tissues and compartments. MC methods also provide insights into the accumulation of NPs in specific tissues, such as tumors, by simulating enhanced permeability and retention (EPR) effects [33]. These predictions are vital for designing NPs that achieve targeted drug delivery and imaging with minimal off-target effects. MC simulations also offer a multi-scale modeling framework, connecting molecular-level interactions with macroscopic bio-distribution patterns [34].

MC simulations play a crucial role in predicting DNA dosimetry and understanding the mechanisms of DNA damage caused by NPs in NP-enhanced radiotherapy [35,36] and imaging applications [37]. When NPs, particularly high atomic number (Z) materials such as gold or platinum, are introduced into biological tissues and exposed to ionizing radiation, they enhance the local radiation dose through a phenomenon known as the NP dose enhancement effect [38]. This increased dose can lead to a higher likelihood of DNA damage, including single-strand and double-strand breaks. MC simulations provide a detailed framework to model the complex interactions between radiation and matter at the molecular level [39]. Figure 3 shows the Monte Carlo simulation models of irradiated single AuNPs of different radii interacting with a DNA molecule. The simulation models include NPs with radii of 5 nm, 3.97 nm, and 3.15 nm. The red tracks in each subfigures represent the paths of secondary electrons generated during the simulation, highlighting the interaction dynamics between the NPs and the DNA molecule. Figure 4 shows the relationship between electron beam energy (50–200 keV) and the total number of strand breaks, influenced by the number of AuNPs in the simulation. The volume of gold was kept constant across all simulations, maintaining a consistent NP concentration. The results indicate that a higher number of AuNPs leads to more strand breaks, and lowering the electron beam energy further increases the strand break count [39].

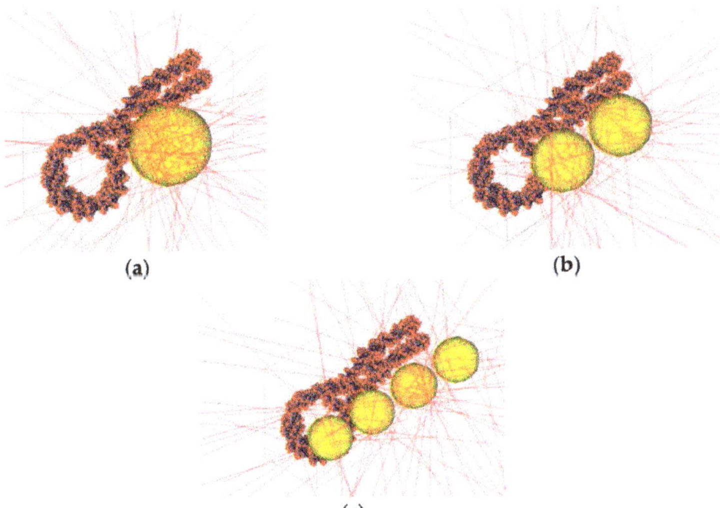

Figure 3. MC simulation models of irradiated single AuNPs of varying radii interacting with a DNA molecule. (**a**) shows a NP with a 5 nm radius, (**b**) shows a NP with a 3.97 nm radius, and (**c**) shows a NP with a 3.15 nm radius. The red tracks in each subfigure represent the paths of secondary electrons generated during the simulation. Reproduced from reference [39] under the Creative Commons Attribution 4.0 International License (https://creativecommons.org/licenses/by/4.0/ (accessed on 10 December 2024)).

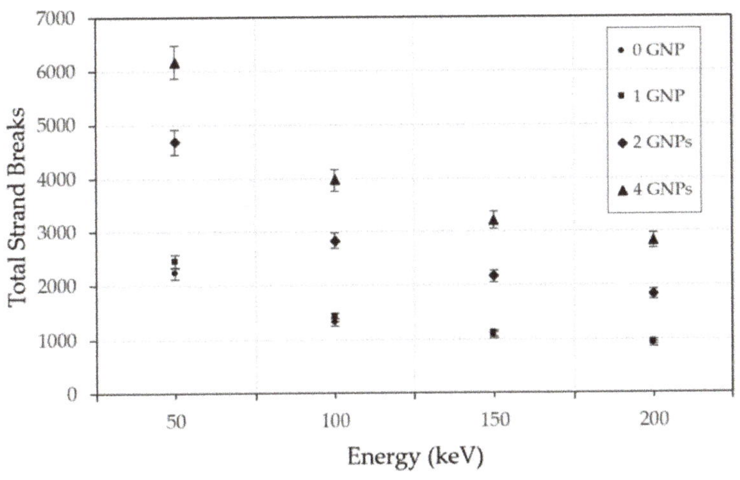

Figure 4. Relationship between total strand breaks and electron energy (keV) is influenced by the number of AuNPs present. Reproduced from reference [39] under the Creative Commons Attribution 4.0 International License (https://creativecommons.org/licenses/by/4.0/ (accessed on 9 December 2024)).

By simulating the transport of radiation particles (e.g., photons, electrons) and their interactions with NPs and surrounding biological material, MC methods can accurately predict the spatial distribution of energy deposition within the cell nucleus [40]. This enables precise dosimetric calculations at the DNA scale, allowing researchers to evaluate the extent and patterns of DNA damage. Such insights are vital for optimizing NP design and radiation parameters to maximize therapeutic efficacy while minimizing collateral damage to healthy tissues.

In FLASH radiotherapy, which involves ultra-high dose rate radiation delivery [41,42], MC simulations are important in understanding the generation and role of reactive oxygen species (ROS). ROS, including free radicals such as superoxide and hydroxyl radicals, are critical mediators of radiation-induced damage in cells [43]. When NPs are present, they can further enhance ROS production through secondary electron generation, leading to increased oxidative stress and cellular damage. FLASH radiotherapy has shown potential for reducing normal tissue toxicity while maintaining tumor control, largely due to its unique radiobiological effects [44]. However, the underlying mechanisms, including the interplay between ROS dynamics and ultra-high dose rates, are still under investigation. MC simulations enable researchers to model the rapid radiation-matter interactions and the subsequent ROS generation at different spatial and temporal scales. By incorporating NP-mediated effects into these models, researchers can predict how NPs influence ROS production and distribution during FLASH radiotherapy. These predictions are critical for understanding the enhanced therapeutic effects of FLASH in combination with NPs and for designing optimized treatment strategies. Figure 5 shows the role of MC simulations in NP research, highlighting their applications in modeling NP interactions, transport, and radiotherapy effects. The diagram illustrates how MC simulations provide insights into biological processes like protein corona formation, cellular uptake, and DNA damage, enabling the optimization of NP design for targeted drug delivery, imaging, and enhanced therapeutic efficacy.

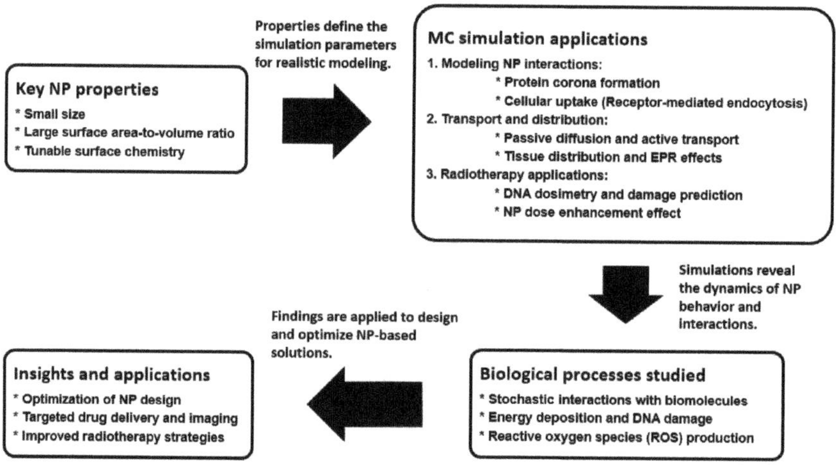

Figure 5. Role of MC Simulations in NP Research.

2.3. MC NP Model

Accurate modeling of NPs in MC simulations is crucial for understanding their behavior in complex biological environments. These models must account for the physical, chemical, and biological characteristics of NPs and their interactions with cells, tissues, and biological fluids. Robust MC NP models are essential for applications in imaging, therapy, and drug delivery, enhancing efficacy and safety [7].

Key aspects of MC NP modeling include incorporating NP-specific properties like size, shape, density, atomic composition, and surface functionalization. High-Z materials like gold and bismuth require detailed atomic structure modeling to simulate interactions with ionizing radiation. The size and shape of NPs influence their transport properties and interaction cross-sections, critical for dose enhancement in radiotherapy or contrast generation in imaging [39]. Surface properties, including charge and functional groups, determine NP interactions with biological components, such as protein corona formation or cellular uptake [28,29].

Modeling NP interactions with biological media is critical, as these environments are highly heterogeneous. MC NP models integrate multiple layers of interaction, simulating how NPs amplify energy deposition in tissues via photoelectric and Compton effects, and cellular uptake mechanisms like receptor-mediated endocytosis and passive diffusion. Advanced MC NP models address the complexity of the tumor microenvironment, including hypoxia, vascular density, and extracellular matrix stiffness, to simulate NP distribution and accumulation in tumors accurately.

A significant advancement in NP modeling is the development of multi-scale MC models, bridging molecular, cellular, and tissue scales [32]. These models simulate radiation interactions with NPs and biomolecules at the molecular level, NP uptake and intracellular distribution at the cellular level, and NP transport and interactions within the tumor microenvironment at the tissue level. Multi-scale MC models predict therapeutic outcomes by simulating how enhanced energy deposition at the molecular scale translates into cellular damage and tumor control at the macroscopic level. They also incorporate ROS production dynamics for accurate predictions of oxidative stress-induced damage.

Despite advancements, challenges remain in modeling the dynamic and heterogeneous tumor microenvironment and incorporating biological processes like NP aggregation, clearance, and immune responses. Hybrid models combining MC methods with agent-based modeling and machine learning hold promise for addressing these challenges [37].

3. NPs in Biomedical Imaging

3.1. Imaging Modalities Enhanced by NPs

NPs have innovated biomedical imaging by enhancing the sensitivity, resolution, and specificity of various imaging modalities. Their unique physical and chemical properties, such as high Z, magnetic susceptibility, and tunable optical characteristics, make them ideal contrast agents in imaging techniques like MRI, Magnetic Particle Imaging (MPI), CT, PET, and optical imaging [45]. These advancements have significantly improved disease diagnosis, monitoring, and treatment planning.

In MRI, NPs, particularly superparamagnetic iron oxide nanoparticles (SPIONs), are employed as contrast agents to improve image contrast in soft tissues [46]. Their magnetic properties influence the relaxation times of nearby hydrogen protons, enhancing the visibility of specific tissues or pathological regions [47]. MPI, another imaging modality, leverages the unique magnetic properties of SPIONs to directly image their spatial distribution with high sensitivity and resolution, providing quantitative and real-time imaging capabilities without background noise from biological tissues. Similarly, in CT, high-Z NPs, such as gold or bismuth, serve as contrast agents by amplifying X-ray attenuation. This results in higher-contrast images with improved differentiation between tissues, enabling more precise detection of tumors and vascular abnormalities [48]. PET imaging benefits from NPs functionalized with positron-emitting isotopes. These radiolabeled NPs not only enhance the imaging signal but also provide a platform for targeted imaging, where NPs are directed to specific cells or tissues using ligands or antibodies [49]. Figure 6 shows three radiolabeling techniques using metallic radionuclides like Cu-64, Ga-68, and Zr-89 with NPs. These techniques are post-radiolabeling, pre-radiolabeling, and direct radiolabeling. This specificity improves the detection of cancerous lesions and metastases. In optical imaging, quantum dots and other fluorescent NPs offer superior brightness and photostability compared to traditional dyes, enabling the real-time imaging of molecular and cellular processes with high spatial resolution [50].

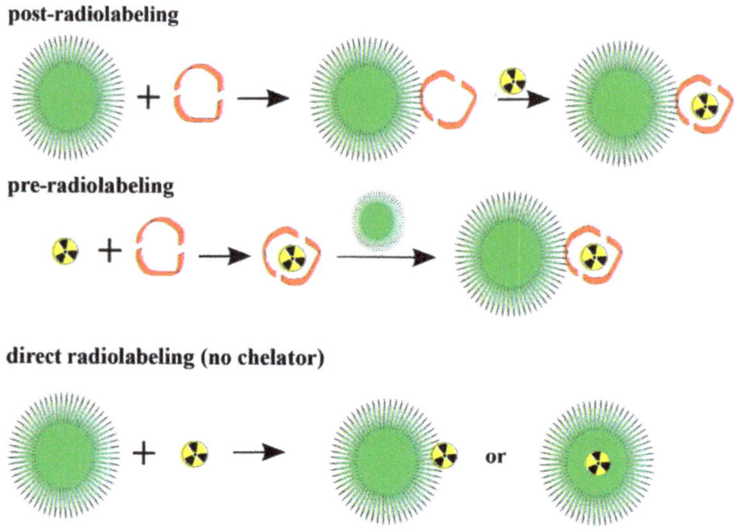

Figure 6. Three common radiolabeling methods involving metallic radionuclides and NPs. Reproduced from reference [49] under the Creative Commons Attribution 4.0 International License (https://creativecommons.org/licenses/by/4.0/ (accessed on 10 December 2024)).

MC simulations have been pivotal in optimizing the use of NPs across these imaging modalities. By simulating complex interactions between particles, tissues, and imaging systems, MC methods enable the evaluation of imaging performance under various conditions. For example, in CT imaging, MC simulations help optimize NP concentration and distribution to maximize contrast while minimizing radiation dose [51,52]. In PET, simulations aid in determining optimal isotope loading on NPs for improved signal-to-noise ratios and reduced patient exposure to radioactivity [53]. Case studies have demonstrated the power of MC simulations in refining imaging strategies. For example, in MRI, simulations have been used to model the distribution of SPIONs within tumors, providing insights into their impact on image contrast and guiding the development of more efficient NP formulations [54,55]. Similarly, in optical imaging, MC simulations have helped optimize light–tissue interactions, improving the accuracy of fluorescence-based imaging techniques [56,57].

3.2. MC Simulation in Image Reconstruction

NPs have enhanced biomedical imaging by offering unique optical, magnetic, and radiative properties that innovate imaging modalities such as fluorescence imaging, MRI, and CT. However, the accurate reconstruction of images involving NPs requires sophisticated modeling to account for their complex interactions with light, radiation, and biological tissues. MC simulations have emerged as a tool for tackling these challenges [58]. Figure 7 shows the outcomes of various MC methods in image reconstruction. Two MC techniques, namely, common MC dropout and the innovative MC-Arbitrary Scan Masking (ASM), enhanced image reconstructions of MRI, CT, and PET compared to the control groups. The MC-ASM algorithm is a novel approach that combines MC sampling with a dynamic scan-masking strategy to selectively mask input data, thereby simulating arbitrary sampling patterns. This method integrates with deep learning models to account for variability in scan data, allowing for more robust and accurate reconstructions even under sparse or incomplete sampling conditions. Notably, the MC-ASM method, leveraging its ability to mimic real-world noise and data irregularities, delivered the best performance, achieving superior image quality and resolution across imaging modalities. The Structural Similarity Index Measure (SSIM) in Figure 7B is a metric that evaluates the quality of an image by comparing its structural, luminance, and contrast similarities to a reference image. The dashed line in Figure 7B indicates the maximum SSIM value in the plot. Unlike traditional metrics like Mean Squared Error (MSE), SSIM aligns closely with human visual perception, with values ranging from −1 to 1, where 1 indicates perfect similarity.

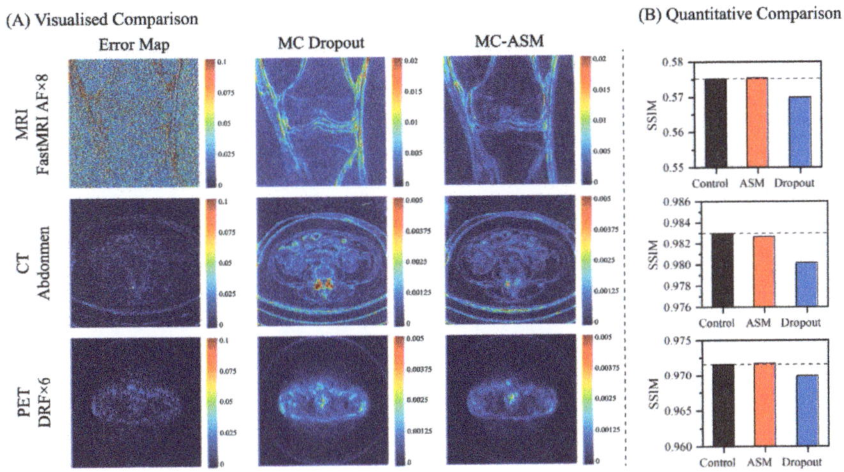

Figure 7. (**A**) Visualized samples of uncertainty maps generated by the MC dropout and MC-ASM, along with the corresponding error maps. (**B**) Quantitative comparison between (1) Arbitrary-Masked Mamba-based model (MambaMIR) without MC-ASM or MC dropout (control group), (2) MambaMIR with MC-ASM, and (3) MambaMIR with MC dropout across three datasets. Reproduced from reference [58] under the Creative Commons Attribution 4.0 International License (https://creativecommons.org/licenses/by/4.0/ (accessed on 10 December 2024)).

NPs can enhance imaging signals, but noise from scattering, absorption, and background interference often reduces the clarity of reconstructed images. MC simulations provide a robust framework for modeling these interactions at a microscopic level, enabling the separation of NP-specific signals from background noise. For example, in fluorescence imaging, MC simulations accurately model the scattering of emitted photons in tissues, improving signal detection and localization [59]. When coupled with NP-based fluorescent probes, this approach achieves a higher signal-to-noise ratio, critical for detecting tumors at early stages [60]. In photoacoustic imaging, NPs such as gold nanorods serve as contrast agents by efficiently converting absorbed light into acoustic waves [61]. MC simulations help optimize the laser parameters and NP distribution to maximize signal strength while minimizing noise, resulting in clearer reconstructed images [62]. These simulations are particularly valuable in scenarios where noise dominates due to tissue heterogeneity or low NP concentrations.

The unique physical properties of NPs, such as their ability to selectively accumulate in tumor tissues (via enhanced permeability and retention effects), inherently enhance imaging contrast. MC simulations amplify this advantage by providing precise models of NP behavior in biological environments. For example, in MRI, SPIONs serve as contrast agents by altering the local magnetic field [63]. MC-based models simulate these magnetic perturbations to refine image reconstruction algorithms, thereby improving contrast between tumor and healthy tissues. Similarly, in CT imaging, NPs like gold exhibit high X-ray attenuation coefficients, offering superior contrast compared to traditional agents. MC simulations model the intricate interactions between X-rays, NPs, and tissues, enabling enhanced resolution and structural delineation in reconstructed images [64]. These advancements allow for the visualization of microvascular details and early-stage tumor boundaries that would otherwise be undetectable.

4. NPs in Cancer Therapy

4.1. Radiotherapy and Radiosensitization

Radiotherapy is a cornerstone of cancer treatment, using ionizing radiation to damage the DNA of cancer cells and inhibit their proliferation [65]. However, the efficacy of radiotherapy can be limited by the inherent radioresistance of tumors and the collateral damage to surrounding healthy tissues [66]. The introduction of NPs, particularly gold nanoparticles (AuNPs), has transformed radiotherapy by acting as potent radiosensitizers. These NPs amplify radiation effects within tumors, offering enhanced therapeutic outcomes [67]. MC simulations play a pivotal role in optimizing the use of AuNPs in radiotherapy, particularly for dosimetry and treatment planning.

AuNPs have emerged as highly effective radiosensitizers due to their unique physical and chemical properties. Their high atomic number ($Z = 79$) results in a significant photoelectric absorption cross-section, particularly under low-energy X-rays, which leads to the generation of secondary electrons, Auger electrons, and free radicals [68]. These reactive species cause localized damage to cancer cells, enhancing the therapeutic efficacy of radiation treatments. In addition to their physical properties, AuNPs can be functionalized with targeting ligands, such as antibodies or peptides, to selectively accumulate in tumors via active targeting mechanisms [5,69]. This selective accumulation minimizes radiation exposure to healthy tissues, reducing side effects. Studies have shown that AuNPs enhance the radiation dose delivered to tumors by up to several fold, particularly in photon-based therapies [70].

The precise application of AuNPs in radiotherapy relies heavily on accurate dosimetry and treatment planning. MC simulations are vital in this context, as they provide a detailed stochastic modeling of radiation transport and interactions within tissues containing NPs [71]. These simulations account for the physical properties of AuNPs, including their size, shape, concentration, and spatial distribution, as well as their interaction with different radiation types, such as photons, electrons, or protons. MC-based nanodosimetry enables the quantification of enhanced dose deposition in tumor tissues due to the presence of AuNPs [72]. For example, MC simulations have demonstrated that NP clusters amplify dose heterogeneity within tumors, offering insights into optimizing radiation delivery [73]. Furthermore, these simulations assist in identifying optimal radiation energies that maximize the NP-induced dose enhancement while minimizing damage to surrounding healthy tissues [74]. In treatment planning, MC methods integrate biological factors, such as the radiosensitivity of tumors and the oxygenation status, with NP distributions to develop personalized radiotherapy protocols [75]. This capability allows for precise tailoring of radiation dose and beam configuration to maximize therapeutic benefits. Figure 8 shows the relationship between the dose enhancement ratio (DER) and skin target thickness, taking into account different AuNP concentrations, with 105 and 220 kVp photon beams. The AuNP concentration is chosen based on the levels used in small-animal experiments. It reveals that, for 220 kVp photon beams, the dose enhancement factor remains largely unaffected by skin target thickness when AuNP concentrations range from 3 to 40 mg/mL.

Conversely, for 105 kVp photon beams, the enhancement factor increases as the target thickness decreases, especially when the NP concentration is above 18 mg/mL [74].

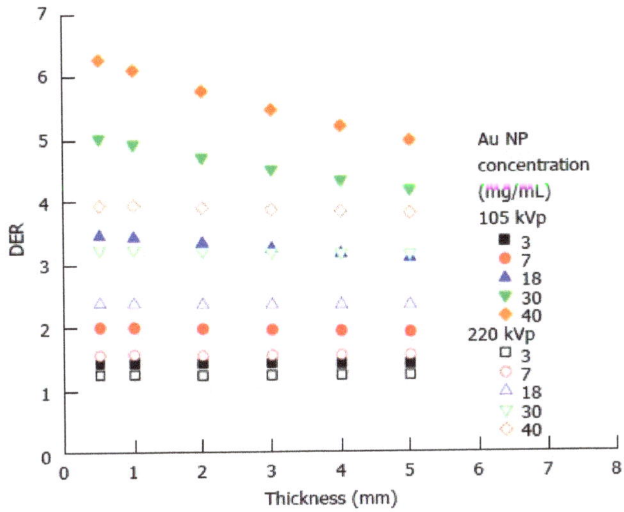

Figure 8. Relationship between the dose enhancement ratio (DER) and skin target thickness with varying concentrations of AuNPs using 105 and 220 kVp photon beams. Reproduced from reference [74] under the Creative Commons Attribution 4.0 International License (https://creativecommons.org/licenses/by/4.0/ (accessed on 10 December 2024)).

4.2. Photothermal and Photodynamic Therapy

Photothermal therapy (PTT) and photodynamic therapy (PDT) are cutting-edge cancer treatments that utilize the unique interactions between light and NPs to achieve localized tumor destruction. In PTT, NPs absorb near-infrared (NIR) light and convert it into heat, leading to thermal ablation of cancer cells. In PDT, light activation of NPs or photosensitizers generates ROS, which cause oxidative damage to tumor tissues. Both therapies benefit immensely from computational modeling and simulation, particularly for optimizing thermal effects, ROS generation, and dose delivery [76]. MC simulations play a critical role in these processes by providing insights into light–tissue interactions and the spatial–temporal dynamics of treatment.

The effectiveness of PTT hinges on the ability to generate sufficient heat to cause irreversible tumor cell damage while sparing surrounding healthy tissues. AuNPs, carbon-based nanostructures, and other plasmonic materials are widely employed in PTT due to their efficient light absorption and photothermal conversion properties [77]. To predict and optimize thermal effects, MC simulations model the transport of light in heterogeneous biological tissues, considering scattering, absorption, and NP distribution. Simulations also account for heat diffusion dynamics, enabling precise predictions of temperature elevation in the targeted region [78]. These models incorporate factors such as laser wavelength, power density, NP concentration, and tumor geometry to achieve controlled and uniform thermal ablation. In PDT, the generation of ROS, particularly singlet oxygen, is the primary mechanism for inducing tumor cell apoptosis. MC simulations model the propagation and absorption of light within tissues, ensuring the sufficient activation of photosensitizers localized in tumors [79]. Moreover, these simulations predict the spatial distribution of ROS and their diffusion in tissue microenvironments, which is critical for maximizing therapeutic effects while minimizing off-target oxidative damage [80].

The success of PTT and PDT depends on precise dose and temperature optimization. MC simulations assist in determining the ideal light dosimetry for NP activation. In PTT, these simulations guide the selection of laser parameters, such as fluence rate and exposure duration, to achieve tumor-specific heating thresholds without causing thermal damage to adjacent healthy tissues [81]. Figure 9 shows the temperature distribution in the YZ plane of normal and tumor tissue at 200 and 600 s after laser irradiation, with the distribution radius ratio of AuNPs (φ_{drr}) = 1, three injections, and a laser power of 50 mW at 1064 nm. The black box indicates tumor tissue. After 600 s, the temperature was higher than at 200 s. Approximately 96% of the tumor volume reached the apoptosis temperature range at 200 s, and 94% at 600 s. However, normal tissue beneath the tumor reached around 48.5 °C after 600 s, risking thermal damage [81].

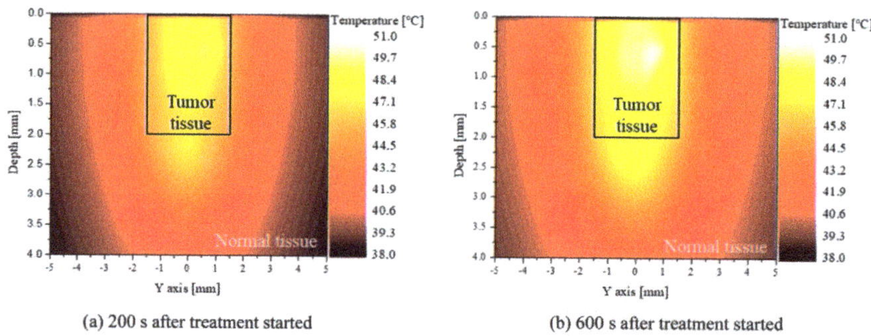

Figure 9. Temperature distribution of the medium (φ_{drr} = 1, 3 AuNP injections, Pl = 50 mW). Reproduced from reference [81] under the Creative Commons Attribution 4.0 International License (https://creativecommons.org/licenses/by/4.0/ (accessed on 10 December 2024)).

For PDT, MC-based models integrate photosensitizer photophysics, tissue oxygenation, and light propagation to optimize the generation and diffusion of ROS. Simulation studies reveal the influence of tumor hypoxia—a common challenge in PDT—on treatment efficacy and help design strategies such as fractionated light delivery or oxygen supplementation to overcome these limitations [82]. Furthermore, hybrid PTT-PDT approaches, which combine the thermal and oxidative effects of NPs, have gained attention for synergistic cancer therapy [83]. MC simulations are instrumental in evaluating the interplay between heat generation and ROS dynamics, enabling comprehensive treatment planning that advantages the strengths of both modalities.

5. Synergistic Applications

The integration of imaging and therapeutic functionalities into single NP systems, known as theranostics, represents a transformative approach in nanomedicine [84]. Theranostic NPs enable simultaneous cancer diagnosis, the real-time monitoring of therapeutic delivery, and effective treatment, creating opportunities for personalized medicine [5]. Figure 10 shows that various functional groups, including PEG, ssDNA, antibodies, peptides, drugs, fluorescence markers, and siRNA, can be attached to the particle surface to enable both imaging and therapeutic functions in cancer nanotheranostics.

MC simulations play an important role in optimizing the design and application of dual-function NP systems, ensuring their efficacy in both imaging and therapy [85]. Theranostic NPs combine diagnostic imaging modalities such as MRI, CT, or fluorescence imaging with therapeutic functions like PTT, PDT, or drug delivery. For example, AuNPs functionalized with fluorescent dyes can enable high-resolution tumor imaging while simultaneously inducing photothermal ablation upon NIR light exposure [86]. Similarly, iron oxide NPs can serve as contrast agents for MRI and be employed for hyperthermia-based cancer therapy [47,63]. One key advantage of theranostic NPs is their ability to localize and act specifically within tumor tissues through passive mechanisms (e.g., EPR effect) or active targeting via ligand-receptor interactions. This targeted dual functionality enhances therapeutic efficacy while minimizing off-target effects [87].

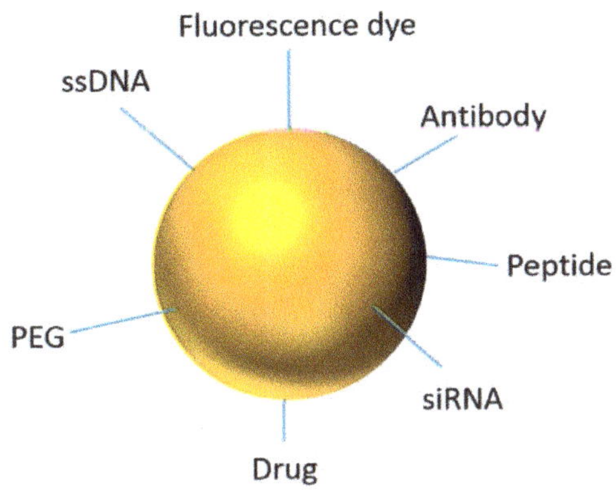

Figure 10. Depiction of an AuNP for theranostic applications. Reproduced from reference [5] under the Creative Commons Attribution 4.0 International License (https://creativecommons.org/licenses/by/4.0/ (accessed on 10 December 2024)).

MC simulations are essential in the development and optimization of dual-function NP systems. By modeling the complex interactions of NPs with biological tissues, radiation, and light, MC simulations provide insights into the behavior of theranostic agents under various physiological and therapeutic conditions. In imaging, MC simulations help evaluate the signal generation capabilities of NPs across different modalities. For example, in fluorescence imaging, MC models account for light scattering, absorption, and NP emission, optimizing the contrast and resolution [88]. In CT or MRI, simulations analyze NP interactions with X-rays or magnetic fields to enhance image quality while minimizing exposure to ionizing radiation or optimizing magnetic field parameters [89]. For therapeutic applications, MC simulations model the deposition of energy within tumor tissues, whether from thermal energy in PTT, ROS in PDT, or radiation in NP-enhanced radiotherapy [90]. In dual-function systems, these simulations address the interplay between imaging and therapeutic mechanisms, ensuring that the NPs perform both roles effectively without compromising one for the other [91]. Furthermore, MC studies enable the optimization of NP properties such as size, shape, composition, and functionalization. They also provide insights into the bio-distribution and clearance of theranostic agents, ensuring safety and efficacy in clinical applications. For example, MC simulations can guide the design of NPs with optimal optical or radiative properties for imaging while maintaining the necessary energy absorption characteristics for therapy [92].

6. Challenges and Future Directions

MC simulations have been instrumental in advancing cancer imaging and therapy by providing detailed insights into the complex interactions of NPs within biological systems. However, several challenges continue to limit their effectiveness, and emerging trends and technologies present opportunities for significant advancements. One of the major challenges in MC simulation lies in modeling the intricate biological complexity of the tumor microenvironment [93]. Tumors exhibit heterogeneous cellular structures, dynamic vasculature, and immune interactions that are difficult to replicate with current simulation capabilities [94]. Simplifications in models can lead to discrepancies between simulations and real-world biological behavior. In addition, the scarcity of reliable biological and physicochemical data at the nanoscale hinders accurate calibration and validation of simulations. Computational intensity remains a significant barrier, as MC simulations require vast numbers of iterations to produce statistically robust results, which is particularly demanding for high-resolution geometries or multiphysics modeling [95]. Bridging the gaps between disciplines, such as computational science, biology, and physics, is another ongoing challenge that limits the practical application of MC simulations in nanomedicine.

Recent advances in NP design, simulation technology, and novel methodologies are beginning to address these challenges. Hybrid models that integrate MC methods with deterministic approaches, such as finite element methods or machine learning algorithms, offer a way to simulate complex systems more efficiently [96]. In parallel, FLASH radiotherapy—a novel technique that delivers ultra-high-dose radiation in extremely short bursts—has emerged as a promising cancer therapy [41,42]. MC simulations are being adapted to model the unique physical and biological effects of FLASH radiotherapy, including its ability to spare normal tissues while maintaining tumor control [97,98]. This requires MC simulations to evolve to capture ultrafast dynamics and high-dose deposition patterns with unprecedented precision.

Another important development is the application of quantum computing to MC simulations. Quantum algorithms have the potential to drastically reduce computational time by efficiently solving complex probabilistic problems inherent in MC methods [99]. By harnessing quantum superposition and entanglement, researchers could simulate larger systems at atomic and molecular scales, enabling more detailed studies of NP interactions within biological environments. Quantum computing could also facilitate the development of patient-specific simulations by processing vast datasets, such as those derived from multi-omics and imaging studies, much faster than classical computers [100].

The integration of big data and high-performance computing further enhances the clinical relevance of MC simulations. Advances in artificial intelligence and machine learning allow simulations to incorporate large-scale experimental datasets for model refinement [101]. This is particularly valuable for designing multifunctional NPs with specific therapeutic and diagnostic properties. Simulations are also increasingly being used to predict the behavior of these NPs in personalized scenarios, optimizing their bio-distribution, dosimetry, and therapeutic efficacy [102].

The translation of these advancements to clinical applications offers exciting opportunities for personalized nanomedicine. MC simulations can guide the design of patient-specific treatments, tailoring NP formulations to optimize outcomes while minimizing side effects. For example, integrating FLASH radiotherapy with personalized dosimetry could revolutionize cancer therapy by providing highly effective treatments with reduced toxicity [103]. Furthermore, MC simulations could streamline clinical trials by predicting outcomes and optimizing protocols, reducing the need for extensive preclinical testing. These simulations also hold potential for aiding regulatory processes by providing robust safety and efficacy data for emerging nanotherapies [104,105]. Table 1 outlines the key limitations of current MC simulation techniques in cancer imaging and therapy, alongside emerging technologies and methodologies aimed at addressing these challenges to enhance clinical translation and personalized nanomedicine.

Table 1. Summary of challenges and future directions for MC simulations in nanomedicine.

Aspect	Challenges	Future Directions
Modeling Biological Complexity	Simplified representations of tumor microenvironments lead to inaccuracies. Limited understanding of dynamic vasculature and immune responses.	Develop high-fidelity models incorporating dynamic and heterogeneous biological interactions. Leverage experimental data from advanced imaging and omics technologies for model refinement.
Data Scarcity and Quality	Lack of reliable physicochemical and biological data at the nanoscale hinders calibration. Incomplete datasets limit validation.	Integrate large-scale datasets (e.g., multi-omics, imaging) for improved accuracy. Use artificial intelligence and machine learning to bridge data gaps and improve prediction accuracy.
Computational Intensity	High computational demands for detailed simulations. Difficulty modeling complex geometries and ultrafast dynamics in a feasible timeframe.	Adopt quantum computing to enhance computational efficiency. Utilize hybrid models combining MC simulations with deterministic methods or artificial intelligence algorithms. Leverage high-performance and cloud computing.
Emerging Treatment Techniques	Lack of established simulation methods for novel therapies like FLASH radiotherapy. Difficulty adapting existing models to capture ultrafast dynamics.	Develop specialized MC simulation frameworks to model FLASH radiotherapy. Incorporate ultrafast radiation dose deposition and biological response dynamics into simulations.
NP Design	Challenges in predicting interactions of multifunctional or novel NPs. Limited capability to simulate complex theranostic agents.	Simulate advanced NP designs, including multifunctional and biodegradable systems. Incorporate molecular-level interactions to refine predictions.
Clinical Translation	Bridging the gap between simulation outcomes and real-world applicability remains difficult. Regulatory hurdles due to limited simulation validation for clinical use.	Use patient-specific data to guide personalized treatment designs. Streamline clinical trials by using MC simulations for prediction and optimization. Support regulatory approval with robust data.
Interdisciplinary Gaps	Limited collaboration among computational scientists, biologists, and physicists. Knowledge silos hinder integrated approaches to complex problems.	Foster interdisciplinary research collaborations. Develop standardized frameworks and protocols for MC simulations in nanomedicine.

7. Conclusions

MC simulations have emerged as powerful tools in advancing NP-based applications for cancer imaging and therapy. This review highlights the pivotal role of MC simulations in unraveling complex biological interactions, optimizing NP design, and enhancing treatment precision. By providing a probabilistic framework to model the stochastic nature of NP behavior in biological systems, MC simulations have become indispensable for simulating processes such as drug delivery, radiation dosimetry, and bio-distribution with remarkable accuracy. Despite current limitations, including challenges in modeling biological complexity, computational intensity, and data scarcity, MC simulations continue to drive progress in nanomedicine. Emerging technologies, such as hybrid modeling approaches, high-performance computing, and quantum algorithms, promise to overcome these barriers. Furthermore, innovations like FLASH radiotherapy and multifunctional theranostic NPs are pushing the boundaries of MC simulations, enabling the exploration of novel therapeutic paradigms and clinical applications. The transformative potential of MC simulations lies in their ability to bridge fundamental research and clinical practice. As these simulations become increasingly integrated with patient-specific data, they offer the possibility of personalized nanomedicine, where treatments are tailored to individual physiological and genetic profiles. Moreover, their role in streamlining clinical trials and supporting regulatory approval processes underscores their importance in accelerating the translation of nanotechnology-based therapies to the clinic.

In conclusion, MC simulations represent a cornerstone of modern nanomedicine, with the capacity to revolutionize cancer diagnostics and therapy. Continued advancements in simulation techniques, interdisciplinary collaboration, and integration with emerging technologies will ensure that MC simulations remain at the forefront of innovation, paving the way for more precise, effective, and accessible healthcare solutions.

Funding: This research was supported by the Government of Canada's New Frontiers in Research. Fund—Exploration (Grant number: NFRFE-2022-00707), through the three federal research funding agencies (CIHR, NSERC and SSHRC).

Data Availability Statement: No new data was created.

Conflicts of Interest: The authors declare no conflict of interest.

References

1. Alexis, F.; Pridgen, E.M.; Langer, R.; Farokhzad, O.C. Nanoparticle technologies for cancer therapy. *Drug Deliv.* **2010**, *197*, 55–86.
2. Subbiah, R.; Veerapandian, M.; Yun, K.S. Nanoparticles: Functionalization and multifunctional applications in biomedical sciences. *Curr. Med. Chem.* **2010**, *17*, 4559–4577. [CrossRef] [PubMed]
3. Mitchell, M.J.; Billingsley, M.M.; Haley, R.M.; Wechsler, M.E.; Peppas, N.A.; Langer, R. Engineering precision nanoparticles for drug delivery. *Nat. Rev. Drug Discov.* **2021**, *20*, 101–124. [CrossRef] [PubMed]
4. Thierry, B. Drug nanocarriers and functional nanoparticles: Applications in cancer therapy. *Curr. Drug Deliv.* **2009**, *6*, 391–403. [CrossRef] [PubMed]
5. Siddique, S.; Chow, J.C.L. Gold nanoparticles for drug delivery and cancer therapy. *Appl. Sci.* **2020**, *10*, 3824. [CrossRef]
6. Al-Thani, A.N.; Jan, A.G.; Abbas, M.; Geetha, M.; Sadasivuni, K.K. Nanoparticles in cancer theragnostic and drug delivery: A comprehensive review. *Life Sci.* **2024**, *352*, 122899. [CrossRef] [PubMed]
7. Chow, J.C.L. Recent progress in Monte Carlo simulation on gold nanoparticle radiosensitization. *AIMS Biophys.* **2018**, *5*, 231–244. [CrossRef]
8. Velikova, T.; Mileva, N.; Naseva, E. Method "Monte Carlo" in healthcare. *World J. Methodol.* **2024**, *14*, 3. [CrossRef] [PubMed]
9. Lee, H. Monte Carlo methods for medical imaging research. *Biomed. Eng. Lett.* **2024**, *14*, 1195–1205. [CrossRef] [PubMed]
10. Li, W.B.; Müllner, M.; Greiter, M.B.; Bissardon, C.; Xie, W.Z.; Schlatll, H.; Oeh, U.; Li, J.L.; Hoeschen, C. Monte Carlo simulations of dose enhancement around gold nanoparticles used as x-ray imaging contrast agents and radiosensitizers. In *Medical Imaging 2014, Physics of Medical Imaging*; SPIE: Bellingham, WA, USA, 2014; Volume 9033, pp. 430–440.
11. Harrison, R.L. Introduction to Monte Carlo simulation. *AIP Conf. Proc.* **2010**, *1204*, 17. [PubMed]
12. Rubinstein, R.Y.; Kroese, D.P. *Simulation and the Monte Carlo Method*; John Wiley & Sons: Hoboken, NJ, USA, 2016.
13. Bonate, P.L. A brief introduction to Monte Carlo simulation. *Clin. Pharmacokinet.* **2001**, *40*, 15–22. [CrossRef] [PubMed]
14. Wu, F.; Dantan, J.Y.; Etienne, A.; Siadat, A.; Martin, P. Improved algorithm for tolerance allocation based on Monte Carlo simulation and discrete optimization. *Comput. Ind. Eng.* **2009**, *56*, 1402–1413. [CrossRef]
15. Alerstam, E.; Svensson, T.; Andersson-Engels, S. Parallel computing with graphics processing units for high-speed Monte Carlo simulation of photon migration. *J. Biomed. Opt.* **2008**, *13*, 060504. [CrossRef]
16. Bielajew, A.F. History of monte carlo. In *Monte Carlo Techniques in Radiation Therapy*; CRC Press: Boca Raton, FL, USA, 2021; pp. 3–15.
17. Los Alamos National Laboratory. Hitting the Jackpot: The Birth of the Monte Carlo Method. Available online: https://www.lanl.gov/media/publications/actinide-research-quarterly/first-quarter-2023/hitting-the-jackpot-the-birth-of-the-monte-carlo-method (accessed on 9 December 2024).
18. Benov, D.M. The Manhattan Project, the first electronic computer and the Monte Carlo method. *Monte Carlo Methods Appl.* **2016**, *22*, 73–79. [CrossRef]
19. Metropolis, N. The beginning of the Monte Carlo method. *Los Alamos Sci.* **1987**, *15*, 125–130.

20. Carlson, J.; Gandolfi, S.; Pederiva, F.; Pieper, S.C.; Schiavilla, R.; Schmidt, K.E.; Wiringa, R.B. Quantum Monte Carlo methods for nuclear physics. *Rev. Mod. Phys.* **2015**, *87*, 1067–1118. [CrossRef]
21. Rogers, D.W. Fifty years of Monte Carlo simulations for medical physics. *Phys. Med. Biol.* **2006**, *51*, R287. [CrossRef] [PubMed]
22. Zaidi, H. Relevance of accurate Monte Carlo modeling in nuclear medical imaging. *Med. Phys.* **1999**, *26*, 574–608. [CrossRef] [PubMed]
23. Brualla, L.; Rodriguez, M.; Lallena, A.M. Monte Carlo systems used for treatment planning and dose verification. *Strahlenther. Onkol.* **2017**, *193*, 243. [CrossRef]
24. Saaidi, R.; Zeghari, A.; El Moursli, R.C. Monte Carlo simulation of two Siemens biograph PET/CT system using gate: Image quality performance. *Radiat. Phys. Chem.* **2024**, *218*, 111653. [CrossRef]
25. Rudramurthy, G.R.; Swamy, M.K. Potential applications of engineered nanoparticles in medicine and biology: An update. *J. Biol. Inorg. Chem.* **2018**, *23*, 1185–1204. [CrossRef] [PubMed]
26. Moradi, F.; Saraee, K.R.; Sani, S.A.; Bradley, D.A. Metallic nanoparticle radiosensitization: The role of Monte Carlo simulations towards progress. *Radiat. Phys. Chem.* **2021**, *180*, 109294. [CrossRef]
27. Mailander, V.; Landfester, K. Interaction of nanoparticles with cells. *Biomacromolecules* **2009**, *10*, 2379–2400. [CrossRef] [PubMed]
28. Zhdanov, V.P. Formation of a protein corona around nanoparticles. *Curr. Opin. Colloid. Interface Sci.* **2019**, *41*, 95–103. [CrossRef]
29. Carnal, F.; Clavier, A.; Stoll, S. Polypeptide-nanoparticle interactions and corona formation investigated by monte carlo simulations. *Polymers* **2016**, *8*, 203. [CrossRef] [PubMed]
30. Deng, H.; Dutta, P.; Liu, J. Stochastic modeling of nanoparticle internalization and expulsion through receptor-mediated transcytosis. *Nanoscale* **2019**, *11*, 11227–11235. [CrossRef] [PubMed]
31. Liu, H.H.; Surawanvijit, S.; Rallo, R.; Orkoulas, G.; Cohen, Y. Analysis of nanoparticle agglomeration in aqueous suspensions via constant-number Monte Carlo simulation. *Environ. Sci. Technol.* **2011**, *45*, 9284–9292. [CrossRef]
32. Shi, X.; Tian, F. Multiscale modeling and simulation of nano-carriers delivery through biological barriers—A review. *Adv. Theory Simul.* **2019**, *2*, 1800105. [CrossRef]
33. Peukert, D.; Kempson, I.; Douglass, M.; Bezak, E. Gold nanoparticle enhanced proton therapy: Monte Carlo modeling of reactive species' distributions around a gold nanoparticle and the effects of nanoparticle proximity and clustering. *Int. J. Mol. Sci.* **2019**, *20*, 4280. [CrossRef] [PubMed]
34. Jiang, L.; Guo, Y.; Liu, Z.; Chen, S. Computational understanding of the coalescence of metallic nanoparticles: A mini review. *Nanoscale* **2024**, *16*, 5521–5536. [CrossRef] [PubMed]
35. Chow, J.C.L. Biophysical insights into nanomaterial-induced DNA damage: Mechanisms, challenges, and future directions. *AIMS Biophys.* **2024**, *11*, 340–369. [CrossRef]
36. Berbeco, R.I.; Korideck, H.; Ngwa, W.; Kumar, R.; Patel, J.; Sridhar, S.; Johnson, S.; Price, B.D.; Kimmelman, A.; Makrigiorgos, G.M. DNA damage enhancement from gold nanoparticles for clinical MV photon beams. *Radiat. Res.* **2012**, *178*, 604–608. [CrossRef] [PubMed]
37. Kirillin, M.; Shirmanova, M.; Sirotkina, M.; Bugrova, M.; Khlebtsov, B.; Zagaynova, E. Contrasting properties of gold nanoshells and titanium dioxide nanoparticles for optical coherence tomography imaging of skin: Monte Carlo simulations and in vivo study. *J. Biomed. Opt.* **2009**, *14*, 021017. [CrossRef]
38. Díaz-Galindo, C.A.; Garnica-Garza, H.M. Gold nanoparticle-enhanced radiotherapy: Dependence of the macroscopic dose enhancement on the microscopic localization of the nanoparticles within the tumor vasculature. *PLoS ONE* **2024**, *19*, e0304670. [CrossRef] [PubMed]
39. Santiago, C.A.; Chow, J.C.L. Variations in Gold Nanoparticle Size on DNA Damage: A Monte Carlo Study Based on a Multiple-Particle Model Using Electron Beams. *Appl. Sci.* **2023**, *13*, 4916. [CrossRef]
40. Antunes, J.; Pinto, C.I.; Campello, M.P.; Santos, P.; Mendes, F.; Paulo, A.; Sampaio, J.M. Utility of realistic microscopy-based cell models in simulation studies of nanoparticle-enhanced photon radiotherapy. *Biomed. Phys. Eng. Express* **2024**, *10*, 025015. [CrossRef] [PubMed]
41. Chow, J.C.L.; Ruda, H.E. Flash Radiotherapy: Innovative Cancer Treatment. *Encyclopedia* **2023**, *3*, 808–823. [CrossRef]
42. Siddique, S.; Ruda, H.E.; Chow, J.C.L. FLASH Radiotherapy and the Use of Radiation Dosimeters. *Cancers* **2023**, *15*, 3883. [CrossRef] [PubMed]
43. Shiraishi, Y.; Matsuya, Y.; Fukunaga, H. Possible mechanisms and simulation modeling of FLASH radiotherapy. *Radiol. Phys. Technol.* **2024**, *17*, 11–23. [CrossRef]
44. Chow, J.C.L.; Ruda, H.E. Mechanisms of Action in FLASH Radiotherapy: A Comprehensive Review of Physicochemical and Biological Processes on Cancerous and Normal Cells. *Cells* **2024**, *13*, 835. [CrossRef]
45. Padmanabhan, P.; Kumar, A.; Kumar, S.; Chaudhary, R.K.; Gulyás, B. Nanoparticles in practice for molecular-imaging applications: An overview. *Acta Biomater.* **2016**, *41*, 1–6. [CrossRef]
46. Antonelli, A.; Magnani, M. SPIO nanoparticles and magnetic erythrocytes as contrast agents for biomedical and diagnostic applications. *J. Magn. Magn. Mater.* **2022**, *541*, 168520. [CrossRef]

47. Chow, J.C.L. Magnetic Nanoparticles in Magnetic Resonance Imaging: Principles and Applications. In *Magnetic Nanoparticles in Nanomedicine*; Wu, K., Wang, J.-P., Eds.; Elsevier: Amsterdam, The Netherlands, 2024; Chapter 13; pp. 371–394. [CrossRef]
48. Cormode, D.P.; Naha, P.C.; Fayad, Z.A. Nanoparticle contrast agents for computed tomography: A focus on micelles. *Contrast Media Mol. Imaging* **2014**, *9*, 37–52. [CrossRef]
49. Stockhofe, K.; Postema, J.M.; Schieferstein, H.; Ross, T.L. Radiolabeling of nanoparticles and polymers for PET imaging. *Pharmaceuticals* **2014**, *7*, 392–418. [CrossRef] [PubMed]
50. Wu, Y.; Ali, M.R.; Chen, K.; Fang, N.; El-Sayed, M.A. Gold nanoparticles in biological optical imaging. *Nano Today* **2019**, *24*, 120–140. [CrossRef]
51. Jones, B.L.; Cho, S.H. The feasibility of polychromatic cone-beam x-ray fluorescence computed tomography (XFCT) imaging of gold nanoparticle-loaded objects: A Monte Carlo study. *Phys. Med. Biol.* **2011**, *56*, 3719. [CrossRef] [PubMed]
52. Dou, Y.; Guo, Y.; Li, X.; Li, X.; Wang, S.; Wang, L.; Lv, G.; Zhang, X.; Wang, H.; Gong, X.; et al. Size-tuning ionization to optimize gold nanoparticles for simultaneous enhanced CT imaging and radiotherapy. *ACS Nano* **2016**, *10*, 2536–2548. [CrossRef]
53. Cho, J.; Wang, M.; Gonzalez-Lepera, C.; Mawlawi, O.; Cho, S.H. Development of bimetallic (Zn@ Au) nanoparticles as potential PET-imageable radiosensitizers. *Med. Phys.* **2016**, *43*, 4775–4788. [CrossRef]
54. Maschmeyer, R.T.; Gholami, Y.H.; Kuncic, Z. Clustering effects in nanoparticle-enhanced β− emitting internal radionuclide therapy: A Monte Carlo study. *Phys. Med. Biol.* **2020**, *65*, 125007. [CrossRef]
55. Martin, É.; Gossuin, Y.; Bals, S.; Kavak, S.; Vuong, Q.L. Monte Carlo simulations of the magnetic behaviour of iron oxide nanoparticle ensembles: Taking size dispersion, particle anisotropy, and dipolar interactions into account. *Eur. Phys. J. B* **2022**, *95*, 201. [CrossRef]
56. Lopushenko, I.; Sieryi, O.; Bykov, A.; Meglinski, I. Exploring the evolution of circular polarized light backscattered from turbid tissue-like disperse medium utilizing generalized Monte Carlo modeling approach with a combined use of Jones and Stokes–Mueller formalisms. *J. Biomed. Opt.* **2024**, *29*, 052913. [CrossRef] [PubMed]
57. Yang, X.; Chai, C.; Zuo, H.; Chen, Y.H.; Shi, J.; Ma, C.; Sawan, M. Monte Carlo-Based Optical Simulation of Optical Distribution in Deep Brain Tissues Using Sixteen Optical Sources. *Bioengineering* **2024**, *11*, 260. [CrossRef]
58. Huang, J.; Yang, L.; Wang, F.; Wu, Y.; Nan, Y.; Wu, W.; Wang, C.; Shi, K.; Aviles-Rivero, A.I.; Schönlieb, C.B.; et al. Enhancing global sensitivity and uncertainty quantification in medical image reconstruction with Monte Carlo arbitrary-masked mamba. *Med. Image Anal.* **2025**, *99*, 103334. [CrossRef]
59. Cong, A.X.; Hofmann, M.C.; Cong, W.; Xu, Y.; Wang, G. Monte Carlo fluorescence microtomography. *J. Biomed. Opt.* **2011**, *16*, 070501. [CrossRef] [PubMed]
60. Mar Blanca, C.; Saloma, C. Monte Carlo analysis of two-photon fluorescence imaging through a scattering medium. *Appl. Opt.* **1998**, *37*, 8092–8102. [CrossRef] [PubMed]
61. Fales, A.M.; Vogt, W.C.; Wear, K.A.; Ilev, I.K.; Pfefer, T.J. Pulsed laser damage of gold nanorods in turbid media and its impact on multi-spectral photoacoustic imaging. *Biomed. Opt. Express* **2019**, *10*, 1919–1934. [CrossRef] [PubMed]
62. Okawa, S.; Hirasawa, T.; Sato, R.; Kushibiki, T.; Ishihara, M.; Teranishi, T. Numerical and experimental investigations of dependence of photoacoustic signals from gold nanoparticles on the optical properties. *Opt. Rev.* **2018**, *25*, 365–374. [CrossRef]
63. Neuwelt, A.; Sidhu, N.; Hu, C.A.; Mlady, G.; Eberhardt, S.C.; Sillerud, L.O. Iron-based superparamagnetic nanoparticle contrast agents for MRI of infection and inflammation. *Am. J. Roentgenol.* **2015**, *204*, W302–W313. [CrossRef]
64. Albayedh, F.; Chow, J.C. Monte Carlo simulation on the imaging contrast enhancement in nanoparticle-enhanced radiotherapy. *J. Med. Phys.* **2018**, *43*, 195–199.
65. Ross, G.M. Induction of cell death by radiotherapy. *Endocr. Relat. Cancer* **1999**, *6*, 41–44. [CrossRef] [PubMed]
66. Xing, J.L.; Stea, B. Molecular mechanisms of sensitivity and resistance to radiotherapy. *Clin. Exp. Metastasis* **2024**, *41*, 1–8. [CrossRef] [PubMed]
67. Chen, Y.; Yang, J.; Fu, S.; Wu, J. Gold nanoparticles as radiosensitizers in cancer radiotherapy. *Int. J. Nanomed.* **2020**, *15*, 9407–9430. [CrossRef] [PubMed]
68. Incerti, S.; Suerfu, B.; Xu, J.; Ivantchenko, V.; Mantero, A.; Brown, J.M.; Bernal, M.A.; Francis, Z.; Karamitros, M.; Tran, H.N. Simulation of Auger electron emission from nanometer-size gold targets using the Geant4 Monte Carlo simulation toolkit. *Nucl. Instrum. Methods Phys. Res. Sect. B Beam Interact. Mater. At.* **2016**, *372*, 91–101. [CrossRef]
69. Rosero, W.A.; Barbezan, A.B.; de Souza, C.D.; Rostelato, M.E. Review of advances in coating and functionalization of gold nanoparticles: From theory to biomedical application. *Pharmaceutics* **2024**, *16*, 255. [CrossRef] [PubMed]
70. Cho, S.H. Estimation of tumour dose enhancement due to gold nanoparticles during typical radiation treatments: A preliminary Monte Carlo study. *Phys. Med. Biol.* **2005**, *50*, N163. [CrossRef] [PubMed]
71. Dos Santos, M.; Delorme, R.; Salmon, R.; Prezado, Y. Minibeam radiation therapy: A micro- and nano-dosimetry Monte Carlo study. *Med. Phys.* **2020**, *47*, 1379–1390. [CrossRef] [PubMed]
72. Sheeraz, Z.; Chow, J.C.L. Evaluation of dose enhancement with gold nanoparticles in kilovoltage radiotherapy using the new EGS geometry library in Monte Carlo simulation. *AIMS Biophys.* **2021**, *8*, 337–345. [CrossRef]

73. Martinov, M.P.; Thomson, R.M. Heterogeneous multiscale Monte Carlo simulations for gold nanoparticle radiosensitization. *Med. Phys.* **2017**, *44*, 644–653. [CrossRef] [PubMed]
74. Zheng, X.J.; Chow, J.C. Radiation dose enhancement in skin therapy with nanoparticle addition: A Monte Carlo study on kilovoltage photon and megavoltage electron beams. *World J. Radiol.* **2017**, *9*, 63. [CrossRef] [PubMed]
75. Harriss-Phillips, W.M.; Bezak, E.; Yeoh, E.K. Monte Carlo radiotherapy simulations of accelerated repopulation and reoxygenation for hypoxic head and neck cancer. *Br. J. Radiol.* **2011**, *84*, 903–918. [CrossRef]
76. Zerakni, F.; Dib, A.S.; Attili, A. Enhancing tumor's skin photothermal therapy using Gold nanoparticles: A Monte Carlo simulation. *Lasers Med. Sci.* **2024**, *39*, 1–7. [CrossRef] [PubMed]
77. Ma, S.; Jiang, L.; Yang, W.; Liu, F.; Wang, D.; Wang, F.; Huang, J. Advances of Nanomaterials in Cancer Photocatalysis Therapy. *Mater. Today Sustain.* **2024**, *29*, 101023. [CrossRef]
78. Maurente, A.; de Sousa, A.N. Anisotropic scattering in photothermal therapy and beyond: A detailed evaluation of transport approximation for gold nanoshells and nanospheres. *J. Braz. Soc. Mech. Sci. Eng.* **2024**, *46*, 1–20. [CrossRef]
79. Periyasamy, V.; Pramanik, M. Advances in Monte Carlo simulation for light propagation in tissue. *IEEE Rev. Biomed. Eng.* **2017**, *10*, 122–135. [CrossRef] [PubMed]
80. Jin, W.; Shi, X.; Yin, H.; Zhang, H.; Wang, Z.; Chen, Q.; Wu, H.; Han, Y.; Li, Y. Comparison of actual and simulated tumoricidal effects induced by photodynamic therapy. *Photodiagnosis Photodyn. Ther.* **2020**, *32*, 102060. [CrossRef]
81. Kim, D.; Paik, J.; Kim, H. Effect of gold nanoparticles distribution radius on photothermal therapy efficacy. *Sci. Rep.* **2023**, *13*, 12135. [CrossRef]
82. Hosseini, F.S.; Naghavi, N.; Sazgarnia, A. A physicochemical model of X-ray induced photodynamic therapy (X-PDT) with an emphasis on tissue oxygen concentration and oxygenation. *Sci. Rep.* **2023**, *13*, 17882. [CrossRef] [PubMed]
83. Bucharskaya, A.B.; Khlebtsov, N.G.; Khlebtsov, B.N.; Maslyakova, G.N.; Navolokin, N.A.; Genin, V.D.; Genina, E.A.; Tuchin, V.V. Photothermal and photodynamic therapy of tumors with plasmonic nanoparticles: Challenges and prospects. *Materials* **2022**, *15*, 1606. [CrossRef] [PubMed]
84. Siddique, S.; Chow, J.C.L. Recent advances in functionalized nanoparticles in cancer theranostics. *Nanomaterials* **2022**, *12*, 2826. [CrossRef]
85. Montenegro, M.; Nahar, S.N.; Pradhan, A.K.; Huang, K.; Yu, Y. Monte Carlo simulations and atomic calculations for Auger processes in biomedical nanotheranostics. *J. Phys. Chem. A* **2009**, *113*, 12364–12369. [CrossRef]
86. Lindstaedt, A.; Doroszuk, J.; Machnikowska, A.; Dziadosz, A.; Barski, P.; Raffa, V.; Witt, D. Effects Induced by the Temperature and Chemical Environment on the Fluorescence of Water-Soluble Gold Nanoparticles Functionalized with a Perylene-Derivative Dye. *Materials* **2024**, *17*, 1097. [CrossRef] [PubMed]
87. Tian, Y.; Carrillo-Malani, N.; Feng, K.; Miller, J.; Busch, T.M.; Sundaram, K.M.; Cheng, Z.; Amirshaghaghi, A.; Tsourkas, A. Theranostic phthalocyanine and naphthalocyanine Nanoparticles for photoacoustic imaging and photothermal therapy of tumors. *Nanotheranostics* **2024**, *8*, 100. [CrossRef] [PubMed]
88. Herzog, J.M.; Sick, V. MCRAD: A Monte Carlo photon transport code for analysis of fluorescence and elastic scattering diagnostics. *SoftwareX* **2024**, *27*, 101672. [CrossRef]
89. Díaz-Galindo, C.A.; Garnica-Garza, H.M. Radiation source personalization for nanoparticle-enhanced radiotherapy using dynamic contrast-enhanced MRI in the treatment planning process. *Radiat. Phys. Chem.* **2024**, *217*, 111518. [CrossRef]
90. Chow, J.C.L.; Jubran, S. Depth Dose Enhancement in Orthovoltage Nanoparticle-enhanced Radiotherapy: A Monte Carlo Phantom Study. *Micromachines* **2023**, *14*, 1230. [CrossRef] [PubMed]
91. Cogno, N.; Bauer, R.; Durante, M. Mechanistic model of radiotherapy-induced lung fibrosis using coupled 3D agent-based and Monte Carlo simulations. *Commun. Med.* **2024**, *4*, 16. [CrossRef] [PubMed]
92. Taheri, A.; Khandaker, M.U.; Moradi, F.; Bradley, D.A. A simulation study on the radiosensitization properties of gold nanorods. *Phys. Med. Biol.* **2024**, *69*, 045029. [CrossRef]
93. Rojas-Domínguez, A.; Arroyo-Duarte, R.; Rincón-Vieyra, F.; Alvarado-Mentado, M. Modeling cancer immunoediting in tumor microenvironment with system characterization through the ising-model Hamiltonian. *BMC Bioinform.* **2022**, *23*, 200. [CrossRef] [PubMed]
94. Metzcar, J.; Wang, Y.; Heiland, R.; Macklin, P. A review of cell-based computational modeling in cancer biology. *JCO Clin. Cancer Inf.* **2019**, *2*, 1–3. [CrossRef]
95. Mountain, R.D.; Thirumalai, D. Quantative measure of efficiency of Monte Carlo simulations. *Phys. A Stat. Mech. Its Appl.* **1994**, *210*, 453–460. [CrossRef]
96. Luo, C.; Keshtegar, B.; Zhu, S.P.; Taylan, O.; Niu, X.P. Hybrid enhanced Monte Carlo simulation coupled with advanced machine learning approach for accurate and efficient structural reliability analysis. *Comput. Methods Appl. Mech. Eng.* **2022**, *388*, 114218. [CrossRef]
97. Chow, J.C.L.; Ruda, H.E. Impact of Scattering Foil Composition on Electron Energy Distribution in a Clinical Linear Accelerator Modified for FLASH Radiotherapy: A Monte Carlo Study. *Materials* **2024**, *17*, 3355. [CrossRef]

98. Lai, Y.; Jia, X.; Chi, Y. Modeling the effect of oxygen on the chemical stage of water radiolysis using GPU-based microscopic Monte Carlo simulations, with an application in FLASH radiotherapy. *Phys. Med. Biol.* **2021**, *66*, 025004. [CrossRef] [PubMed]
99. Mazzola, G. Quantum computing for chemistry and physics applications from a Monte Carlo perspective. *J. Chem. Phys.* **2024**, *160*, 010901. [CrossRef] [PubMed]
100. Chow, J.C.L. Quantum Computing in Medicine. *Med. Sci.* **2024**, *12*, 67. [CrossRef]
101. Neph, R.; Lyu, Q.; Huang, Y.; Yang, Y.M.; Sheng, K. DeepMC: A deep learning method for efficient Monte Carlo beamlet dose calculation by predictive denoising in magnetic resonance-guided radiotherapy. *Phys. Med. Biol.* **2021**, *66*, 035022. [CrossRef] [PubMed]
102. Gong, F.Q.; Xiong, K.; Cheng, J. Constructing machine learning potential for metal nanoparticles of varying sizes via basin-hoping Monte Carlo and active learning. *Natl. Sci. Open* **2024**, *3*, 20230088. [CrossRef]
103. Holmes, J.; Feng, H.; Zhang, L.; Fix, M.K.; Jiang, S.B.; Liu, W. Fast Monte Carlo dose calculation in proton therapy. *Phys. Med. Biol.* **2024**, *69*, 17TR01. [CrossRef]
104. O'Quigley, J.; Chevret, S. Methods for dose finding studies in cancer clinical trials: A review and results of a Monte Carlo study. *Stat. Med.* **1991**, *10*, 1647–1664. [CrossRef] [PubMed]
105. Goldenholz, D.M.; Tharayil, J.; Moss, R.; Myers, E.; Theodore, W.H. Monte Carlo simulations of randomized clinical trials in epilepsy. *Ann. Clin. Transl. Neurol.* **2017**, *4*, 544–552. [CrossRef]

MDPI AG
Grosspeteranlage 5
4052 Basel
Switzerland
Tel.: +41 61 683 77 34

Nanomaterials Editorial Office
E-mail: nanomaterials@mdpi.com
www.mdpi.com/journal/nanomaterials

www.ingramcontent.com/pod-product-compliance
Lightning Source LLC
Chambersburg PA
CBHW061557210926
51850CB00009B/2126
9783725863006